GLIDER FLYING HANDBOOK

2024

U.S. Department of Transportation
FEDERAL AVIATION ADMINISTRATION
Flight Standards Service

Skyhorse Publishing

First published in December 2024
First Skyhorse Publishing edition 2025

Skyhorse Publishing books may be purchased in bulk at special discounts for sales promotion, corporate gifts, fund-raising, or educational purposes. Special editions can also be created to specifications. For details, contact the Special Sales Department, Skyhorse Publishing, 307 West 36th Street, 11th Floor, New York, NY 10018 or info@skyhorsepublishing.com.

Skyhorse® and Skyhorse Publishing® are registered trademarks of Skyhorse Publishing, Inc.®, a Delaware corporation.

Visit our website at www.skyhorsepublishing.com.
Please follow our publisher Tony Lyons on Instagram @tonylyonsisuncertain.

10 9 8 7 6 5 4 3 2 1

Library of Congress Cataloging-in-Publication Data is available on file.

Cover design by Federal Aviation Administration

ISBN: 978-1-5107-8441-3
eBook ISBN: 978-1-5107-8442-0

Printed in China

GLIDER FLYING HANDBOOK

Preface

The Glider Flying Handbook is designed as a technical manual for applicants who are preparing for glider category rating and for currently certificated glider pilots who wish to improve their knowledge. Certificated flight instructors will find this handbook a valuable training aid, since detailed coverage of aeronautical decision-making, components and systems, aerodynamics, flight instruments, performance limitations, ground operations, flight maneuvers, traffic patterns, emergencies, soaring weather, soaring techniques, and cross-country flight is included. Topics such as radio navigation and communication, use of flight information publications, and regulations are available in other Federal Aviation Administration (FAA) publications.

The discussion and explanations reflect the most commonly used practices and principles. Occasionally, the word "must" or similar language is used where the desired action is deemed critical. The use of such language is not intended to add to, interpret, or relieve a duty imposed by Title 14 of the Code of Federal Regulations (14 CFR). Persons working towards a glider rating are advised to review the references from the applicable practical test standards. Resources for study include FAA-H-8083-25, Pilot's Handbook of Aeronautical Knowledge; FAA-H-8083-2, Risk Management Handbook; and FAA-H-8083-28, Aviation Weather Handbook, as these documents contain basic material not duplicated herein. All beginning applicants should refer to FAA-H-8083-25, Pilot's Handbook of Aeronautical Knowledge, for study and basic library reference.

It is essential for persons using this handbook to become familiar with and apply the pertinent parts of 14 CFR and the Aeronautical Information Manual (AIM). The AIM is available on the FAA website. The current Flight Standards Service airman training and testing material and learning statements for all airman certificates and ratings can be obtained from the FAA website.

This handbook supersedes FAA-H-8083-13A, Glider Flying Handbook, dated 2013.

This handbook is available for download, in PDF format, from the FAA website.

This handbook is published by the United States Department of Transportation, Federal Aviation Administration, General Aviation & Commercial Division, Training & Certification Group (AFS-810), Testing Standards Section, P.O. Box 25082, Oklahoma City, OK 73125.

Comments regarding this publication should be emailed to AFS630comments@faa.gov.

Acknowledgments

The Glider Flying Handbook was produced by the Federal Aviation Administration (FAA). The FAA wishes to acknowledge the following contributors:

Sue Telford of Telford Fishing & Hunting Services for images used in Chapter 1.

JerryZieba (www.dianasailplanes.com) for images used in Chapter 2.

Tim Mara (www.wingsandwheels.com) for images used in Chapters 2 and 12.

Uli Kremer of Alexander Schleicher GmbH & Co for images used in Chapter 2.

Richard Lancaster (www.carrotworks.com) for images and content used in Chapter 3.

Dave Nadler of Nadler & Associates for images used in Chapter 6.

Dave McConeghey for images used in Chapter 6.

John Brandon (www.raa.asn.au) for images and content used in Chapter 7.

Patrick Panzera (www.contactmagazine.com) for images used in Chapter 8.

Jeff Haby (www.theweatherprediction) for images used in Chapter 8.

National Soaring Museum (www.soaringmuseum.org) for content used in Chapter 9.

Bill Elliot (www.soaringcafe.com) for images used in Chapter 12.

Tiffany Fidler for images used in Chapter 12.

Phil Klauder, Philadelphia Glider Council, for assistance with graphics revisions and review input.

Additional appreciation is extended to the Soaring Society of America, Inc. (www.ssa.com), the Soaring Safety Foundation, and Mr. Brad Temeyer and Mr. Bill Martin from the National Oceanic and Atmospheric Administration (NOAA) for their technical support and input.

Table of Contents

Chapter 4: Flight Instruments

Chapter 5: Glider Performance

Chapter 6: Preflight & Ground Operations

Chapter 7: Launch, Flight Maneuvers, Landing, & Recovery Procedures

Chapter 8: Abnormal & Emergency Procedures

Chapter 9: Glider Flight & Weather

Chapter 10: Soaring Techniques

Chapter 13: Human Factors

Chapter 1: Gliders & Sailplanes

Introduction

A modern glider, such as the one pictured below [*Figure 1-1*] can fly high and make long cross-country flights with a skilled pilot at the controls.

Figure 1-1. *A DG Flugzeugbau GmbH 800B-series glider.*

The Code of Federal Regulations (14 CFR part 1, section 1.1) states, "glider means a heavier-than-air aircraft, that is supported in flight by the dynamic reaction of the air against its lifting surfaces and whose free flight does not depend principally on an engine." The term "glider" also designates the rating placed on a pilot certificate once an applicant successfully completes required glider training, has the requisite experience, passes any required knowledge test, and passes the appropriate practical test.

For a glider to fly, it needs a means to become airborne. Early gliders could only launch from the top of a hill. [*Figure 1-2*] and [*Figure 1-3*].

Figure 1-2. *Otto Lilienthal (the Glider King) in flight during the mid-1890s.*

Figure 1-3. *Orville Wright (left) and Dan Tate (right) launched the Wright 1902 glider off the east slope of the Big Hill, Kill Devil Hills, North Carolina, on October 17, 1902. Wilbur Wright was flying the glider.*

After development of powered flight, an airplane could tow a glider to altitude, and this became a common means to launch a glider. While early glider designs would only descend after tow and release, later designs could release and take advantage of natural rising air to continue to gain altitude. Some gliders that require a tow to altitude also have a sustainer engine for use in flight. The pilot can start and stop the powerplant while in flight, and in some models, the pilot may retract the propeller system into the body of the glider for increased aerodynamic efficiency. A self-launching motor glider can takeoff and climb to soaring altitudes without a tow. [*Figure 1-4*]

Figure 1-4. *An ASH 26 E self-launching motor glider with the propeller extended.*

Glider Pilot Training

How does a person obtain glider flight training? With a general location in mind, an individual may consider several options, including an FAA-approved glider school, privately owned commercial glider school, or college, university, or private soaring club. These will have FAA-certified flight instructors who can provide instruction. Published articles, soaring-related websites, and discussions with other pilots may help a prospective student create a list of items to look for in a training provider.

An interested person should consider the quality of training provided. Good instruction follows a structured syllabus and a building block approach. Prior to picking a school, visiting the training provider and talking with management, instructors, and other students can reveal the pros and cons of choosing a particular club or commercial school. Before making a commitment, the prospective student should take an introductory lesson. After deciding on a provider and making the necessary arrangements, training can begin. Individual commitment to a regular training schedule maximizes student progress and retention.

To be eligible to fly solo in a glider, unrated pilots need to obtain a student pilot certificate, be at least 14 years of age, and demonstrate satisfactory aeronautical knowledge on a pre-solo written test administered by their instructor. Before solo, a student pilot also receives ground and flight training for the maneuvers and procedures listed in Title 14 of the Code of Federal Regulations (14 CFR) part 61, section 61.87(i). After a student pilot meets the administrative requirements and demonstrates satisfactory proficiency, the instructor may endorse the student's logbook for solo flight.

Note that rated airplane pilots can increase their overall knowledge, skill, and understanding of safety of flight by adding a glider rating. The addition of a glider rating enhances an airplane pilot's ability to manage flight without power should an engine malfunction occur.

Rating Eligibility

A student pilot 16 years of age or older or an existing FAA-rated pilot who meets the flight time requirements may take the practical test for a sport pilot certificate with a glider endorsement or the practical test for a private pilot certificate with a glider rating after accomplishing the training requirements listed in 14 CFR part 61 for the desired level of certification.

To be eligible for a commercial or flight instructor glider certificate, an individual must be 18 years of age and complete the specific training requirements described in 14 CFR part 61.

The applicable FAA Airman Certification Standards (ACS) or Practical Test Standards (PTS) contain the knowledge and skills required for pilot certification and describe the testing process. Applicants may also refer to FAA-G-ACS- 2, the ACS Companion Guide for Pilots, FAA Advisory Circular (AC) 60-22, Aeronautical Decision Making; the Pilot's Handbook of Aeronautical Knowledge (FAA-H-8083-25); the Risk Management Handbook (FAA-H-8083-2); and the Aviation Weather Handbook (FAA-H-8083-28) to gain additional aviation-related information. For more information on the certification of gliders, refer to 14 CFR part 21, the European Aviation Safety Agency (EASA) Certification Specifications (CS) 22.221, and the Weight and Balance Handbook (FAA-H-8083-1).

Medical Eligibility

A person may exercise the privileges of a glider rating or those of an authorized instructor in a glider without holding a medical certificate. However, 14 CFR part 61, section 61.53 states, "...a person shall not act as pilot in command, or in any other capacity as a required pilot flight crewmember, while that person knows or has reason to know of any medical condition that would make the person unable to operate the aircraft in a safe manner."

FAA Wings Program

Rated pilots should compare continuous training and practice to 14 CFR part 61, section 61.56(c)(1) and (2), which allow for training and a sign-off within the previous 24 calendar months in order to act as a pilot in command. Many astute pilots realize that this regulation specifies a minimum requirement, and the path to enhanced proficiency, safety, and enjoyment of flying takes a higher degree of commitment such as available using 14 CFR part 61, section 61.56(e). For this reason, many pilots keep their flight review up to date using the FAA WINGS program. The program provides continuing pilot education and contains interesting and relevant study materials that pilots can use all year round.

A pilot may create a WINGS account to obtain current information concerning risk mitigation. The program provides a means to improve risk management skill as a means to increase safety. As an added bonus, completion of a phase of the Wings Program can count for a flight review and participants may receive a discount on certain flight insurance policies. The link to create an account is www.faasafety.gov.

Chapter Summary

Gliders include heavier-than-air aircraft that need a means to become airborne. In a modern glider and once aloft, pilots have a variety of means to sustain flight. To become a glider pilot, a prospective student should investigate training options and pick a suitable training provider.

Chapter 2: Components & Systems

Introduction

Glider Design

Glider airframes include a fuselage, wings, and empennage or tail section. [*Figure 2-1*]

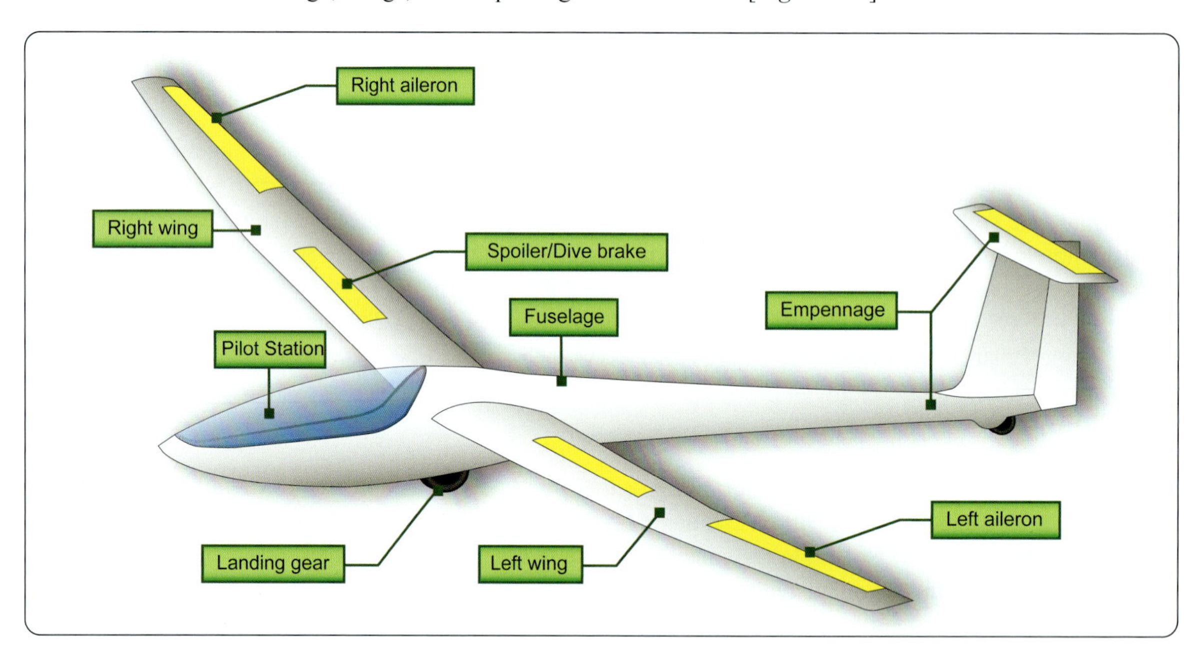

Figure 2-1. *Components of a glider.*

The Fuselage

The fuselage contains the controls for the glider, as well as a seat for each occupant. The wings and empennage attach to the fuselage. Manufacturers typically use composites, fiberglass, or carbon fiber, however in the past manufacturers used wood, fabric over steel tubing, aluminum, or a combination of these materials to build a fuselage.

Introduction

When air flows over the wings of a glider, the wings produce lift that allows the aircraft to stay aloft. Glider wing designs produce maximum lift with minimum drag.

Glider wings incorporate several components that help the pilot maintain the attitude of the glider and control lift and drag. These include ailerons and other lift and drag devices.

Ailerons

The ailerons attach to the outboard trailing edge of each wing. When the pilot moves the aileron control to the right of center, the right aileron deflects upward [*Figure 2-2*] and the left aileron deflects downward. In flight and with air flowing over the wings, these deflections result in increased lift on the left wing and decreased lift on the right wing. The increased lift on the left wing and decreased lift on the right wing apply a force to roll the glider to the right. Moving the aileron control to the left of center deflects the right aileron down and the left aileron up. This applies a force to roll the glider to the left.

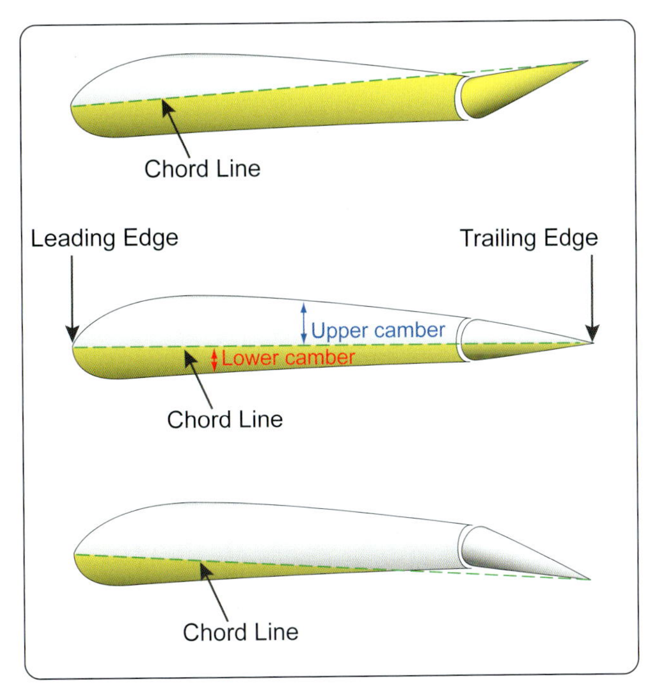

Figure 2-2. *The ailerons change the camber or curvature of the wing and increase or decrease lift.*

Lift/Drag Devices

Gliders may use other devices that modify the lift/drag of the wing. These high drag devices include spoilers, dive brakes, and flaps. [*Figure 2-3*] Spoilers extend from the upper surface of the wings, alter the airflow, and cause the glider to descend more rapidly. Dive brakes extend from both the upper and lower surfaces of the wing and increase drag. Some high-performance gliders have dive brake speed limitations to prevent structural damage.

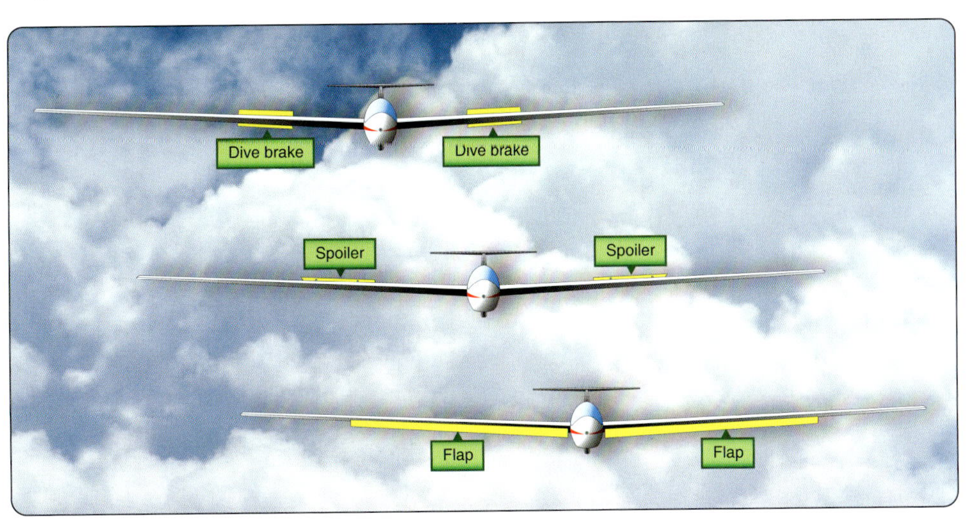

Figure 2-3. *Types of lift/drag devices.*

Some gliders have flaps installed on the trailing edge of each wing inboard of the ailerons, which can change lift, drag, and descent rate. The pilot can generally set flaps in three different positions, which are trail, down, or negative. [*Figure 2-4*] When the pilot sets the flaps to deflect downward in flight, the wing produces more lift and drag. On the other hand, a negative flap position results in reduced lift and drag.

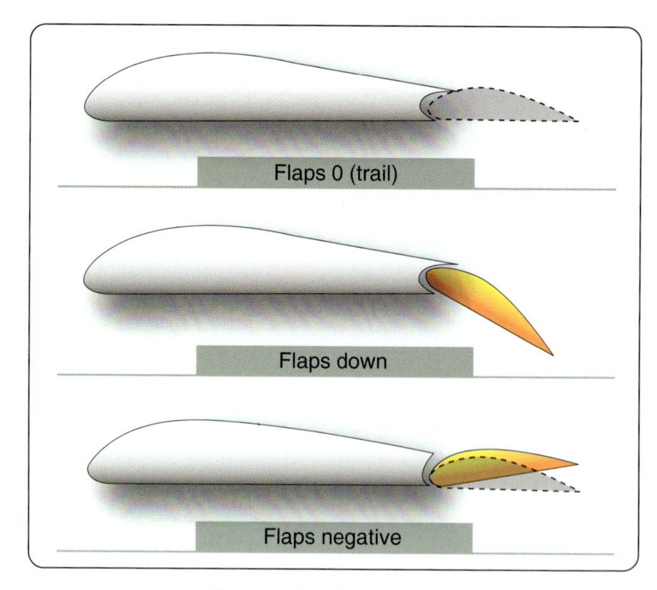

Figure 2-4. *Flap positions.*

Empennage

The empennage includes the entire tail section, consisting of the fixed surfaces, such as the horizontal stabilizer and vertical fin, and movable surfaces, such as the elevator or stabilator, rudder, and any trim tabs. The two fixed surfaces act like the feathers on an arrow to steady the glider and help maintain a straight path through the air. [*Figure 2-5*]

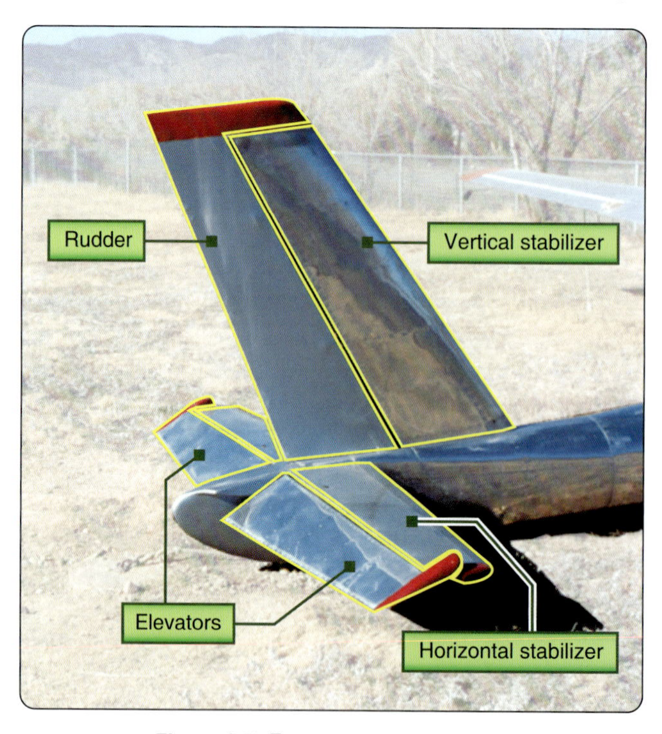

Figure 2-5. *Empennage components.*

The rudder attaches to the back of the vertical stabilizer, and the pilot deflects the rudder using foot pedals. The rudder controls yaw and turn coordination during flight.

During straight-and-level flight, pilot-controlled deflection of the elevator applies a force to move the glider's nose up and down relative to the horizon. Raising the nose results in lower airspeed while lowering the nose increases airspeed. Instead of a horizontal stabilizer and elevator, some gliders use a stabilator, where the entire horizontal tail surface pivots up and down on a central hinge point. Pilots refer to the movement of the elevator or stabilator as controlling the pitch attitude of the glider.

When the pilot deflects the pitch controls from the neutral position, the airflow pushes back against the controls. This opposing force provides feedback to the pilot but also adds to the pilot's workload. The pilot may relieve the elevator control pressure using elevator trim. A common trim system consists of a small, pilot-controlled adjustable tab on the trailing edge of the elevator. [*Figure 2-6*] Trim tabs come in servo and anti-servo designs. [*Figure 2-7*] A servo tab relieves pressure. The pilot can adjust an anti-servo tab, usually installed on stabilators, to relieve pressure, but it also self-adjusts to increase opposing pressure when the pilot increases stabilator displacement.

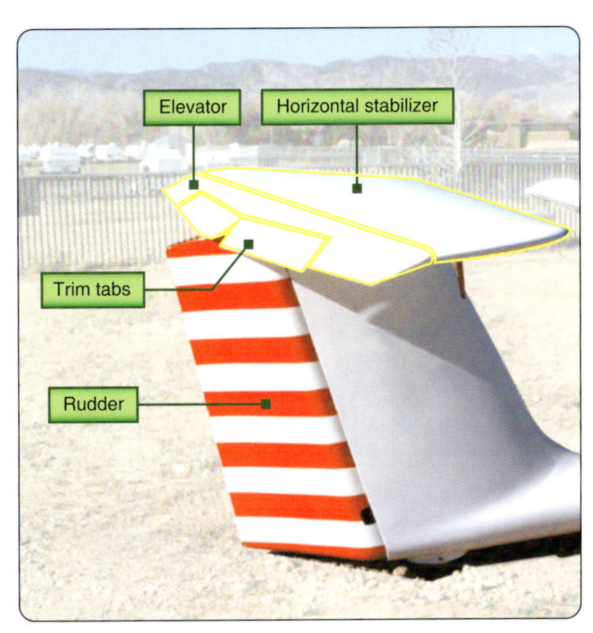

Figure 2-6. *Additional empennage components.*

Figure 2-7. *An anti-servo tab.*

Glider designs with a conventional tail incorporate the horizontal stabilizer mounted at the bottom of the vertical stabilizer. T-tail gliders have the horizontal stabilizer mounted on the top of the vertical stabilizer forming a "T" shape. V-tails have two tail surfaces mounted to form a "V" that combines elevator and rudder movements.

Towhook Devices

Gliders that launch using a tow have an approved towhook. For an aerotow, the crew normally connects the tow line to a towhook located on or just under the nose of the glider. For a tow using a winch or ground vehicle, the crew attaches the tow line to a towhook positioned below the glider center of gravity (CG)—the point where the glider would balance on the ground if lifted from above or below that position. [*Figure 2-8*]. Both towhook designs allow for quick release of the rope or cable when the glider pilot pulls the release handle. An aerotow launch of a glider using a specific towhook may only occur if the Glider Flight Manual/Pilot's Operating Handbook approves the procedure.

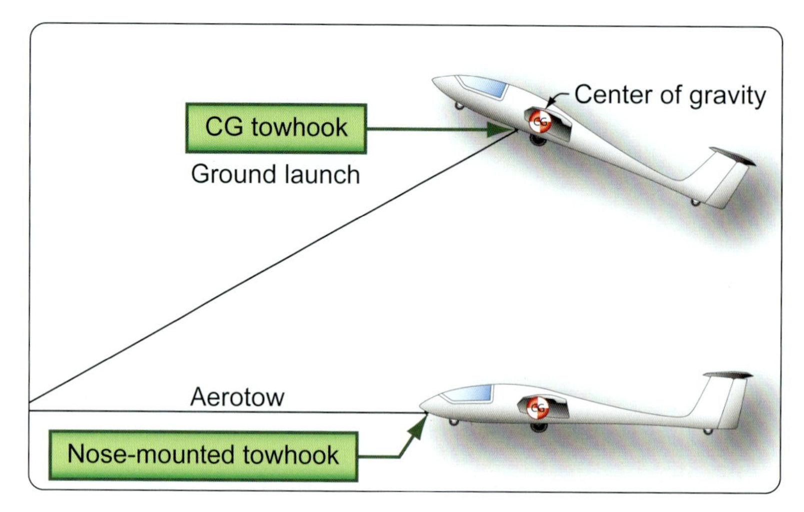

Figure 2-8. *Tow hook locations.*

Powerplant

While tow planes provide the most common means to launch a glider in the United States, self-launching gliders with built-in engines have become more commonplace.

Self-Launching Gliders

There are two types of self-launching gliders: touring motor gliders and high-performance self-launching gliders.

Touring Motor Gliders

Touring motor gliders have a nose-mounted engine and a full feathering propeller. [*Figure 2-9*] Although touring motor gliders have some basic airplane characteristics, they are certified in the glider category.

Figure 2-9. *A Grob G109B touring motor glider.*

High-Performance Gliders

High-performance self-launching gliders have engines that retract into the fuselage for minimal drag. [*Figure 2-10*] The propeller may fold, or simply align with the engine. This configuration preserves a smooth low drag configuration.

Figure 2-10. *A DG-808B 18-meter high-performance glider in self-launch.*

Gliders with Sustainer Engines

Some gliders have sustainer engines powered by either electricity or gasoline. A pilot can use a sustainer engine to remain aloft; however, sustainer engines do not provide sufficient power to launch the glider and may not have enough power to compensate for sinking air. [*Figure 2-11*]

Figure 2-11. *A Schleicher ASG-29 E with gasoline sustainer engine mast extended.*

Landing Gear

A glider landing gear system usually includes a main wheel. Gliders designed for high speed and low drag often feature a retractable main landing gear. Other components may include a front skid or wheel, a tail wheel or skid, or wing tip wheels or skid plates. [*Figure 2-12*]

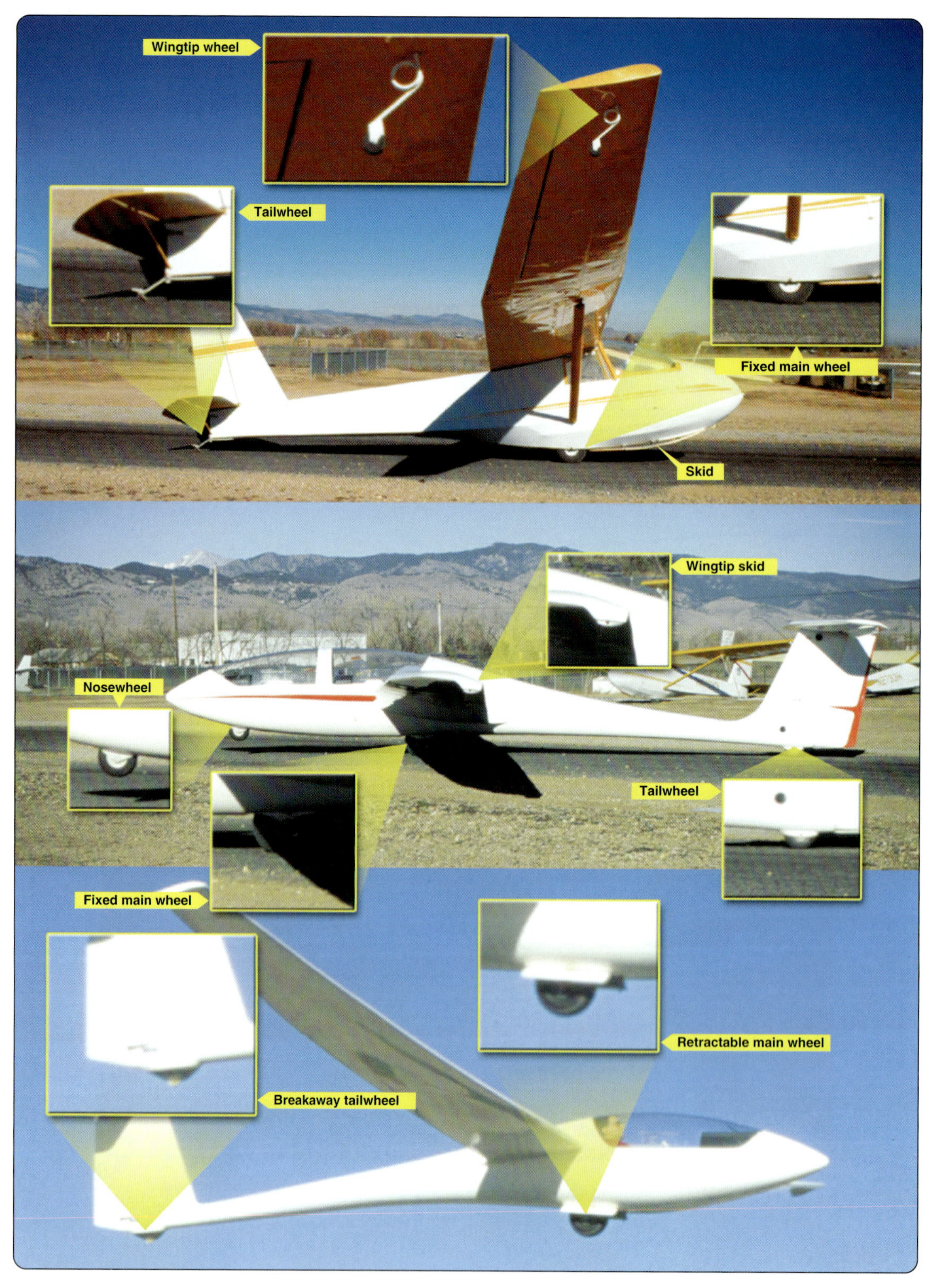

Figure 2-12. *Landing gear wheels on a glider.*

Almost all high-performance gliders have retractable landing gear, so pilots should include landing gear down in their prelanding checklist. Most landing gear handles are on the right side of the pilot station. However, a few models have gear handles on the left side, and pilots should use caution when reaching for a gear handle on the left to make sure it controls the gear and not flaps or airbrakes. A common error includes neglecting to retract the landing gear after takeoff, and then mistakenly retracting it as part of the prelanding checklist.

Some high-performance gliders have only one center of gravity (CG) tow hook either ahead of the landing gear or in the landing gear well. With a CG hook within the landing gear well, retracting the gear on tow interferes with the tow line. Even if the glider has a nose hook, retracting the gear should wait. A CG hook, as compared to a nose hook, makes a crosswind takeoff more difficult since the glider can weathervane into the wind more easily. In addition, a CG hook makes the glider more susceptible to kiting (climbing above the tow plane) on takeoff, which threatens the safety of the tow pilot.

Wheel Brakes

Early gliders often relied on friction between the nose skid and the ground to come to a stop. Later models use a wheel brake mounted on the main landing gear wheel, which helps the glider slow down or stop after touchdown. Modern gliders commonly use a hydraulic disk brake, which provides substantial braking capability.

Chapter Summary

Although gliders come in an array of shapes and sizes, most gliders share basic design features. These include a fuselage, wings and components, lift/drag devices, and empennage. Depending on the launch method used, a glider may have a towhook or an engine.

Chapter 3: Aerodynamics of Flight

Introduction

This chapter discusses glider-related aerodynamics. A pilot who understands how forces affect a glider can operate safely while maximizing performance. To obtain a more detailed description of general aerodynamics, see the Pilot's Handbook of Aeronautical Knowledge (FAA-H-8083-25).

A glider maneuvers around three axes of rotation: vertical, lateral, and longitudinal. Each axis is perpendicular to the other two, and all three axes intersect at one central point called the center of gravity (CG), which varies with the loading of the glider. Any object on the ground will balance in any orientation if supported from the CG or a point directly above or below the CG.

Yaw describes movement around the vertical axis, represented by an imaginary straight line drawn through the CG. [*Figure 3-1*] In flight, moving the rudder left or right causes the glider to yaw. The lateral axis runs parallel to a line from wingtip to wingtip. Pulling the stick back or pushing it forward changes the pitch of the glider and controls its movement around the lateral axis. Roll describes movement around the longitudinal axis caused by displacing the ailerons in opposite directions. This axis runs parallel to a line drawn from the nose to the tail.

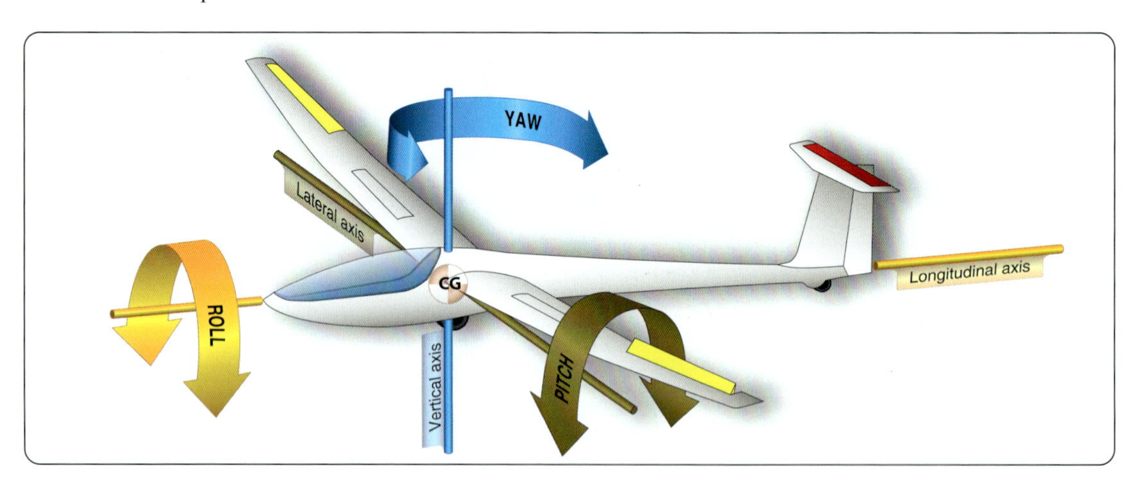

Figure 3-1. *Three axes of rotation with each perpendicular to the other two and all intersecting at the CG.*

Forces of Flight

Three forces act on an unpowered glider while in flight—lift, drag, and weight. Thrust is another force of flight that enables self-launching gliders to launch on their own and stay aloft when soaring conditions subside.

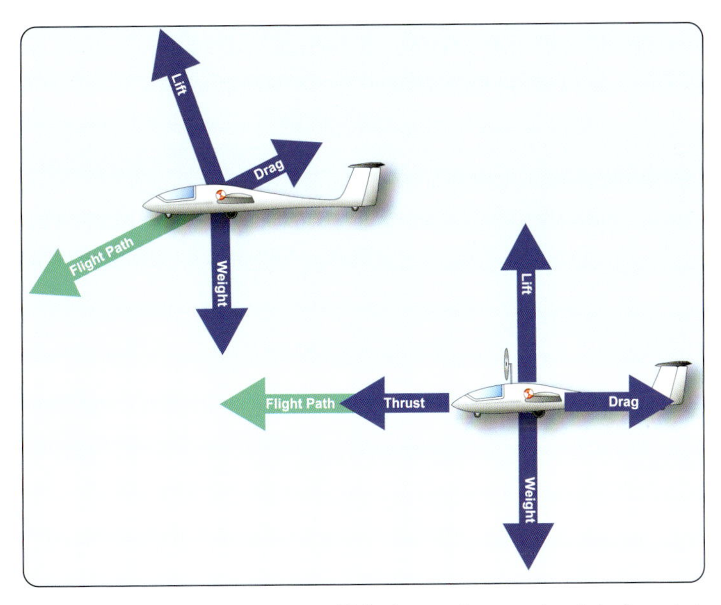

Figure 3-2. *Vector components of lift, thrust, drag, and weight (gravity).*

Lift

Newton's laws and Bernoulli's principle explain lift from different perspectives. Newton's third law describes the overall interaction between atmosphere and wing, while Bernoulli's principle looks at the effect of air changing speed as it moves past the wing. Together, these models provide a valid explanation of lift.

Newton's Third Law

According to Newton's Third Law of Motion, for every action there is an equal and opposite reaction. As air deflects downward because of interaction with the wing, the wing experiences an upward (lifting) reaction.

Bernoulli's Principle

Bernoulli's Principle states that as the velocity of a moving fluid (liquid or gas) increases, the pressure within the fluid decreases. This principle explains what happens when air moves faster to pass over the top of a wing positioned at an angle to the relative wind, i.e., the wind resulting from the forward motion of the glider. The increase in speed of the air as it travels over the top of the wing produces a drop in pressure against the wing, and the higher air pressure below the wing results in a net lifting force.

Lift Formula

A mathematical relationship exists between lift, the coefficient of lift, airspeed, air density, and the size of the wing. *Figure 3-3* shows the relationship.

$$L = C_L V^2 \frac{\rho}{2} S$$

L = Lift

C_L = Coefficient of lift
(This dimensionless number is the ratio of lift pressure to dynamic pressure and area. It is specific to a particular airfoil shape, and, below the stall, it is proportional to angle of attack.)

V = Velocity (feet per second)

ρ = Air density (slugs per cubic foot)

S = Wing surface area (square feet)

Figure 3-3. *Lift Equation.*

The lift equation shows that total lift changes as the factors on the right side of the equation change. For example, lift varies directly with the coefficient of lift. A pilot should understand the concept of angle of attack (AOA) to understand how the coefficient of lift varies. The angle of attack is the acute angle between the chord line of the wing and the relative wind developed by the motion of the glider through the air. [*Figure 3-4*] The coefficient of lift increases linearly until reaching a critical angle, at which point lift decreases even though the angle of attack increases. Once reaching the critical angle, any further increase in the angle of attack disturbs the smooth airflow over the top of the wing and causes a decrease in lift or a stall. Lift also varies with the square of velocity or airspeed. Doubling airspeed quadruples the amount of lift. As air density decreases with increasing altitude or rising temperature, lift decreases.

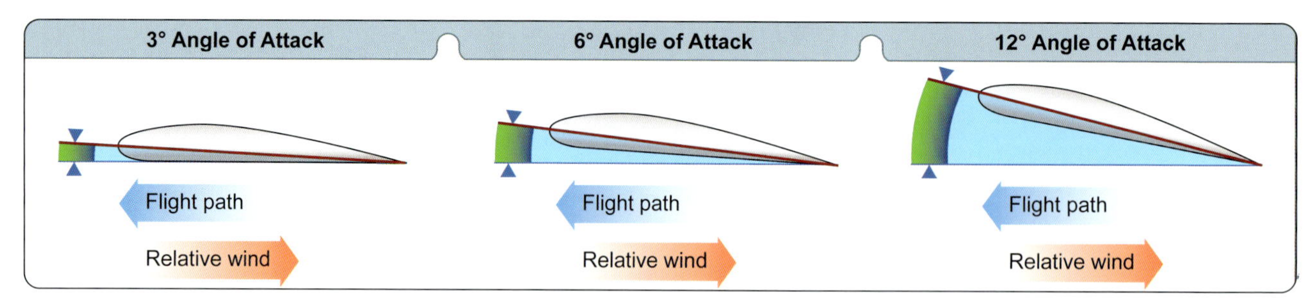

Figure 3-4. *Angle of attack.*

When describing lift as a vector, the total lift acts through a point known as the center of lift (CL). This point occurs aft of the center of gravity. Therefore, lift creates a pitching moment on the glider, which the tail must counteract. The total lift vector acts perpendicular to the flightpath through the CL and perpendicular to the lateral axis.

Drag

The force that resists the movement of the glider consists of parasite and induced drag, which combine to form total drag.

Parasite Drag

Parasite drag includes the resistance of the air to any object moving through it created by skin friction, the shape of the object, and interference patterns within the airflow around the object. While glider wing designs generate minimal induced and parasite drag, other parts of the glider may create significant parasite drag. Parasite drag increases with the square of speed. Simply put, if the speed of the glider doubles, parasite drag increases four times. [*Figure 3-5*]

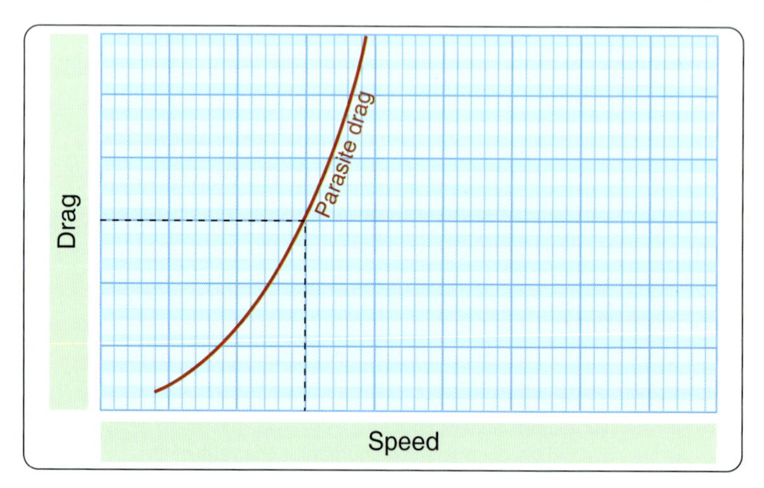

Figure 3-5. *Parasite drag versus speed.*

Induced Drag

As the angle of attack increases, more air flows around the wingtip from the lower to the upper surface, which creates larger wingtip vortices. [*Figure 3-6*]. Panel 4 of *Figure 3-6*, depicts the wing moving horizontally in level flight. The vertical lift vector develops perpendicular to the flight path and oncoming relative wind. Since wingtip vortices cause a downward divergence of the average relative wind from the flightpath, the total lift vector, which is perpendicular to the average relative wind, tilts back. This backward tilt of the total lift vector creates induced drag as a byproduct of lift. Factors that increase the angle between the total lift and vertical lift vectors, such as low airspeed and high angle of attack, increase induced drag.

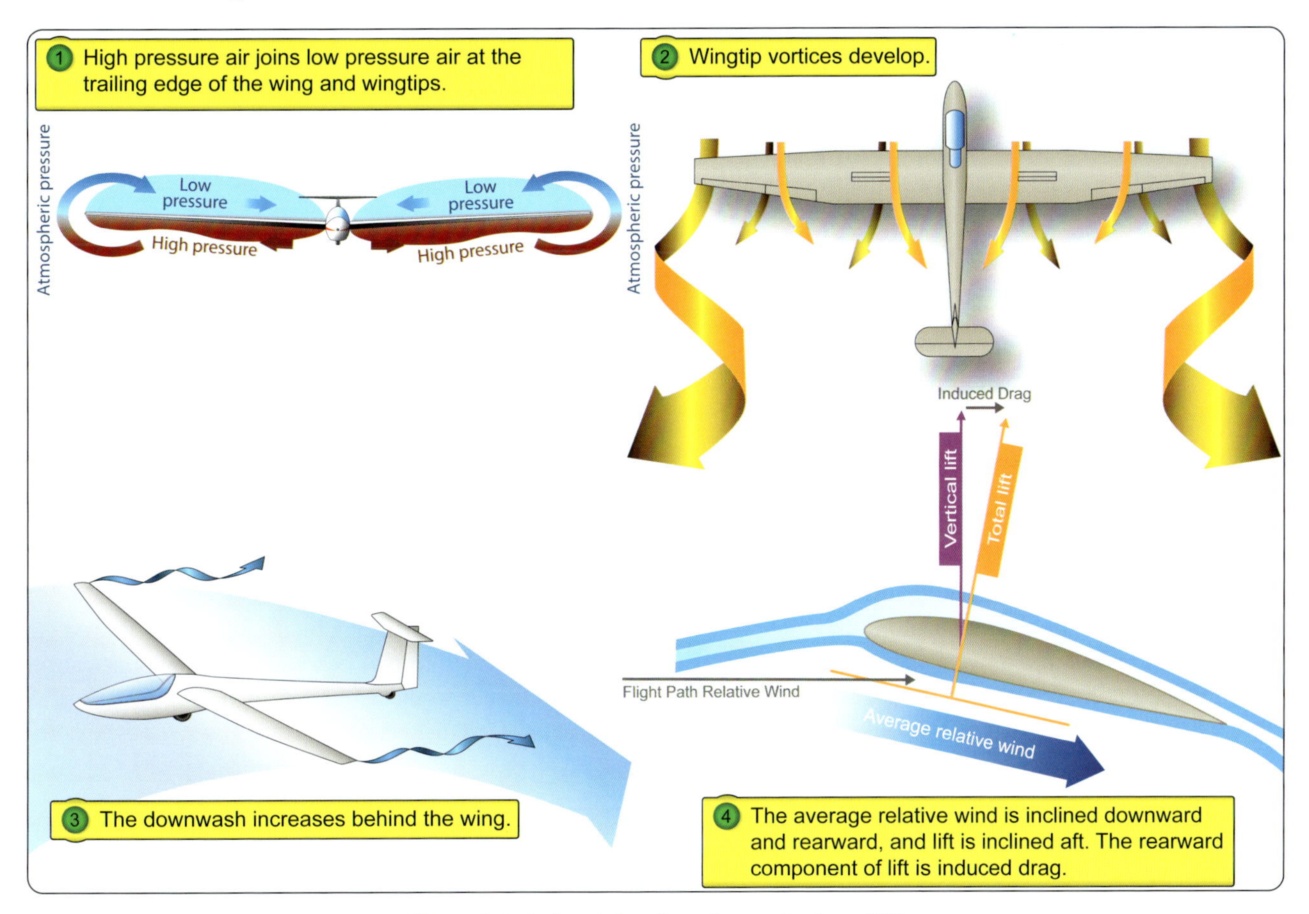

Figure 3-6. *Induced drag from the production of lift.*

As a glider flies faster, the wings can generate the same amount of lift with a reduced angle of attack. Increased speed with a smaller angle of attack reduces the backward slant of the total lift vector and reduces induced drag.

Total Drag

The total drag curve represents the combination of parasite and induced drag and varies with airspeed. [*Figure 3-7*]

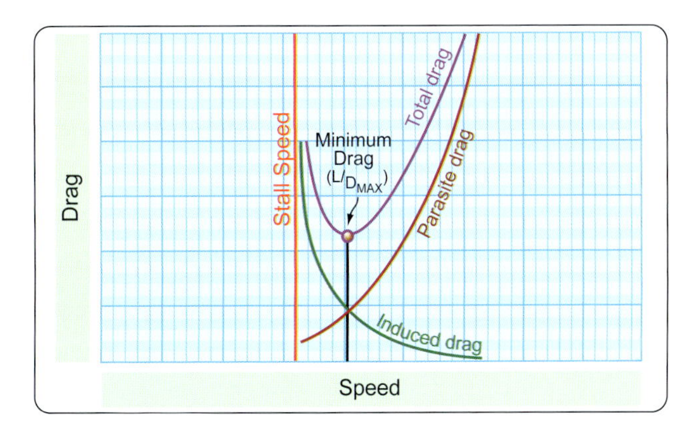

Figure 3-7. *Parasite drag, induced drag, and total drag versus airspeed.*

Ground Effect

Operating within one wingspan above the ground modifies the three-dimensional airflow pattern around the glider. This ground effect decreases downwash, reduces the size and effect of wingtip vortices, reduces induced drag, and results in more efficient flight. This effect allows the glider to fly at a lower airspeed during takeoff and reduces the sink rate of a glider on landing.

Weight

Weight results from the force of gravity acting on the mass of the glider. The vertical components of both lift and drag act in opposition to the weight vector, which acts vertically downward through the center of gravity.

Thrust

Unpowered gliders use an outside source, such as a tow plane, winch, or vehicle, to launch. Once released, these gliders maintain forward motion from conversion of potential energy to kinetic energy. A glider descends through the surrounding air to make this conversion. Powered gliders have engines, which can provide thrust for launch or to sustain flight.

Unpowered Glide Vector Analysis

What propels an unpowered glider in a continuous descent in the surrounding airmass? A vector diagram with the flight path as one axis and a second axis perpendicular to the flight path depicts how the forces of weight, lift, and drag balance each other during an unpowered straight-line descent. [*Figure 3-8*] Weight (W) always points to the ground (center of the Earth). Lift (L) develops perpendicular to the flight path. Drag always acts backward along the flight path. While weight does not align with either axis on the diagram, the weight vector can resolve into two perpendicular components, one forward along the flight path opposing drag (Wf) and the other perpendicular to the flight path opposing lift (Wp). As shown in the figure, Wp balances lift, while Wf balances drag during an unaccelerated descent. Thus, gravity is the external engine that pulls the glider forward by acting on Wf.

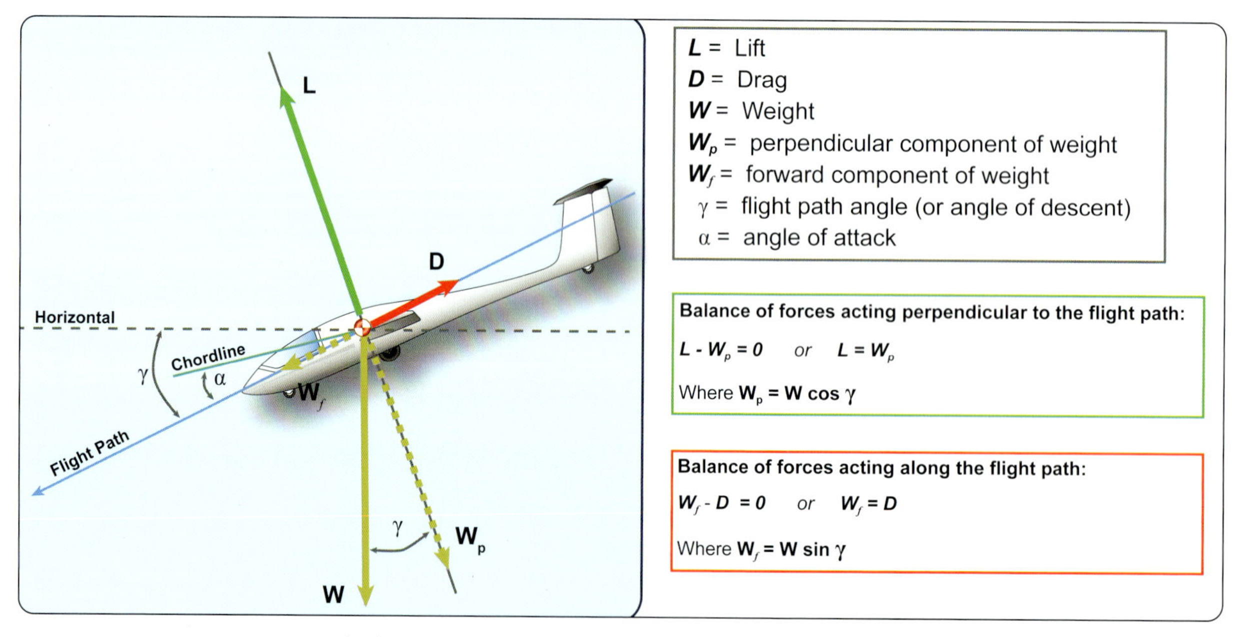

Figure 3-8. *The forces along an unpowered glider's flight path and its perpendicular.*

An unpowered descent converts the glider's potential energy of height above the ground into kinetic energy of motion on a continuous basis. The flight path angle or angle of descent (γ) is the same as the angle between Wp and W. Wf, Wp, and W could form three sides of a right triangle. Trigonometry gives the Wf and Wp components of weight using the formulas shown within the balance of forces boxes in *Figure 3-8*. A steeper flight path angle (γ) increases the forward component of weight (Wf) and decreases the perpendicular component of weight (Wp).

Glide Ratio & Wing Design

One specific point appears in *Figure 3-7* above. The point displayed, (L/D_{MAX}), corresponds to a speed where the total lift capacity of the glider, when compared to the total drag reaches a maximum value. In calm air, this speed yields maximum glide distance and the published glide ratio for a glider. The glide ratio gives the distance the glider can travel during a given descent in altitude. For example, a glide ratio of 50:1 means a glider could travel 50 feet forward while losing one foot of altitude.

Wing Planform

The shape (planform) of the wings affects the amount of lift and drag produced. The four most common wing planforms used on gliders are elliptical, rectangular, tapered, and swept forward. [*Figure 3-9*]

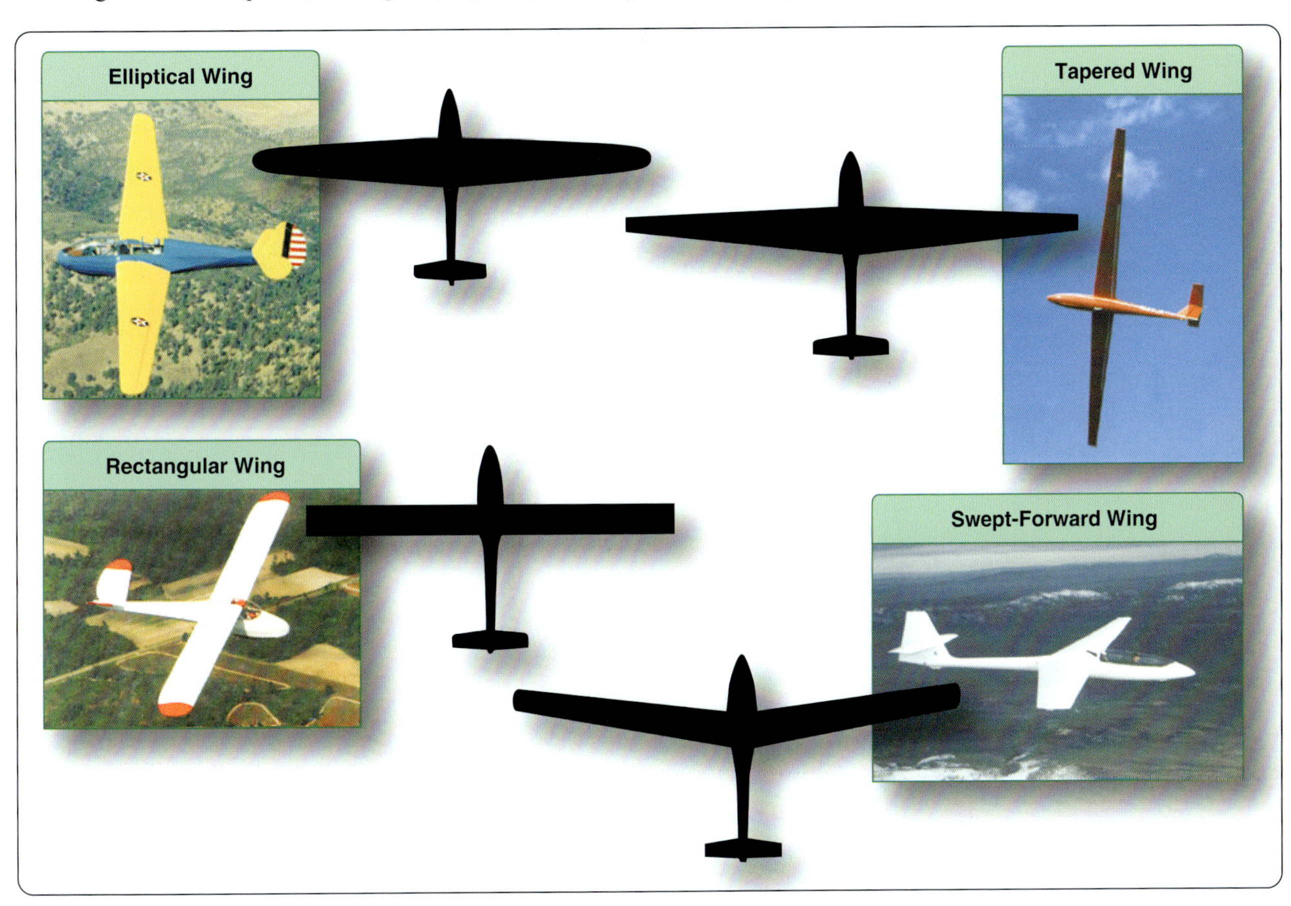

Figure 3-9. *Planforms of glider wings*

Aspect Ratio

Dividing the wingspan (from wingtip to wingtip) by the average wing chord determines the aspect ratio for a glider. Glider wings have a high aspect ratio, as shown in *Figure 3-10*, which generates significant lift at low angles of attack with minimal induced drag.

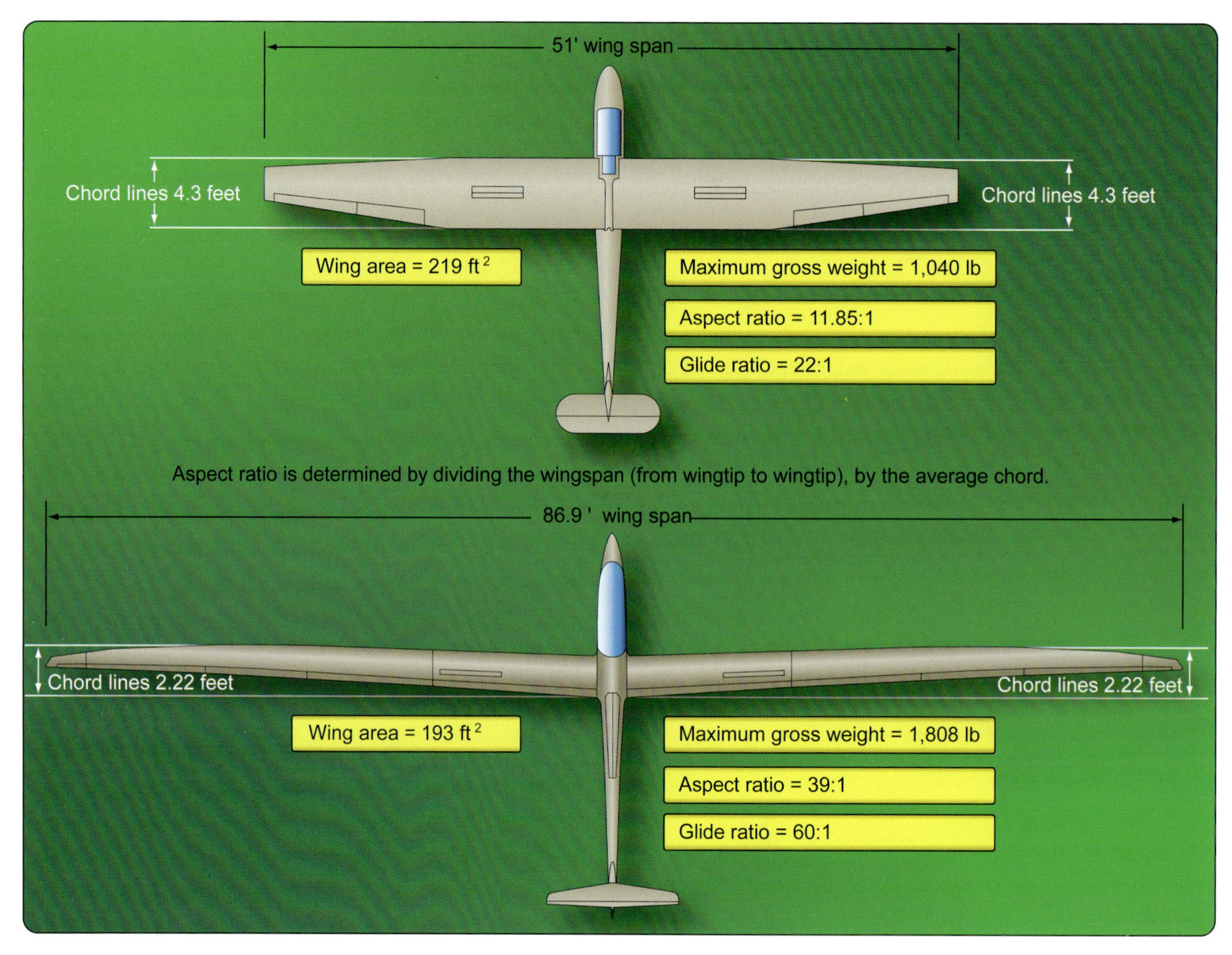

Figure 3-10. *Aspect ratio*

Winglets

Wingtip devices, or winglets, also improve efficiency of the glider by altering the airflow near the wingtips and reducing induced drag.

Washout

The wing root refers to the portion of the wing nearest the fuselage. Washout refers to a slight wing twist between the wing root and wingtip, which causes the wing root to have a greater angle of attack (AOA) than the wing tip. If the AOA becomes excessive, airflow will separate at the wing root before separation occurs at the wing tip. This wing design provides warning of any impending stall or overall separation of air from the wing and allows for continued aileron control at the onset of a stall.

Stability

Vertical gusts, a sudden shift in CG, or deflection of the controls by the pilot can displace the glider from its orientation in flight. Static and dynamic stability define how the glider reacts after a displacement. Static stability describes the initial direction of the response. A glider with positive static stability initially moves back toward its original orientation after a change. A glider with negative static stability would increase displacement after a change. A glider with neutral static stability tends to hold any new orientation. The level of stability about each axis of a glider results from its design and loading.

A glider with positive static stability will swing past its original pitch attitude and undergo a series of oscillations. If that glider has positive dynamic stability, the size of any oscillations will dampen out over time. The same glider with negative dynamic stability would experience oscillations that increase in amplitude over time. That glider with neutral dynamic stability would experience a series of constant oscillations over time. [*Figure 3-11*]

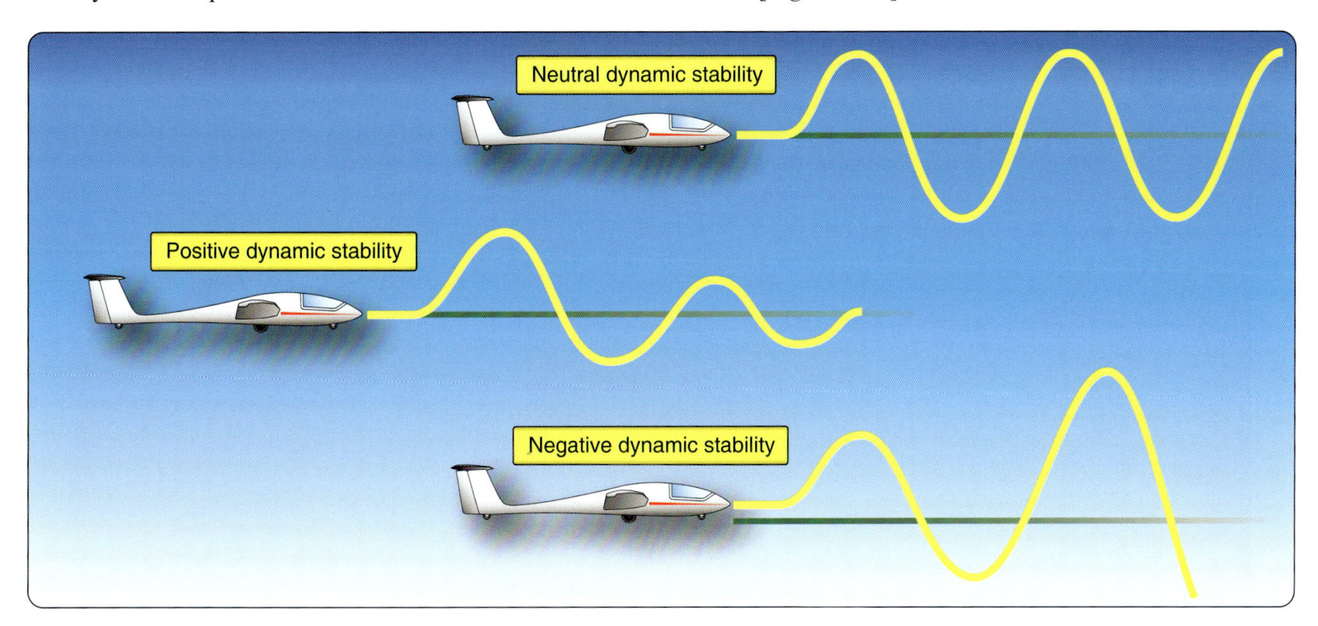

Figure 3-11. *Three types of dynamic stability.*

Gliders have a CG in front of the center of lift—the single point through which the sum of all lift acts. This design requires an opposing tail-down force to maintain control and to create positive static stability for pitch. [Figure 3- 12] For example, if a pilot displaces a glider's nose upward and releases the controls, the glider will lose airspeed. The resulting reduction in down force provided by the tail causes the glider's nose to drop back toward its original position. This design results in positive initial longitudinal stability (stability around the lateral axis). When a positive dynamically stable glider oscillates in pitch, the amplitude of the oscillations diminishes through each cycle and eventually stops at the speed where downward force on the tail offsets the tendency to nose down.

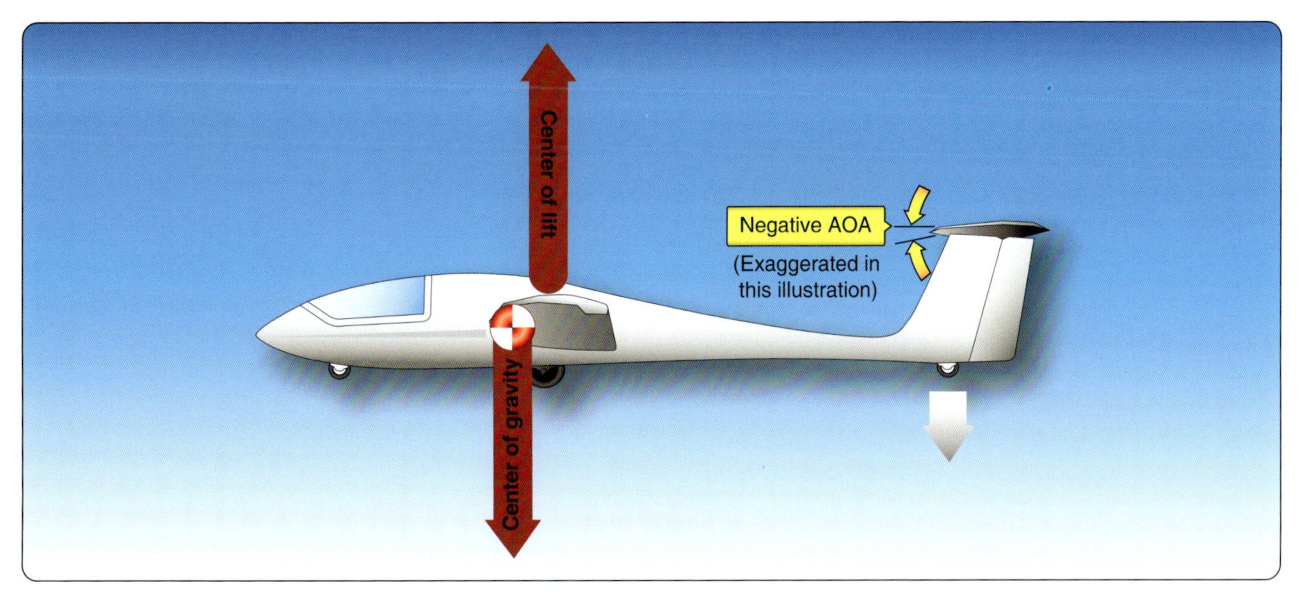

Figure 3-12. *The horizontal stabilizer offsets the natural tendency of a glider to pitch down.*

The glider POH lists an acceptable range for the CG, and the pilot normally computes the CG location before flight to verify the glider will fly as designed. A glider with an aft CG requires less tail-down force, which makes pitch oscillations more difficult to dampen. As airspeed decreases in a glider loaded aft of the permissible CG limit, the nose of the glider might rise uncontrollably and lead to an unrecoverable stall or spin. A forward CG increases induced drag and reduces performance. A glider with a CG ahead of the published forward limit may not provide enough pitch control to raise the nose during a landing. Chapter 5, Glider Performance, contains further discussion of proper loading of a glider and the importance of CG.

Lateral Stability

Lateral stability describes the glider's tendency to return to wings-level flight following a displacement. [*Figure 3-13*] For example, due to a gust of wind, the glider may start to roll. The angle of attack increases slightly on one wing as it moves down, which causes the lift to increase on that wing. On the rising wing, the opposite effect decreases lift. This lift differential tends to dampen the rolling motion but does not bring the glider back to wings-level.

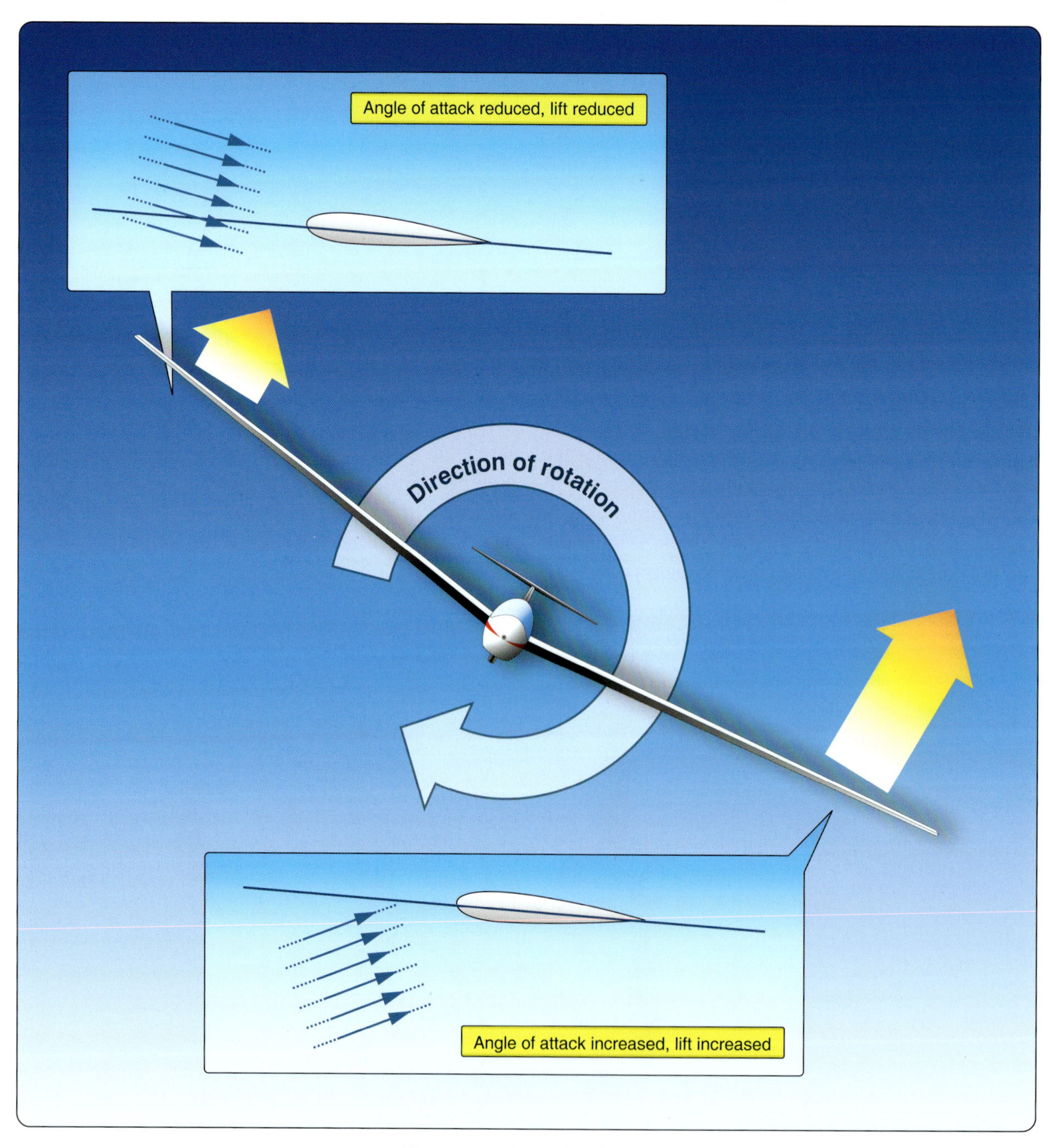

Figure 3-13. *Lateral stability.*

Wing dihedral, the upward angle of the wings from horizontal, adds to lateral stability. As the glider rolls, the descending wing has a larger vertical lift component than the rising wing. This difference in the vertical lift component between each wing tends to roll the glider back toward level flight. [*Figure 3-14*]

Figure 3-14. *Dihedral angle.*

Flutter

Due to the flexibility of a glider wing design, the wing may oscillate rapidly or flutter at high airspeeds. Additionally, looseness in the control surfaces can also result in flutter of flight control surfaces near maximum speed. Improper balance of the control surfaces may also cause flutter. Since prolonged flutter may cause structural failure, the pilot should reduce the airspeed sufficiently to stop any flutter.

Pilot Induced Oscillation (PIO)

Pilot-induced oscillations (PIOs) can occur when the pilot applies excess control pressure that causes an overshoot of the desired flight attitude. If the pilot repeatedly moves the controls back and forth using excess pressure, the glider oscillates continuously past the desired attitude. The oscillations can increase in amplitude and may cause loss of control. While PIOs most likely occur as pitch oscillations, roll and yaw induced PIOs may also occur.

Although PIOs can occur at any time, they often arise during primary training. Experienced pilots may also induce PIOs when flying an unfamiliar make and model glider. For that reason and before flying an unfamiliar glider, all pilots should review and understand the flight characteristics of that glider.

If encountering PIOs, the pilot should remember that changes in flight attitude take time. The pilot should begin to ease flight control pressure as the glider begins to respond in the desired direction. As the glider nears the desired attitude, the pilot centers the appropriate flight control so that overshooting does not occur.

During the first moments of the takeoff roll, as airflow begins to impact the control surfaces, it takes considerable displacement of the flight controls to affect the glider's flightpath. The pilot also experiences a higher control lag time due to reduced control effectiveness at low speed. As the glider accelerates, aerodynamic response improves, lag time decreases, and PIOs become less likely.

Turning Flight

Any moving object continues in a straight line until a force causes a change in direction. A pilot creates turning force by using the ailerons to roll the glider so that the direction of the total lift vector inclines. When a glider rolls away from wings-level, lift divides into two components. The vertical component opposes weight, while the other acts horizontally to oppose centrifugal force. [*Figure 3-15*]

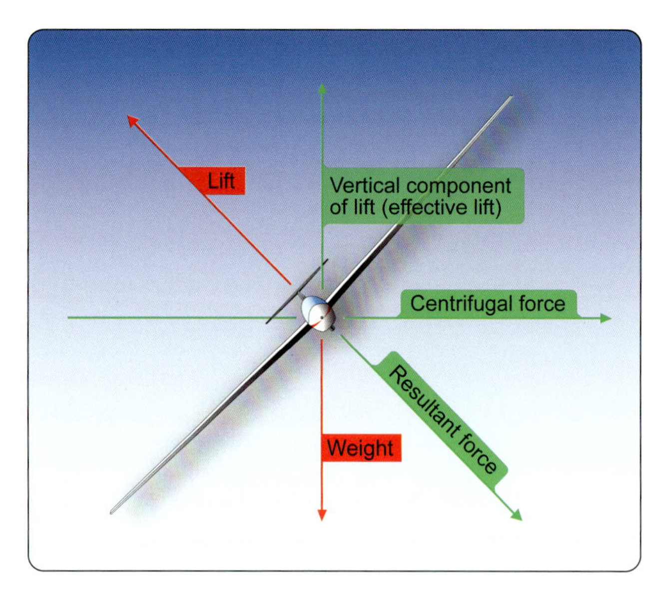

Figure 3-15. *Forces in a banked turn.*

When using the horizontal component of lift to turn the glider, a reduction in the vertical component of lift will occur unless the pilot increases back pressure on the control stick, increasing the angle of attack and total lift produced by the wings. Increasing total lift can restore the vertical component of lift to that required to counteract weight. Otherwise, the descent rate in the airmass increases.

Load Factor

Any force applied to a glider to deflect its flight from a straight line produces a stress on its structure; load factor describes this force. The load factor is the ratio of the total air load acting on the glider to the weight of the glider. Load factor may be positive or negative and depends on the current flightpath. Load factor units often use "G," a gravitational force equivalent. A load factor of one, or 1 G, represents conditions in which the lift is equal to the weight. A glider in flight with a load factor of one does not necessarily mean the glider is in straight-and-level flight, but rather that the total lift equals that of unaccelerated straight-and-level flight.

When subjecting a glider to added Gs in a pull up from a dive, anyone in the glider feels a sensation of pressing into the seat with a force equal to the number of Gs times the person's weight. In addition, the person's extremities require added muscular force to resist the downward force. Added Gs can affect blood flow to the brain, affect cognitive ability, and cause disorientation.

Load factor increases rapidly as the angle of bank increases during a turn when the pilot increases lift to prevent a change in vertical speed. From an aerodynamic perspective, load factors concern the pilot for two distinct reasons:

1. Below a certain airspeed, a stall occurs before the pilot can create a dangerous overload on the glider structure.

2. Above a certain airspeed, known as the maneuvering speed or V*A, a pilot can generate enough lift to create a dangerous overload on the glider structure.

In a turn at constant speed, the pilot pulls back on the stick to furnish the extra lift necessary to maintain a constant vertical speed in the airmass. The load and stall speed increase significantly as bank angles exceed 30 degrees. [*Figure 3-16*] and [*Figure 3-17*]

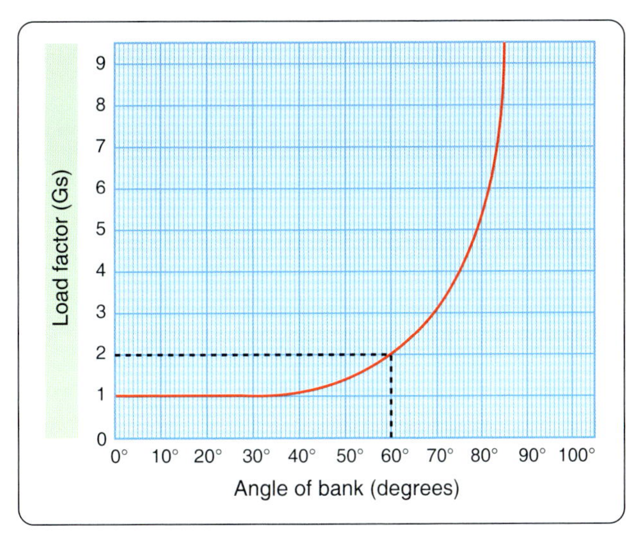

Figure 3-16. *The loads placed on a glider while maintaining a constant rate of descent with respect to the surrounding air increase as the angle of bank increases.*

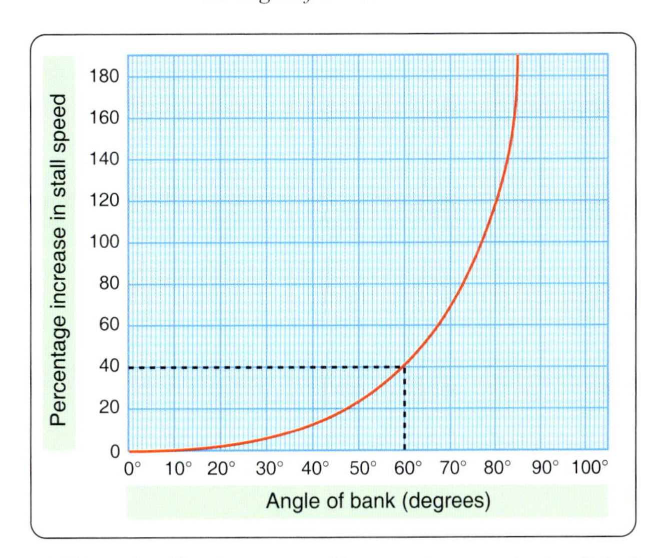

Figure 3-17. *A 60° angle of bank causes a 41 percent increase in the glider's stall speed.*

Rate of Turn

Rate of turn refers to the amount of time it takes for a glider to turn a specified number of degrees. If flown at the same airspeed and angle of bank, every glider turns at the same rate. If airspeed increases while the angle of bank remains the same, the rate of turn decreases. Conversely, a constant airspeed coupled with an increased angle of bank results in an increased rate of turn.

Radius of Turn

The horizontal distance an aircraft uses to complete a turn depends upon the radius of turn. The radius of turn at any given bank angle varies directly with the square of the airspeed. Therefore, if the airspeed of the glider were doubled, the radius of the turn would be four times greater. The radius of turn also depends on a glider's angle of bank. If the angle of bank increases and the airspeed remains the same, the radius of turn decreases.[*Figure 3-18*] When flying in thermals, a smaller turn radius enables a glider to fly closer to the fastest rising core of the thermal and gain altitude more quickly.

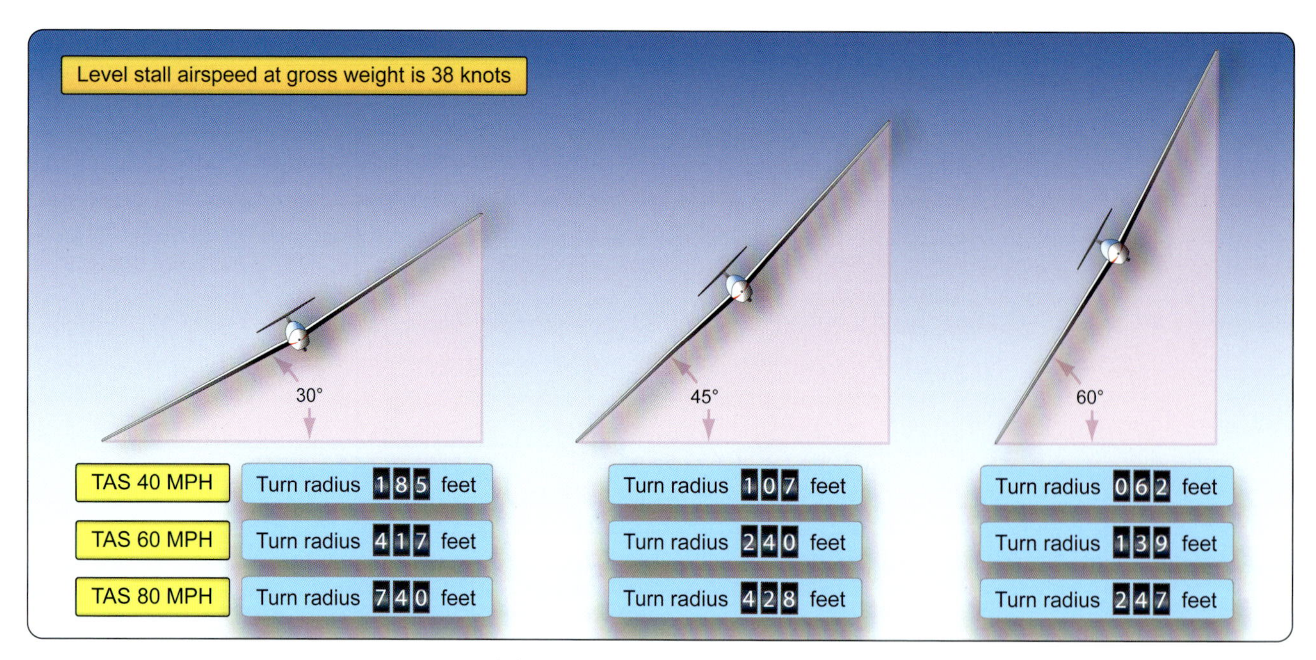

Figure 3-18. *A glider's radius of turn as compared to angle of bank.*

Turn Coordination

When rolling into a turn, the lowered aileron on the outside wing produces more lift for that wing. Since induced drag is a byproduct of lift, the outside wing also experiences more drag than the inside wing. This causes adverse yaw, a yawing tendency toward the outside of the turn. Appropriate use of rudder corrects for any adverse yaw caused by aileron drag. Because glider wings provide a long lever arm for adverse yaw, the pilot may need to use substantial rudder pressure during a coordinated turn. The amount of adverse yaw may surprise a pilot transitioning from an airplane, and the amount of adverse yaw will be much greater than experienced in a typical airplane.

Slips

While uncoordinated flight decreases performance, pilots can use different slipping techniques to steepen the descent angle or to counteract a crosswind during landing.

Because of the location of the pitot tube and static vents, airspeed indicators in some gliders may have considerable error when the pilot places the glider in a slip. The pilot should recognize this error and know how to perform slips based on secondary indications such as the attitude of the glider, the sound of the airflow, and the feel of the flight controls.

A pilot normally coordinates rudder and aileron inputs during a turn. Using too little rudder, or if rudder is applied too late, results in a slip. Too much rudder, or rudder applied before aileron, results in a skid. Both skids and slips expose a side of the fuselage to the relative wind, creating additional parasite drag.

Forward Slip

During a forward slip the glider's horizontal path over the ground remains unchanged. [*Figure 3-19*] This slip uses uncoordinated flight to dissipate energy which increases rate of descent without increasing the glider's forward speed. Pilots sometimes use forward slips during a landing approach over obstacles or for short-field landings when necessary to dissipate altitude no longer needed for a margin of safety during the approach.

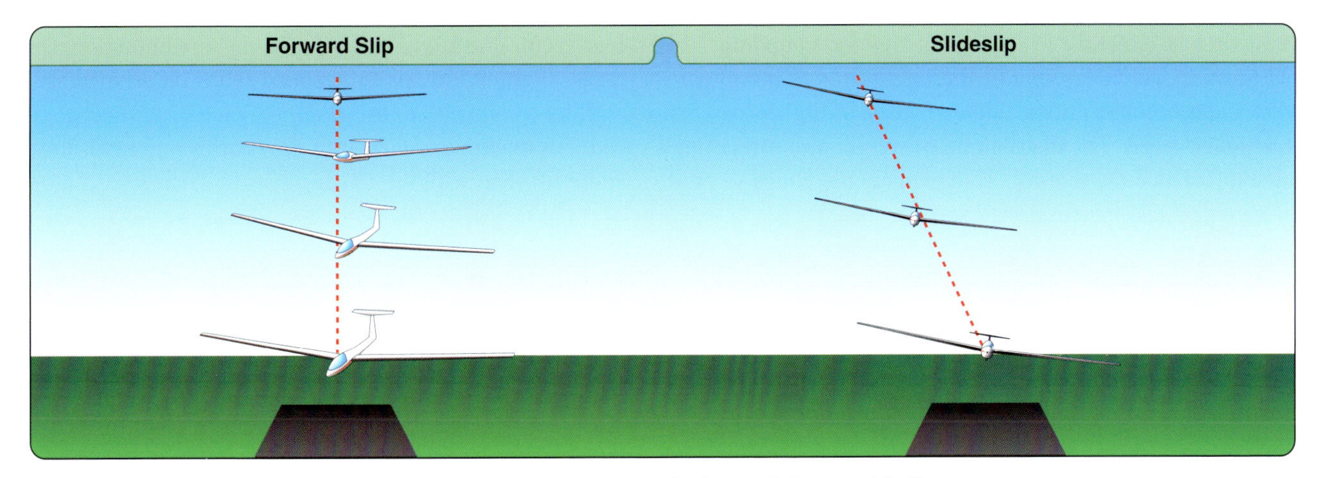

Figure 3-19. *A comparison of a forward slip to a sideslip.*

To enter a slip from straight flight, the pilot lowers the wing on the side toward which the slip occurs using the ailerons. Simultaneously, the pilot yaws the aircraft's nose in the opposite direction by applying enough opposite rudder so that the glider's longitudinal axis no longer aligns with its flightpath. The pilot should yaw the glider such that it maintains the original ground track and raises the nose sufficiently to prevent the airspeed from increasing. In crosswinds, initiating a slip by lowering the wing on windward side of the glider provides more stable path control. The pilot discontinues a slip by leveling the wings and by smoothly and simultaneously releasing the rudder pressure, while readjusting the pitch attitude for a normal glide.

Note: Forward slips with wing flaps extended should not occur if the manufacturer's operating instructions prohibit such operation.

Sideslip

During crosswind landings, a sideslip can counteract wind drift and allows the glider to touch down with its longitudinal axis parallel to the direction of motion. The pilot uses rudder pressure to keep the glider's longitudinal axis parallel to the desired ground track, but the path over the ground can change depending on the amount of bank. To perform a sideslip, the pilot lowers the upwind wing and simultaneously applies sufficient opposite rudder to maintain the nose alignment.

Stalls

A stall occurs whenever the angle between the chord line and relative wind exceeds the critical AOA. [*Figure 3-20*] A stall results in a reduction in lift, although the wings still support some of the aircraft's weight during a stall. Stalls may occur at any airspeed and in any flight attitude.

Figure 3-20. *A stall occurs when the angle of attack exceeds the critical angle of attack.*

Many factors affect the stall speed of a glider, including weight, load factor due to maneuvering, wing contamination, and CG location. As the weight or the load factor of the glider increases, flight at any given airspeed relies on an increased AOA—closer to the critical angle of attack. A higher load factor or glider weight causes the glider to reach the critical angle at a higher speed. The distribution of weight also affects stall speed. For example, a forward CG requires more tail-down force to balance the aircraft. This requires the wings to produce more lift than with the CG further aft. Therefore, a more forward CG also increases stall speed.

Environmental factors also affect stall speed. Snow, ice, or frost accumulation on wing surfaces can increase the weight of the wing and disrupt airflow, both of which increase stall speed. Turbulence has an impact on the stall speed of a glider because the vertical gusts change the direction of the relative wind and abruptly increase the AOA. During landing in gusty conditions, pilots normally increase the approach airspeed by half of the difference between the steady wind and gust value to maintain a safe margin above stall. For example, if the winds were 10 knots gusting to 16 knots, it would be prudent to add 3 knots ((16 – 10) ÷ 2 = 3) to the approach speed.

Spins

A spin develops from an aggravated stall that results in the glider descending in a helical or corkscrew path. A spin may develop as a complex, uncoordinated flight maneuver in which one wing becomes more stalled than the other. Upon entering a spin, the more completely stalled wing usually drops before the other, and the nose of the aircraft yaws in the direction of the low wing. In this spin scenario, the ascending wing experiences more lift and less drag. [*Figure 3-21*] The opposite wing moves down and back due to less lift and increased drag.

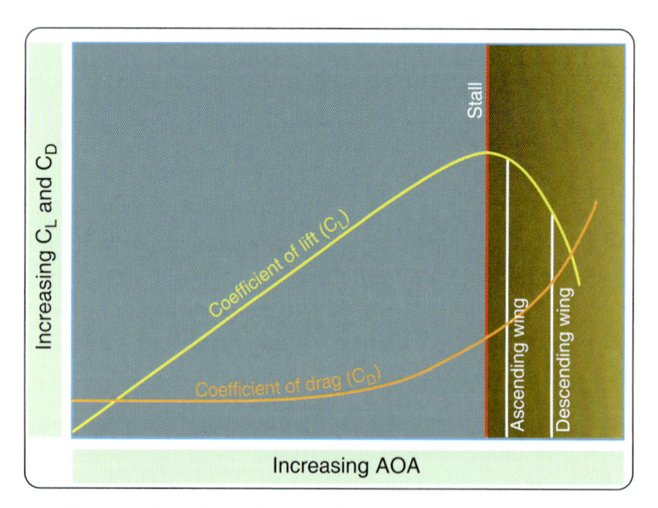

Figure 3-21. *The relative coefficients of lift and drag for each wing during a spin, which generate differential lift and drag and induces roll and yaw.*

Spins may occur after a glider stalls in uncoordinated flight with unequal airflow over the wings. Any resultant spin usually occurs in the direction of rudder application. The entry, wing form, and CG usually determine the type of spin that results from an uncoordinated wing stall. Glider pilots should understand the stall characteristics for any glider flown, and spin recovery techniques as described in either the GFM, if applicable, or in Chapter 8, Abnormal and Emergency Procedures.

Spin classification includes three categories, as shown in *Figure 3-22*. The entry, wing form, and CG usually determine the type of spin that results from an uncoordinated wing stall. The most common type of spin is the upright or erect spin, which maintains a slightly nose-down rolling and yawing motion in the same direction. A second type of spin, an inverted spin, involves the aircraft spinning upside down with the yaw and roll occurring in opposite directions. In a third type of spin, the flat spin, the glider yaws around the vertical axis at a pitch attitude nearly level with the horizon. A flat spin often has a very high rate of rotation with a difficult or impossible recovery. A properly loaded glider should not enter a flat spin. Flat spins can also be inverted.

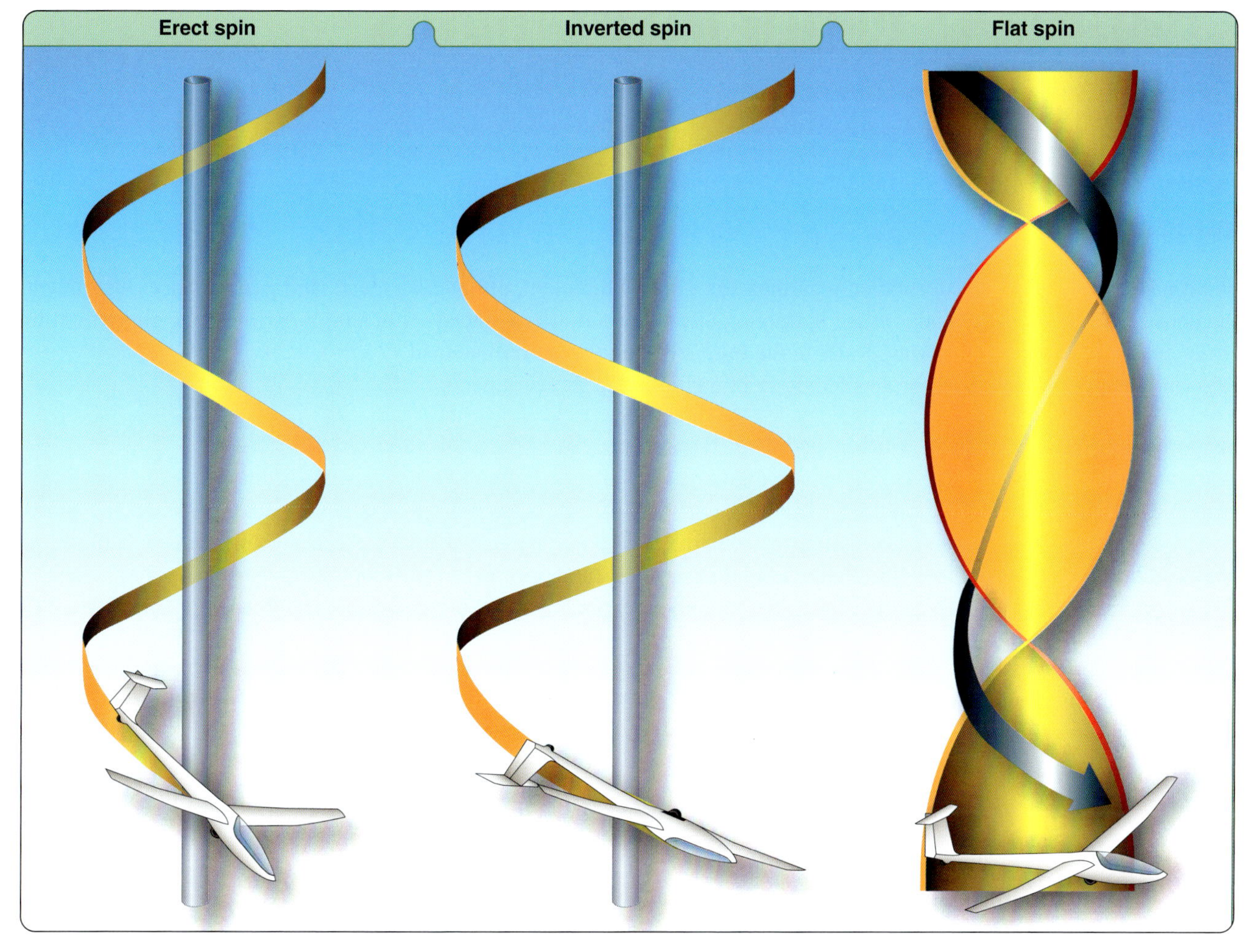

Figure 3-22. *Three types of spins.*

A stall and spin entry near the ground may prove fatal because a pilot may not have adequate altitude to recover. Therefore, pilots should avoid errors that could lead to a stall/spin accident. For example, during the approach and landing phase with a tailwind on the base leg, a pilot might try to tighten the turn to final using rudder, or to make a steep turn to prevent overshooting the final approach course. A skidding turn could lead to the lower wing exceeding its critical AOA before the upper wing and could result in a spin. An excessively steep turn could also result in an accelerated stall and spin.

Chapter Summary

This chapter focuses on the four forces of flight, which include lift, drag, thrust, and weight. Components of one force can contribute to another. When lift offsets gravity and thrust offsets drag, unaccelerated flight occurs. The chapter also introduces the concept of ground effect, which results from a reduction in induced drag near the ground. A glider has 3 axes of rotation and moves about each axis because of natural disturbances and pilot input. Roll occurs around the longitudinal axis, pitch around the lateral axis, and yaw around the vertical axis. Stability refers to the ability of the aircraft to return to its original path and orientation after an upset. Pilots unfamiliar with the effects of control inputs and stability can induce oscillations. This chapter covers turning flight and the effect of forces on rate of turn, radius of turn, turn coordination, and load factors. The chapter also discusses stalls and spins.

Chapter 4: Flight Instruments

Introduction

Flight instruments provide information regarding the glider's direction, altitude, airspeed, and performance. Instruments can consist of a basic set typically found in training aircraft or a more advanced set in a high-performance glider used for cross-country or competition flying. Refer to the Pilot's Handbook of Aeronautical Knowledge (FAA-H-8083-25) for detailed descriptions of different instruments.

Instruments displaying airspeed, altitude, and vertical speed are part of the pitot-static system. Heading instruments display magnetic direction by sensing the earth's magnetic field. Performance instruments, using gyroscopic principles, display the aircraft's attitude, heading, and rate of turn. Electronic instruments using computer and global positioning system (GPS) satellite technology provide pilots with moving map displays, electronic airspeed and altitude, air mass conditions, and other functions relative to flight management. Examples of self-contained instruments and indicators that are useful to the pilot include the yaw string, inclinometer, and outside air temperature (OAT) gauge.

This chapter describes basic glider instruments and systems. Pilots flying gliders with advanced electronic instruments should consult the manufacturer's documentation for a complete description of operation and seek instruction as needed.

Pitot-Static Instruments

The pitot-static system uses two different air pressure measurements:

1. Static ports transport ambient atmospheric pressure to instruments through tubing.
2. The pitot tube transports ambient air pressure plus any ram air pressure resulting from forward motion to instruments through tubing.

Impact & Static Pressure Lines

Impact or ram-air pressure from the forward motion of the glider increases air pressure in the pitot tube, which mounts on either the nose or vertical stabilizer to allow uninterrupted exposure to the oncoming airflow. Glider static ports often mount either on the vertical stabilizer or the side of the fuselage. [*Figure 4-1*]

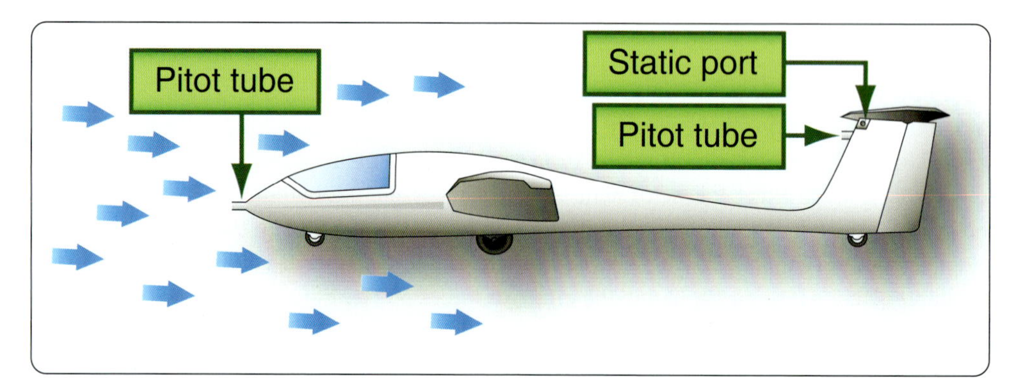

Figure 4-1. *A pitot tube often mounts in the glider's nose or the vertical fin, with the forward-facing open end exposed directly to the oncoming airflow.*

As airspeed increases, pressure builds in the pitot tube and in the connected expandable diaphragm. The pressure rises in the pitot tube and diaphragm until it reaches a state of equilibrium preventing any further rise in pressure. Conversely, pressure in the system decreases as airspeed decreases since air can also flow out of the system. [*Figure 4-2*]

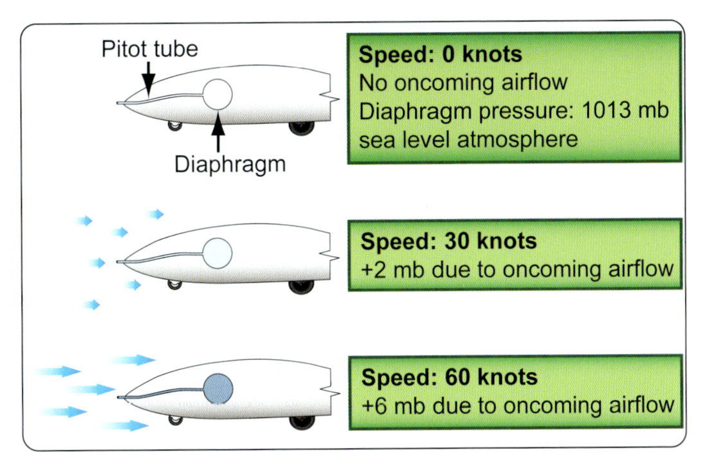

Figure 4-2. *Pressure inside the diaphragm as a function of airspeed.*

The static pressure (pressure of the still air) comes from the movement of air in and out of the static ports and tubing. Gliders using flush mounted static sources often have two vents, one on each side of the fuselage. This compensates for variation of static pressure due to changes in glider attitude and air turbulence.

Pilots check the openings of both the pitot tube and the static port(s) during the preflight inspection to ensure obstructions do not block the free flow of air. A certificated mechanic should clean any blockage. Blowing into these openings could damage flight instruments.

Airspeed Indicator

The airspeed indicator measures the difference between the pitot pressure and static pressure, and displays this difference as the indicated airspeed (IAS) of the glider. [*Figure 4-3*] Color-coded arcs depict airspeed ranges for different phases of flight. The upper and lower limits of the arcs correspond to defined airspeeds. *Figure 4-4* shows the internal structure of an airspeed indicator.

Figure 4-3. *Airspeed indicator.*

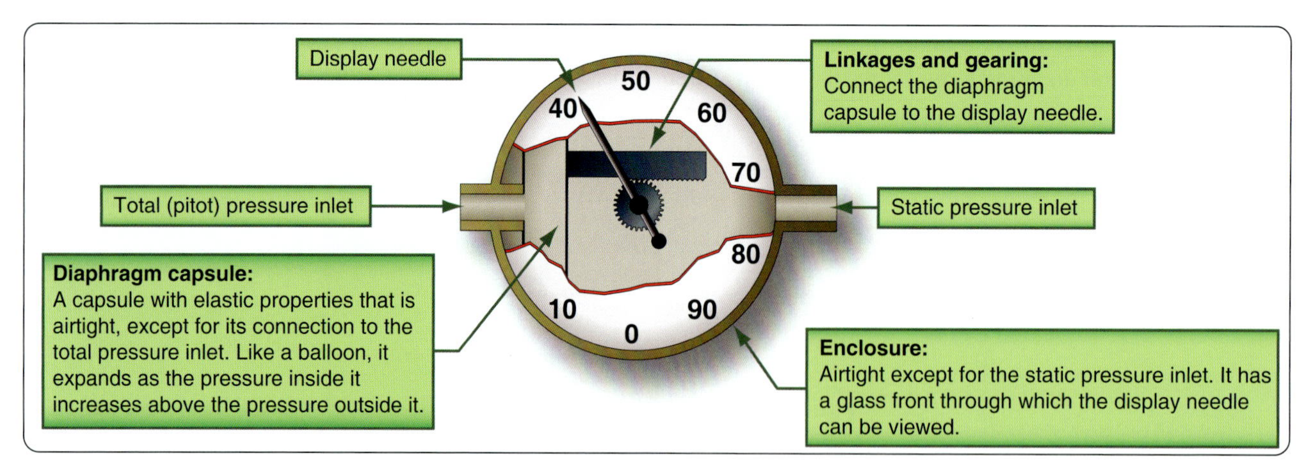

Figure 4-4. *Anatomy of the airspeed indicator.*

As shown in *Figure 4-4* above, the airspeed indicator contains a diaphragm that expands or contracts in response to the difference between pitot and static pressure. This diaphragm movement drives the needle (airspeed needle pointer) on the face of the instrument. When pitot pressure equals static pressure, the indicator reads zero. As pitot pressure becomes progressively greater than static pressure, the needle moves and points to the corresponding airspeed. [*Figure 4-5*]

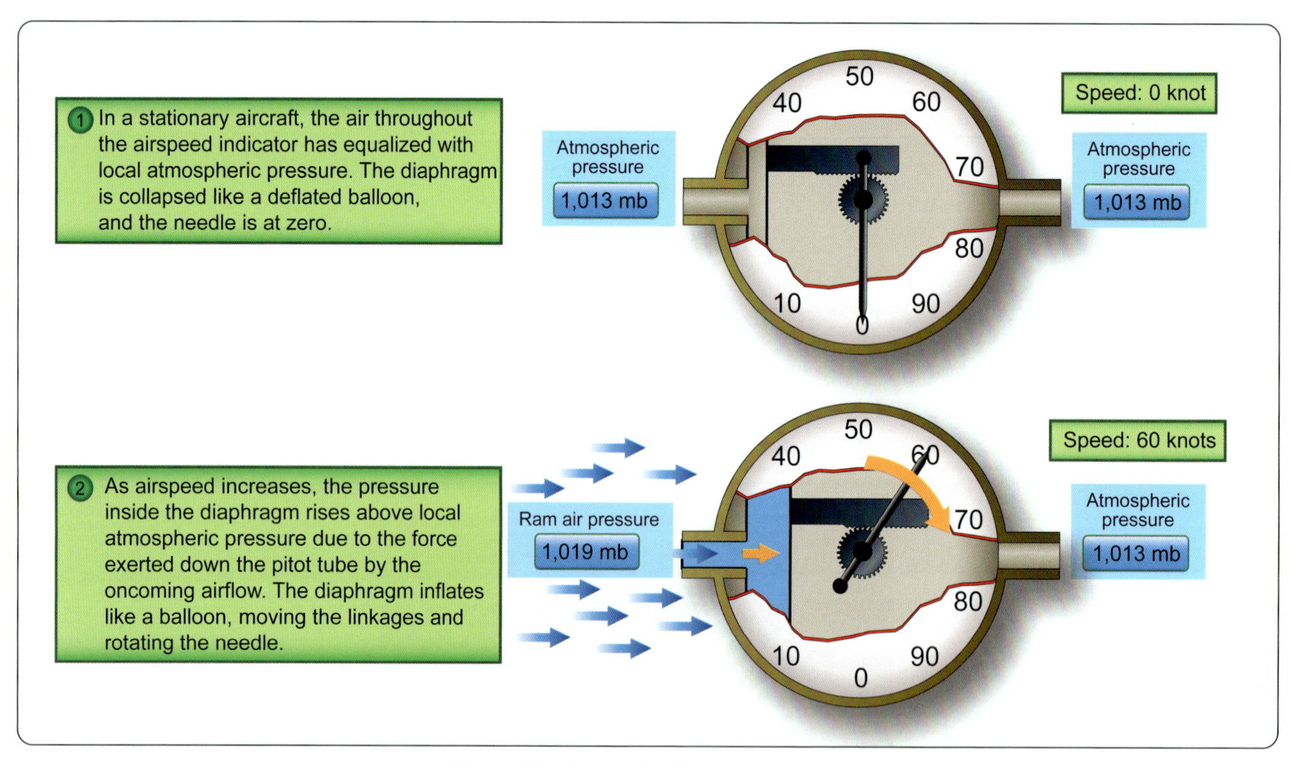

Figure 4-5. *Airspeed indicator operation.*

The Effects of Altitude on the Airspeed Indicator

The airspeed indicator displays dynamic pressure (pitot pressure minus static pressure) calibrated to an airspeed at standard sea-level pressure. Due to lower air density as altitude increases, the buildup of pressure in the diaphragm decreases and the indication becomes lower than at sea level. The error becomes greater as altitude increases. [*Figure 4-6*]

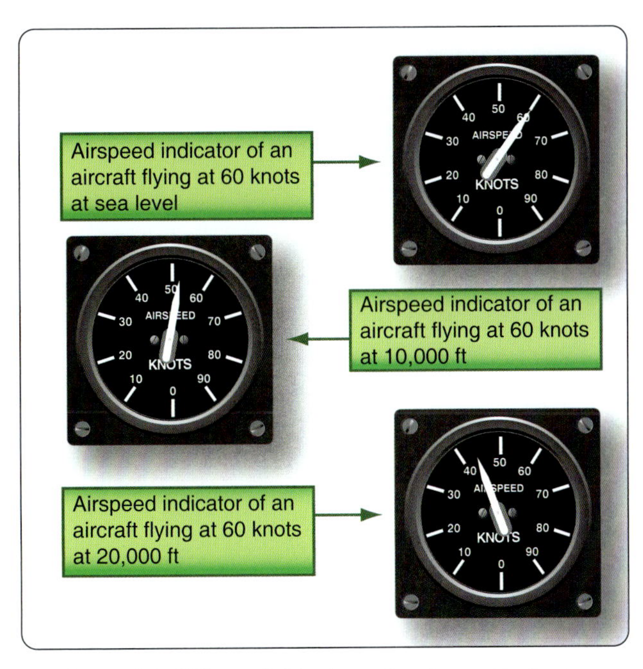

Figure 4-6. *Effects of altitude on the airspeed indicator.*

Types of Airspeed

Pilots work with various airspeed numbers including indicated airspeed (IAS), calibrated airspeed (CAS), and true airspeed (TAS). [*Figure 4-7*]

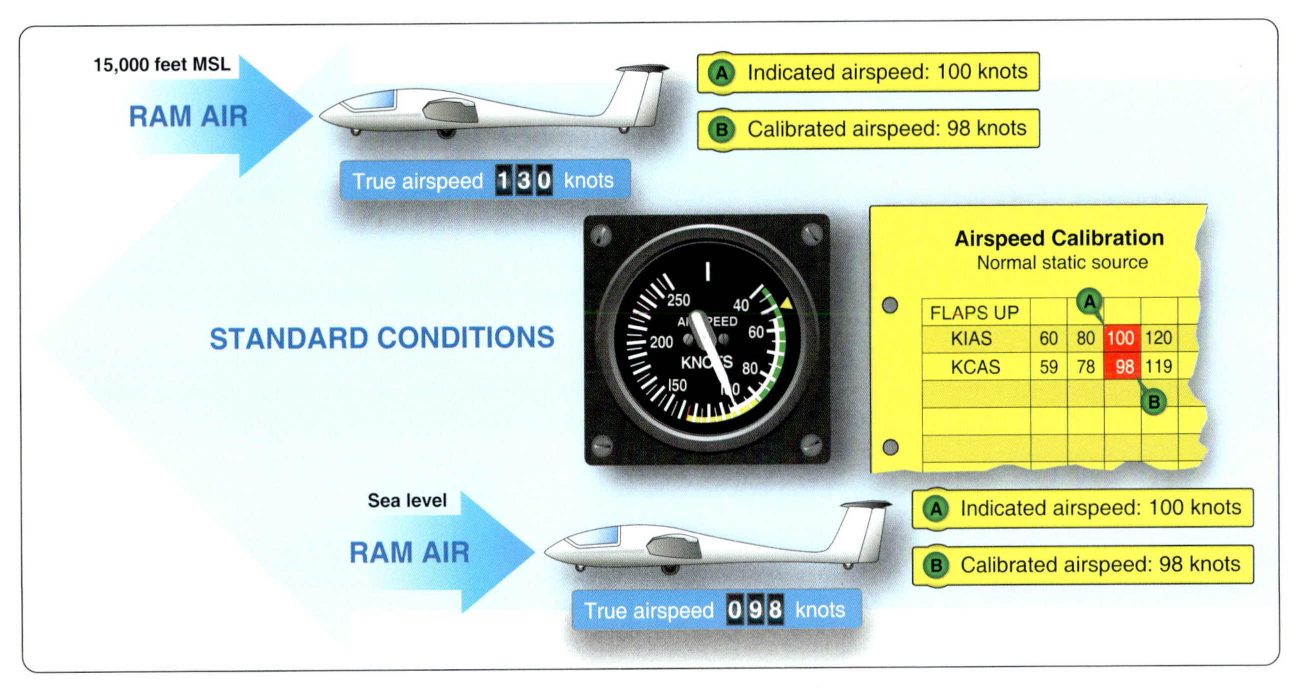

Figure 4-7. *Three types of airspeed.*

Indicated Airspeed (IAS)

The pilot reads the IAS directly from the airspeed indicator, uncorrected for installation error or instrument error. Although air density affects indicated airspeed, a particular glider in steady wings-level flight at a fixed weight stalls at a specific dynamic pressure (indicated airspeed) regardless of the air density. [*Figure 4-8*]

Altitude center vertically	IAS at which stall occurs in steady wings-level flight	True airspeed (TAS)
1,000 feet	40 knots	41 knots
10,000 feet	40 knots	48 knots
20,000 feet	40 knots	56 knots
30,000 feet	40 knots	64 knots
40,000 feet	40 knots	72 knots

Figure 4-8. *Indicated stall airspeeds and true airspeeds at various altitudes.*

Calibrated Airspeed (CAS)

CAS corrects IAS for installation and instrument errors. Significant errors may occur at low airspeeds. At cruising and higher airspeed ranges, IAS and CAS differences become small.

Pilots should refer to the airspeed calibration chart to correct for possible airspeed errors because airspeeds, such as those found on the color-coded face of the airspeed indicator, on placards, or in the Glider Flight Manual or Pilot's Operating Handbook (GFM/POH), usually reflect CAS. Some manufacturers use IAS rather than CAS to denote the airspeeds.

Dirt, dust, ice, or snow collecting at the pitot tube opening may obstruct air passage and prevent correct indications. Degradation due to age and vibration may also affect the sensitivity of the diaphragm. Therefore, airspeed indicators should undergo periodic calibration.

True Airspeed (TAS)

TAS is the actual speed at which the aircraft moves through the air. In still air, TAS equals the actual speed over the ground. However, the airspeed indicator only indicates TAS under standard atmospheric conditions at sea level (29.92 inches of mercury ("Hg) and 15°C). At a constant IAS, TAS increases as the glider climbs because air density decreases with an increase of altitude.

The airspeed indicator provides a convenient means to manage most flight parameters since it indicates dynamic pressure. If the airspeed indicator did display TAS, the pilot would need to use a quick reference card to look up the stall speed, best L/D speed, and minimum sink speed for the current altitude while trying to fly. Fortunately, the pilot only needs to remember one set of numbers for most parameters. For example, *Figure 4-8* illustrates that for each glider, the pilot need only remember one stall speed for all altitudes.

A pilot can determine TAS by two methods. The first and more accurate method involves using a flight computer. In this method, the pilot corrects the CAS for temperature and pressure variation using the airspeed correction scale on the computer. The second method approximates TAS by increasing the current IAS by 2 percent for every 1,000 feet above sea level.

Airspeed Indicator Markings

Gliders manufactured after 1945 have airspeed indicators that conform to a standard color-coded marking system. [*Figure 4-3*] This system enables the pilot to determine important airspeed information quickly. For example, the green arc represents the normal operating range, while the yellow arc represents the caution range. A pilot who notes an airspeed needle in the yellow arc and rapidly approaching the red line while maneuvering should immediately take corrective action to reduce the airspeed. In this case, the pilot should use smooth control pressure at high airspeeds to avoid unsafe stress on the glider structure.

A description of the standard markings on an airspeed indicator follows:

- The white arc—flap operating range.

- The lower limit of the white arc—stalling speed in the landing configuration.

- The top of the white arc—maximum speed for use of full flaps. If flaps are operated at higher airspeeds, severe strain or structural failure could result.

- The lower limit of the green arc—stalling speed with the wing flaps and landing gear retracted.

- The upper limit of the green arc—maximum structural cruising speed. This is the maximum speed for normal operations.

- The yellow arc—caution range. The pilot should avoid speeds within this area unless in smooth air.

- The red line—never-exceed speed. This is the maximum speed at which the glider can be operated in smooth air. This pilot should never intentionally exceed this speed.

Effect of Altitude on V_{NE}

The never-exceed speed (V_{NE}) decreases with increased altitude due to the possibility of flutter at higher true airspeeds. At high altitudes maintaining a speed at or below the red line may exceed actual V_{NE}. Since the decrease in V_{NE} varies by model, the flight manual may include a table, such as the one shown in *Figure 4-9*, that documents the decrease in V_{NE} with altitude. At high true airspeeds during a rapid descent, the glider structure could suddenly flutter and break apart. Glider manufacturers test for flutter and adherence to V_{NE} speeds published in the flight manual for the specific make and model should prevent it.

Altitude (in feet)	V_{NE} (IAS in knots)
Up to 6,500	135
10,000	128
13,000	121
16,500	115

Figure 4-9. *IAS corresponding to V_{NE} decreases with altitude.*

Other Airspeed Limitations

Placards in the view of the pilot may display other important airspeed limitations not marked on the face of the airspeed indicator. [Figure 4-10]

- Maneuvering speed (V~A)— a structural design
- airspeed used in determining the strength requirements for the glider and its control surfaces. The structural design requirements do not cover multiple inputs in one axis or control inputs in more than one axis at a time at any speed, even below V~A. If encountering rough air or severe turbulence during flight, the pilot should reduce airspeed to maneuvering speed or less to prevent exceeding structural limits. Maneuvering speed varies with the weight of the glider and does not appear as a specific airspeed on the airspeed indicator. For gliders with a published rough airspeed limitation (V~B), the pilot should keep below that speed in rough air to accommodate maximum gust intensity.

- Landing gear operating speed (V~LO)—maximum speed for extending or retracting the landing gear if using a glider equipped with retractable landing gear.

- Minimum sink speed—airspeed that results in the least amount of altitude loss over a given time, or
- which maximizes the altitude gain when taking advantage of rising air.

- Best glide speed—airspeed that results in the least amount of altitude loss over a given distance, not considering the effects of wind.

- Maximum aerotow or ground launch speed—maximum airspeed for tow without exceeding design specifications.

Valid when lower or side hook is installed:				
Maximum winch-launching speed	65 KIAS	OR	Maximum winch-launching speed	120 km/hr IAS
Maximum aerotowing speed	81 KIAS		Maximum aerotowing speed	150 km/hr IAS
Maximum maneuvering speed	81 KIAS		Maximum maneuvering speed	150 km/hr IAS
Valid when front hook only is installed:				
Maximum aerotowing speed	81 KIAS	OR	Maximum aerotowing speed	150 km/hr IAS
Maximum maneuvering speed	81 KIAS		Maximum maneuvering speed	150 km/hr IAS

Figure 4-10. *Sample speed limitation placards placed in a glider and in view of the pilot.*

Altimeter

The altimeter measures the static air pressure of the surrounding air. Tubing connects the altimeter static pressure inlet to the static port holes located on the side of the glider. [*Figure 4-11*]

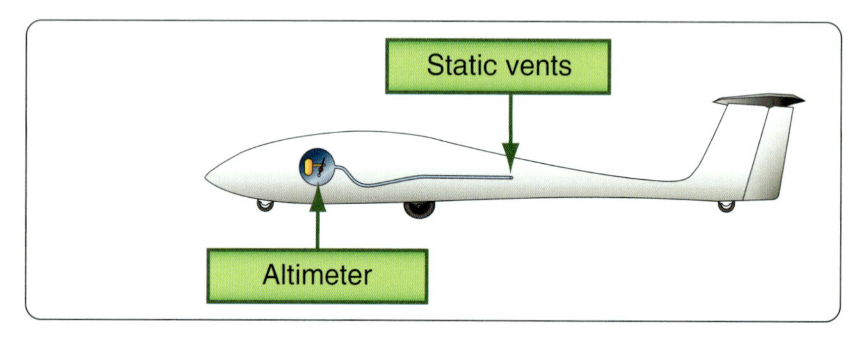

Figure 4-11. *Static vents and altimeter plumbing.*

If using the local altimeter setting (the local pressure corrected to sea level), the altimeter indicates the glider's current height above mean sea level (MSL). [*Figure 4-12*] Subtracting the ground elevation from current MSL altitude gives the height above ground.

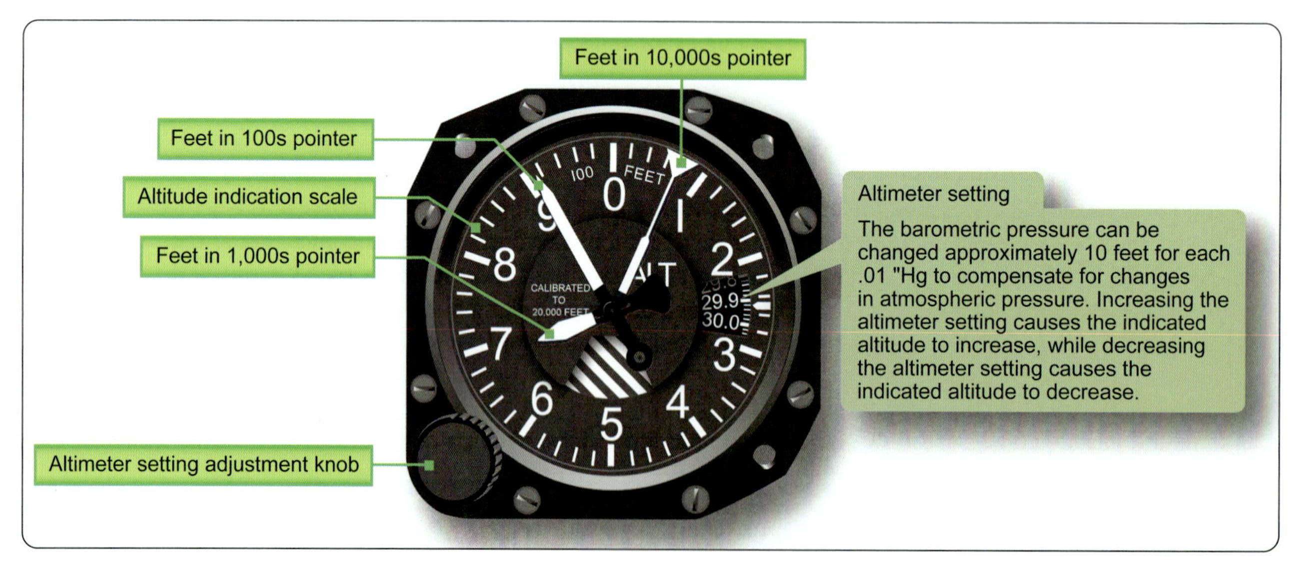

Figure 4-12. *Altimeter.*

The weight of a column of air above a given location creates atmospheric pressure. At sea level, an overlying column of air exerts a force equivalent to 14.7 pounds per square inch, 1013.2 mb, or 29.92 inches of mercury under standard conditions. At a higher altitude, the shorter overlying column of air weighs less and exerts less pressure. Therefore, atmospheric pressure decreases with altitude. At 18,000 feet, atmospheric pressure drops to approximately half that at sea level. [*Figure 4-13*]

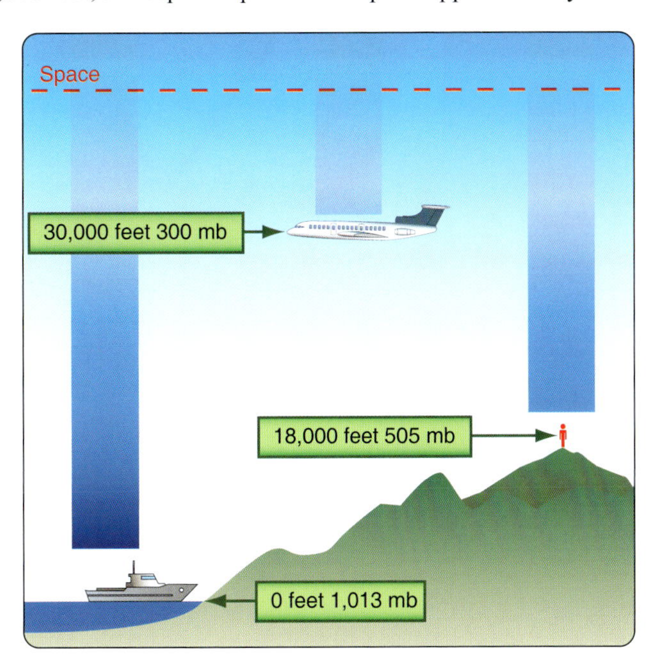

Figure 4-13. *Atmospheric pressure and altitude.*

Principles of Operation

The altimeter works like an aneroid barometer that measures atmospheric pressure at the current elevation except that the altimeter indicates pressure in feet. The altimeter indicates changes in altitude during a climb or descent as atmospheric pressure changes. *Figure 4-14* and *Figure 4-15* illustrate how the altimeter functions. Some altimeters have one pointer while others have more.

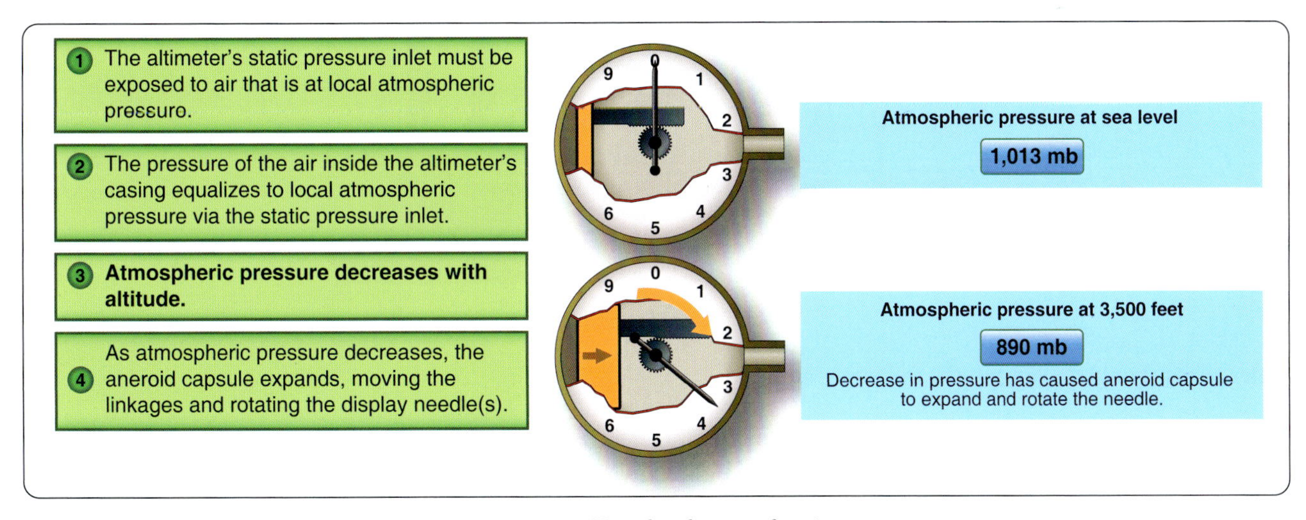

Figure 4-14. *How the altimeter functions.*

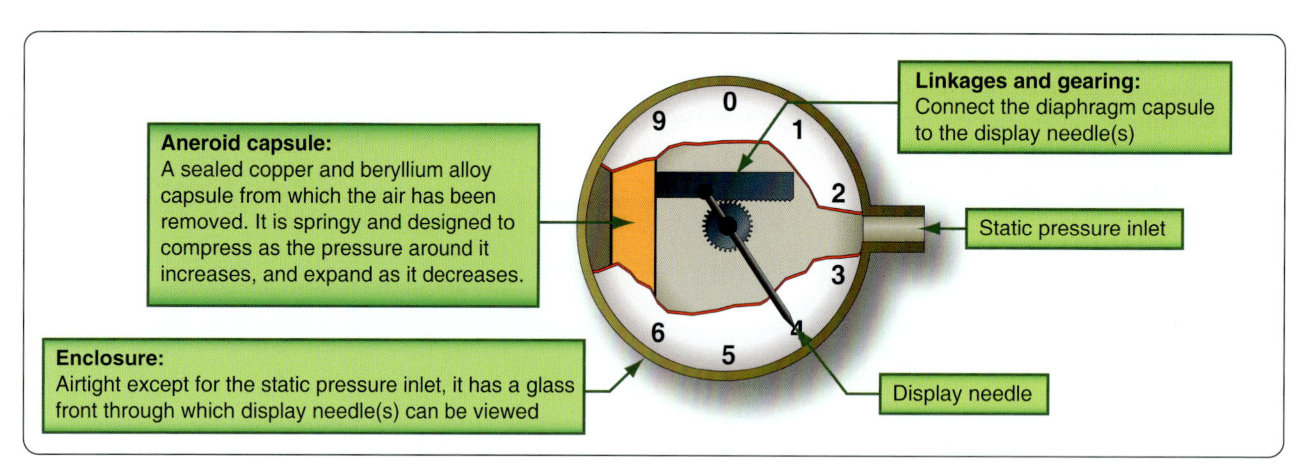

Figure 4-15. *Inside the altimeter.*

The markings on the dial of a typical altimeter include numerals arranged clockwise from 0 to 9 inclusive, as shown in *Figure 4-12*. When the surrounding pressure changes, the expansion or contraction of the aneroid element moves the hands through a gear train. The hands sweep the calibrated dial to indicate altitude. In the altimeter with three hands shown in *Figure 4-12*, the thinnest hand with an end shaped like a triangle indicates altitude in tens of thousands of feet; the shortest pointed hand indicates thousands of feet; and the long pointed hand indicates hundreds of feet, subdivided into 20-foot increments.

Types of Altitude

Altitude corresponds to a vertical distance above some point or level used as a reference. Altitude measured from different reference levels serves different purposes. [*Figure 4-16*]

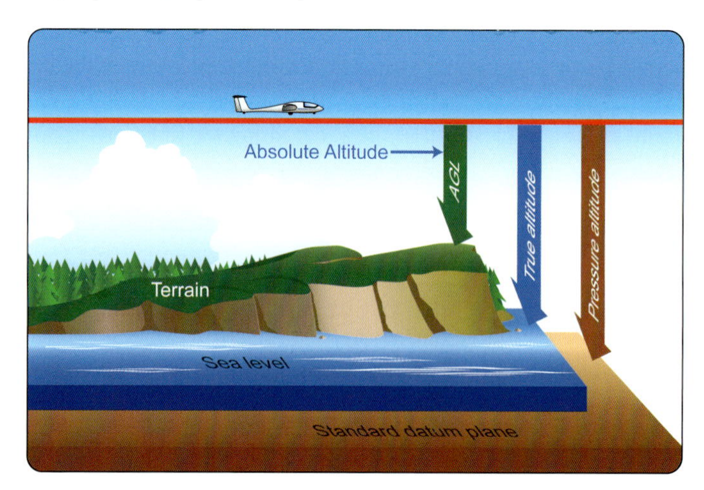

Figure 4-16. *Types of altitude.*

Glider pilots should understand the following altitudes:

- Indicated altitude—altitude read directly from the altimeter. During preflight, the pilot should set the altimeter to the current local altimeter setting. If the indicated altitude deviates from the known field elevation by ±75 feet or more, the pilot should not fly and refer the altimeter to an appropriately rated repair station for evaluation and correction. The pilot can use indicated altitude for terrain and obstacle clearance. However, true altitude and indicated altitude may differ depending on pressure and temperature conditions.

- True altitude—the vertical distance of the glider above sea level in standard atmospheric conditions (known as MSL often expressed in this manner: 10,900 feet MSL, 5,280 feet MSL, or 940 feet MSL). Airport, terrain, and obstacle elevations found on aeronautical charts are expressed as MSL (true altitudes).

- Pressure altitude—altitude indicated with the altimeter setting adjusted to 29.92. The pressure altitude corresponds to the height above the standard datum plane, a theoretical plane where air pressure equals 29.92 inHg. Pilots use pressure altitude for computer solutions to determine density altitude, true altitude, and TAS, etc. When flying in class A airspace, pilots set the altimeter to 29.92.

- Density altitude —pressure altitude corrected for nonstandard temperature variations. In standard conditions, pressure altitude equals density altitude. In temperatures above standard, the density exceeds pressure altitude. With temperatures below standard, the density altitude is less than pressure altitude. The density altitude determines the glider's performance and affects the power output of a tow plane or self-launching glider.

- Absolute altitude—vertical distance above the terrain, above ground level (AGL). The pilot can use absolute altitude to estimate gliding distance over terrain without benefit of lift.

Effect of Nonstandard Pressure

On a flight made from a high-pressure area to a low-pressure area without adjusting the altimeter, the glider descends if the pilot maintains a given indicated altitude. When flying from a low-pressure area to a high-pressure area without adjusting the altimeter, the glider climbs if the pilot maintains a given indicated altitude. The pilot adjusts for this phenomenon by setting the altimeter. A correctly set altimeter provides an appropriate amount of vertical separation between aircraft at different cruising altitudes and can help prevent mid-air collisions. It also allows a more accurate absolute altitude computation, which gives a glider pilot the ability to determine gliding distance more precisely.

Setting the Altimeter

To adjust the altimeter for nonstandard pressure, the pilot sets the pressure scale in the altimeter window (Kollsman window) to the given local altimeter setting or to the field elevation. Altimeter settings correspond to station pressure reduced to sea level, expressed in inches of mercury.

A reporting station takes an hourly measurement of the atmospheric pressure and corrects this value to sea-level pressure. These altimeter settings reflect height above sea level only near the reporting station. When flying below 18,000 feet MSL, the pilot should re-adjust the altimeter as the flight progresses from one station to the next. When flying at or above 18,000 feet MSL, the pilot sets the altimeter to 29.92.

When flying over high mountainous terrain, certain atmospheric conditions can cause the altimeter to indicate an altitude of 1,000 feet or more above the true altitude. For this reason, the pilot should fly with a margin of increased altitude—not only for possible altimeter error, but also for downdrafts, which may occur if encountering high winds.

A cross-country flight from TSA Gliderport, Midlothian, Texas, to Winston Airport, Snyder, Texas, via Stephens County Airport, Breckenridge, Texas, illustrates the use of altimeter settings. Before launch from TSA Gliderport, the pilot receives the current local altimeter setting of 29.85 and adjusts the altimeter to this value. The indication varies slightly from the known airport elevation of 660 feet due to a slight altimeter calibration error.

When over Stephens County Airport, the pilot receives a current area altimeter setting of 29.94 and applies this setting to the altimeter. Before entering the traffic pattern at Winston Airport, the pilot receives a new altimeter setting of 29.69 from the Automated Weather Observing System (AWOS). If the pilot desires to enter the traffic pattern at approximately 1,000 feet above the terrain, and if the field elevation of Winston Airport is 2,430 feet MSL, the pilot should use an indicated altitude of 3,400 feet.

2,430 feet + 1,000 feet = 3,430 feet, rounded to 3,400 feet

For illustration, assume a distraction caused the pilot to neglect the adjustment for the Winston Airport altimeter setting and to continue using the Stephens County Airport setting of 29.94. The pattern entry would occur approximately 250 feet below the Winston Airport's traffic pattern altitude of 3,400 feet and the altimeter would indicate approximately 2,680 feet upon landing or 250 feet higher than the field elevation.

Actual altimeter setting = 29.94

Correct altimeter setting = 29.69

Difference = .25

One inch of pressure is equal to approximately 1,000 feet of altitude.

.25 × 1,000 feet = 250 feet

In this scenario, the pilot, although low, might fly a successful visual approach angle to the landing zone. The pilot should adjust the visual angle to the landing zone to compensate for the lower altitude. However, risk of an accident increases due to the incorrect pattern entry altitude. For example, the glider might strike an obstacle in the flight path if not seen by the pilot. If the pilot does not reset the altimeter, the following memory aid illustrates what can happen "From a high to a low—look out below."

Effect of Nonstandard Temperature

Variations in air temperature also affect the altimeter. On a warm day, air weighs less per unit volume than on a cold day. For example, the pressure level at which the altimeter indicates 10,000 feet occurs at a higher altitude on a warm day than under standard conditions. On a cold day, the 10,000-foot indication moves lower. The adjustment made by the pilot to compensate for nonstandard pressure does not compensate for nonstandard temperature. If considering terrain or obstacle clearance during the selection of a cruising true altitude, particularly at higher altitudes, the pilot should consider this effect. Colder than standard temperature places the glider closer to the ground for a given true altitude, and the pilot should use a higher altitude to provide adequate terrain clearance. [*Figure 4-17*]

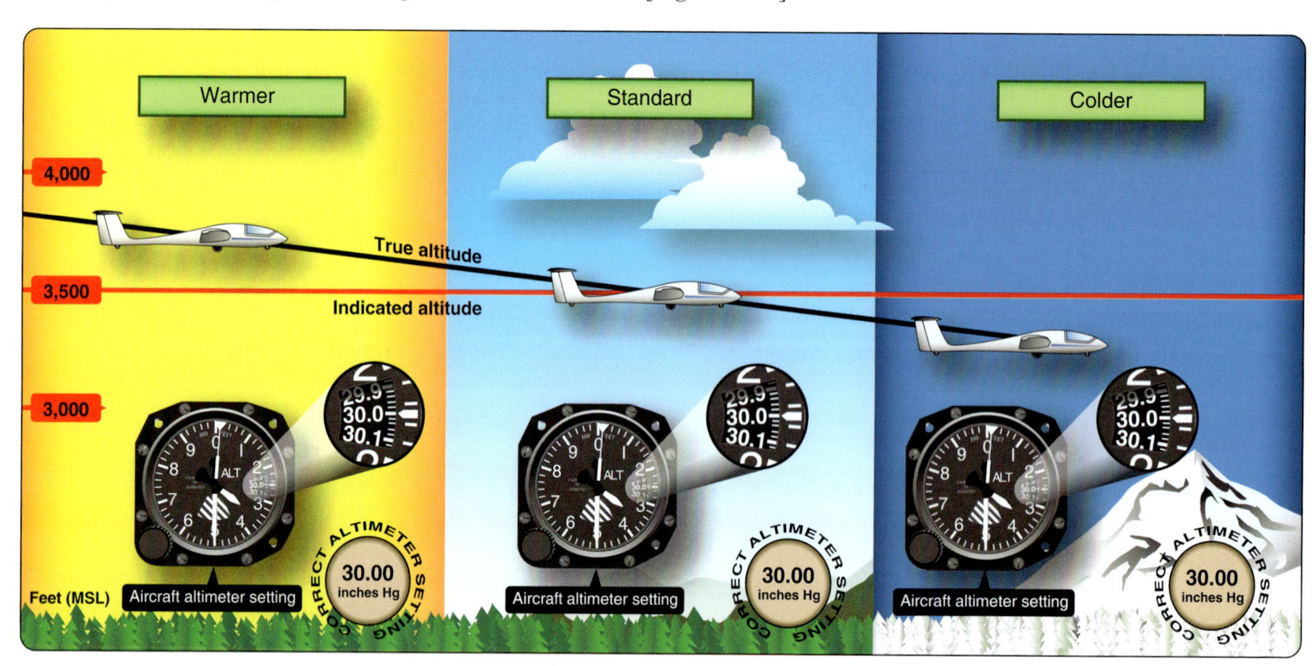

Figure 4-17. *Nonstandard pressure and temperature.*

Variometer

Variometer instruments measure the vertical ascent or descent of the local air mass and glider combined and display that information as vertical speed. The variometer can be considered a simple flow meter measuring air flowing between an outside reference (static or total energy port) and an internal reference flask. The variometer depends upon the pressure lapse rate in the atmosphere to derive information about rate of climb or rate of descent. A non-electric variometer uses a separate insulated tank (thermos or capacity flask) as a reference chamber to increase sensitivity and accuracy of the instrument. The tubing runs from the reference chamber through the variometer instrument to an outside static port in an uncompensated variometer. [*Figure 4-18* and *4-19*]

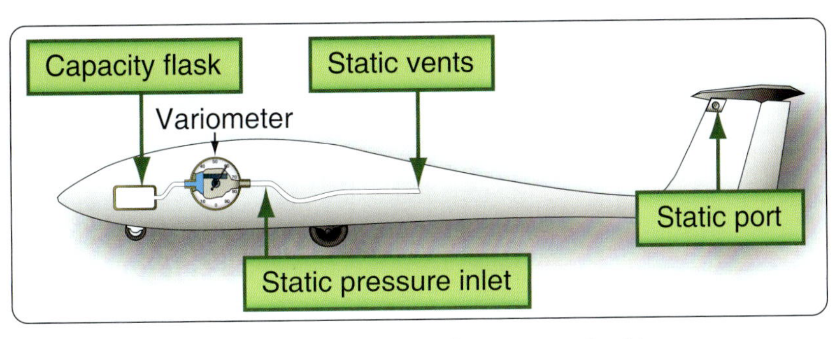

Figure 4-18. *Uncompensated variometer plumbing.*

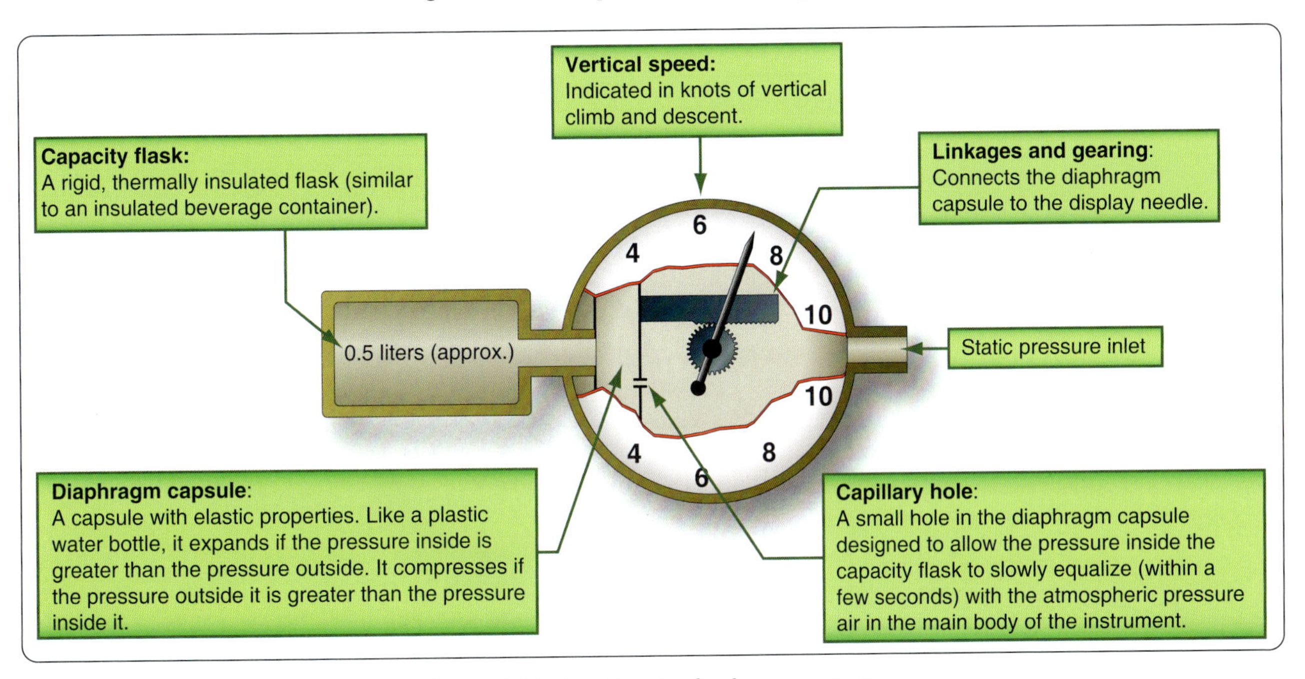

Figure 4-19. *A variometer diaphragm anatomy.*

Pressure differences between the air inside the variometer/reference chamber system and the air outside of the system tend to equalize as air flows from high-pressure areas to low-pressure areas. When pressure inside the reference chamber exceeds the pressure outside, air flows out of the reference chamber through the mechanical variometer to the outside environment, and the variometer indicates a climb. When air pressure outside the reference chamber exceeds the pressure inside, air flows through the variometer and into the reference chamber until pressure equalizes. In this case, the variometer needle indicates conditions that force the glider to lose height, *Figures 4-20* and *4-21* illustrate how the variometer works in level flight and while the glider ascends. In addition, *Figure 4-22* illustrates certain flight maneuvers that cause the variometer to display changes in altitude.

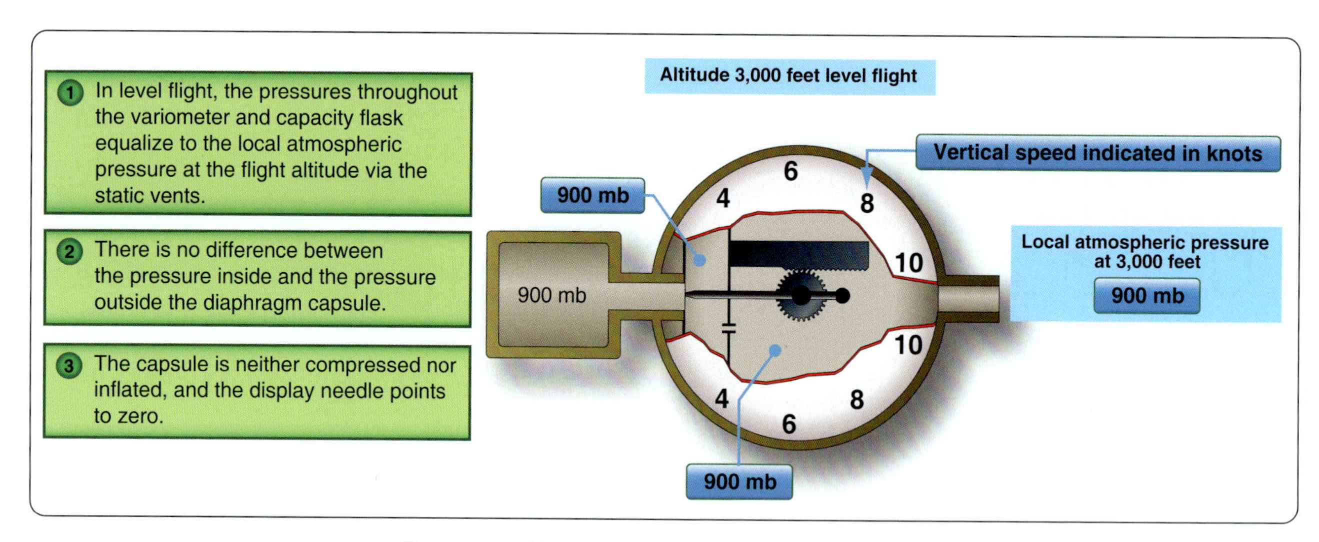

Figure 4-20. *Uncompensated variometer in level flight.*

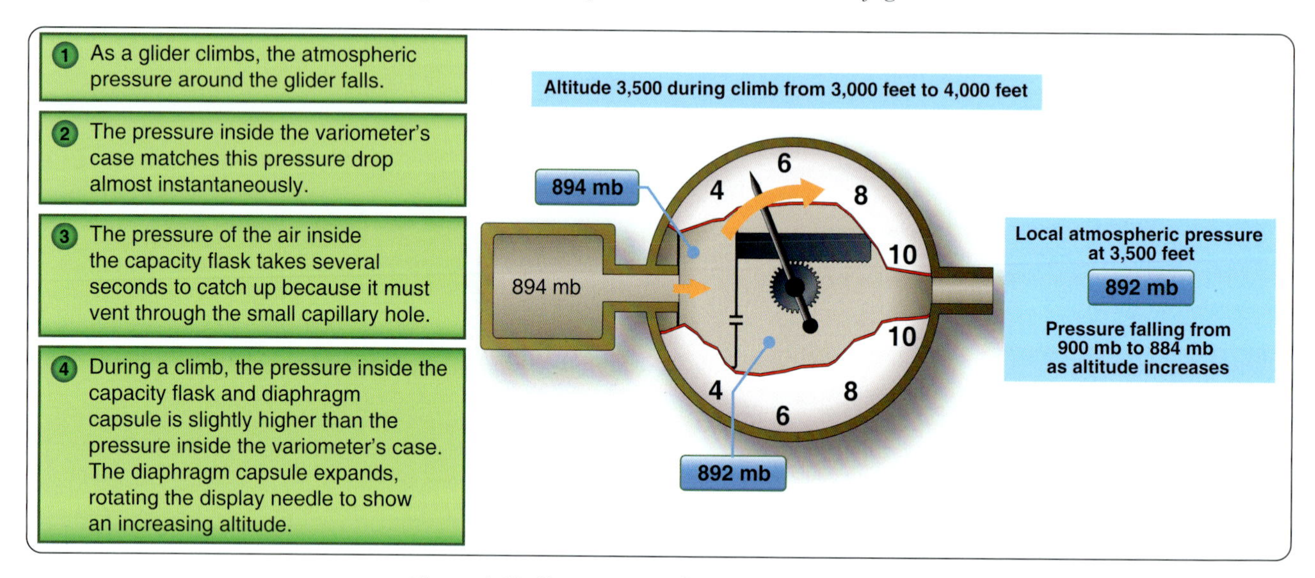

Figure 4-21. *Uncompensated variometer in a climb.*

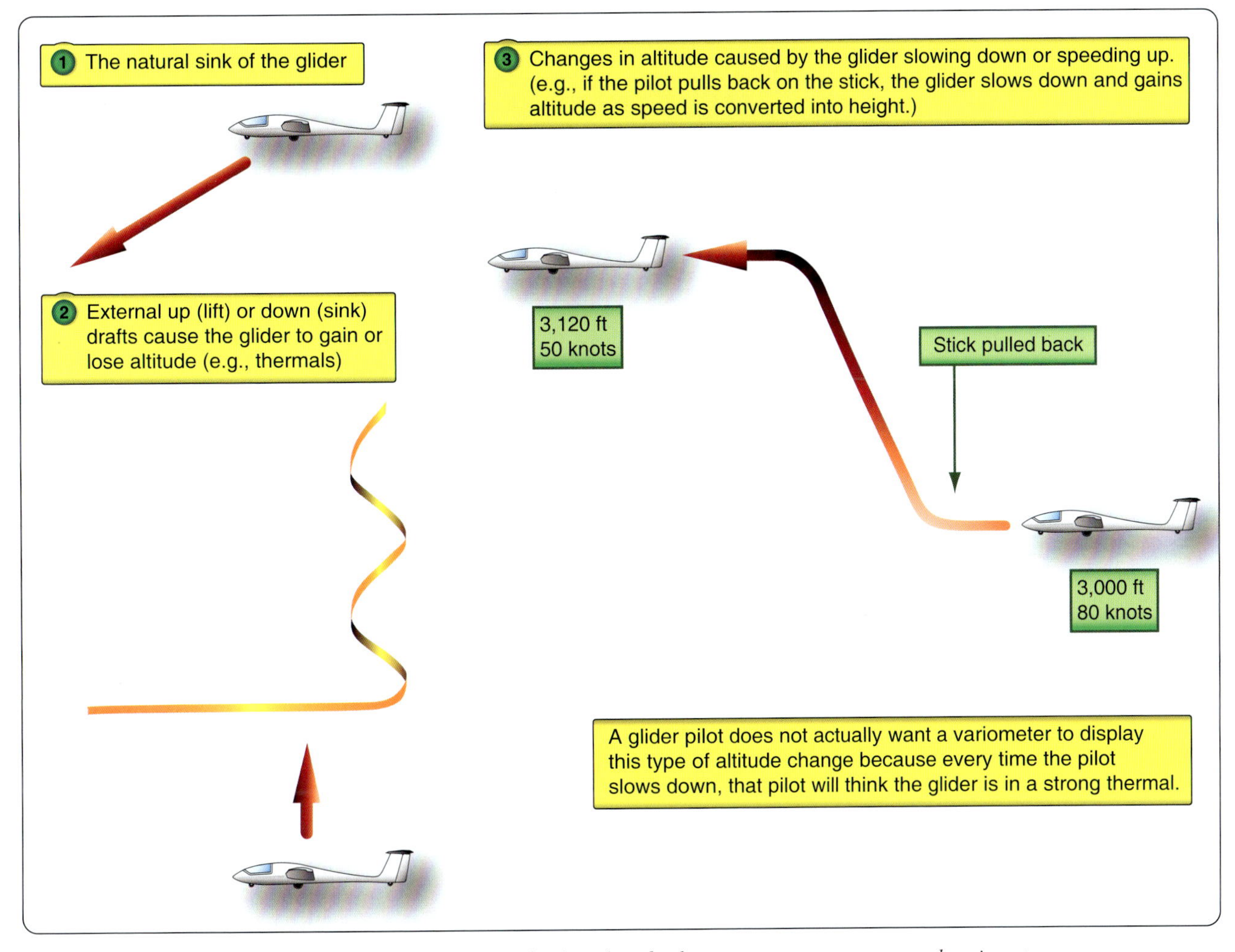

Figure 4-22. *Flight maneuvers that display altitude changes on an uncompensated variometer.*

Electric-powered variometers offer several advantages over the non-electric variety. These advantages include more rapid response rates and separate audible signals for climb and descent.

Some electric variometers use special sensors. As air flows into or out of the reference chamber, it cools sensors in a circuit and alters the electrical resistance measured by the system. The resulting change in resistance corresponds to the rate of climb or descent. The system displays that information on the variometer.

Many electric variometers provide audible tones, or beeps, that indicate the rate of climb or rate of descent of the glider. Pilots using an audio variometer can listen for the rate of climb or descent, which allows more time to focus attention outside the aircraft. [*Figure 4-23*]

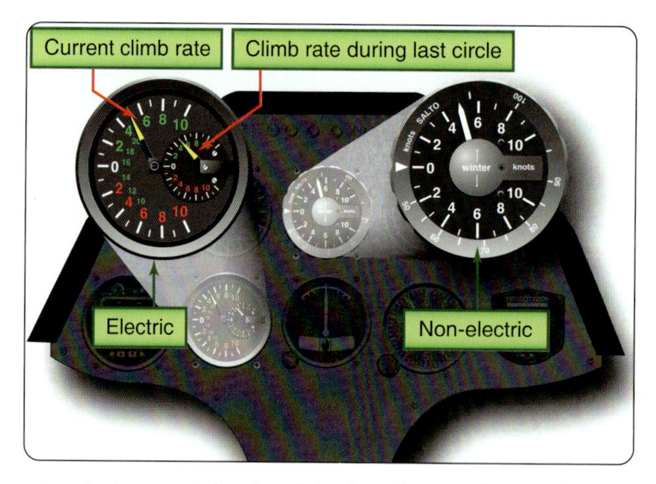

Figure 4-23. *When an electric variometer is mounted to the glider, installation of a non-electric variometer can provide a backup.*

In the past, some variometers had a rotatable rim speed scale called a MacCready ring. This scale indicates the optimum airspeed to fly when traveling between thermals for maximum cross-country performance. During the glide between thermals, the pilot sets the index arrow to the rate of climb expected in the next thermal. On the speed ring, the variometer needle points to the optimum speed to fly between thermals. If the pilot expects a low rate of climb, the instrument selects a lower optimum speed between thermals. When the pilot expects strong lift at the next thermal, the instrument suggests a faster optimum cruise airspeed. [*Figure 4-24*] A MacCready ring has single pilot values, values with a passenger, and may also adjust for ballast since more weight results in higher gliding speeds. Electronic instrument displays have become more common than Macready rings.

Figure 4-24. *The MacCready ring.*

Pilot induced climbs and dives result in changes in airspeed and affect an uncompensated variometer by causing changes in pressure altitude. In still air, when the pilot initiates a dive, the variometer indicates a descent. When the glider pilot pulls out of the dive and initiates a rapid climb, the variometer indicates an ascent. A glider with an uncompensated variometer gives an accurate indication of rising and descending air only if the pilot maintains a constant airspeed.

Total Energy System

A variometer with a total energy system senses changes in airspeed and tends to cancel out the resulting climb and dive indications (stick thermals). This gives a glider pilot an indication of rising or descending air despite changes in airspeed.

A popular type of total energy system consists of a small venturi, a pair of holes, or simply a slot on the back side of a small vertical tube mounted in the air stream and connected to the static outlet of the variometer. When airspeed increases during a dive, more suction from the venturi offsets the increased pressure at the static outlet of the variometer. Similarly, when airspeed decreases during a climb, reduced suction from the venturi offsets the pressure reduction at the static outlet of the variometer. The net effect reduces climb and dive indications caused by airspeed changes. To maximize the precision of this compensation effect, the system can use a total energy probe, which sits in undisturbed airflow ahead of the aircraft nose or tail fin. [*Figure 4-25*]

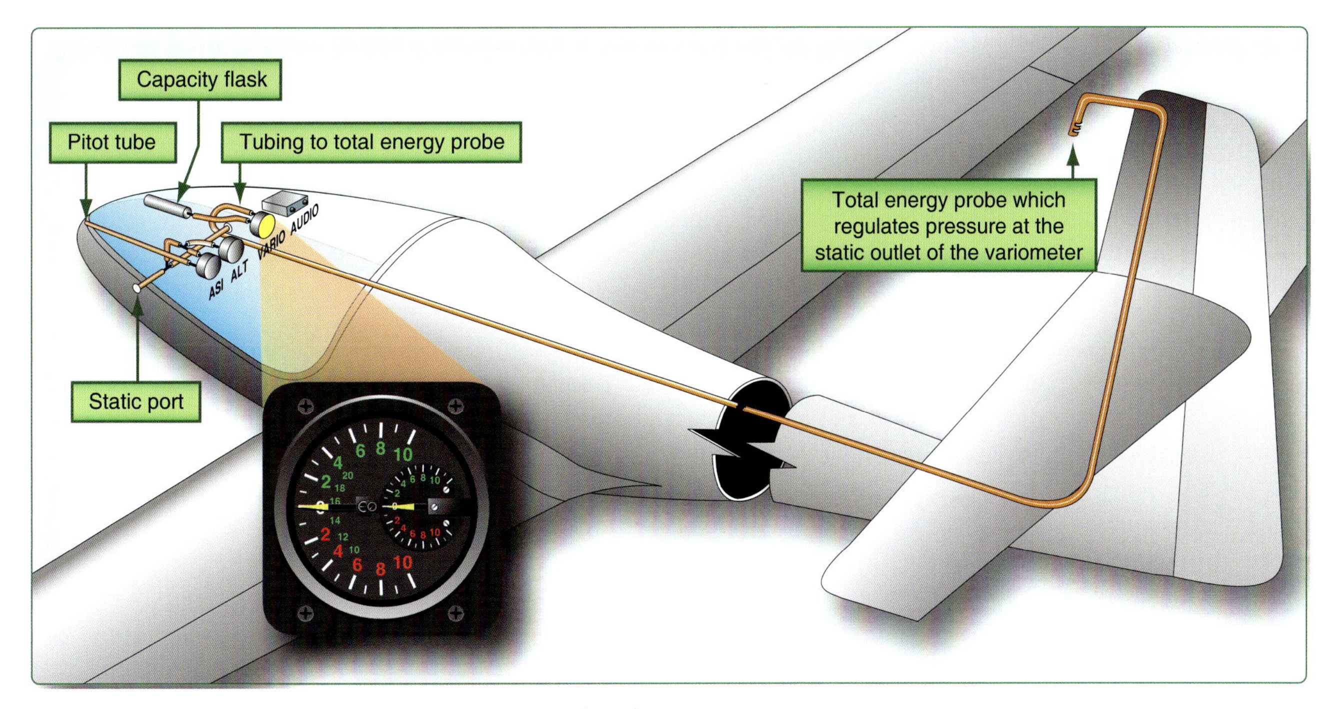

Figure 4-25. *A total energy variometer system.*

Another type of total energy system design uses a diaphragm-type compensator placed in line from the pitot tube to the line coming from the reference chamber (thermos or capacity flask). Deflection of the diaphragm offsets the effect the airspeed change has on pitot pressure. In effect, the diaphragm modulates pressure changes in the capacity flask and masks stick thermals.

Netto

A Netto variometer indicates the vertical movement of the air mass, regardless of the glider's climb or descent rate. Some Netto variometer systems employ a calibrated capillary tube that functions as a tiny valve. Pitot pressure pushes minute quantities of air through the valve and into the reference chamber tubing. This removes the glider's known sink rate at various airspeeds from the variometer indication. [*Figure 4-26*]

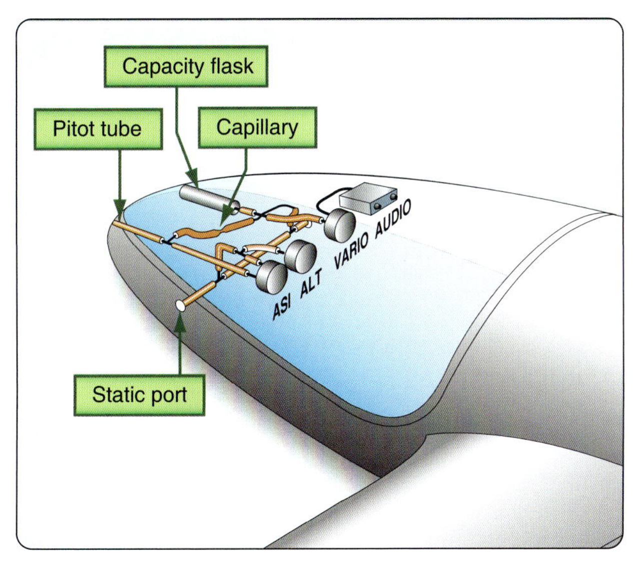

Figure 4-26. *An example of a Netto variometer system.*

Computerized (electronic) Netto variometers employ a different method to remove the glider sink rate. In this type of system, sensors for both pitot pressure and static pressure provide airspeed information to the computer. The computer stores the known sink rate of the glider at every airspeed. At any given airspeed, the computer removes the sink rate of the glider, and the variometer displays the rate of ascent or descent of the air mass itself.

Electronic Flight Computers

Since nonpowered gliders lack a generator or alternator, electrical components, such as the flight computer, if installed, draw power from the glider's rechargeable battery or batteries. Some gliders use solar cells arrayed behind the pilot, or on top of the instrument panel cover, to supply additional power to the electrical system during flight in sunny conditions.

The primary components of most flight computer systems include an electric variometer, a coupled GPS receiver, and a microprocessor. The variometer measures rate of climb and descent. The GPS provides position information. The microprocessor interprets altitude, speed, and position information. The microprocessor output aids the pilot in cross-country decision-making by suggesting a speed to fly. *Figure 4-27* shows a glider flight computer.

Figure 4-27. *Glider flight computer display.*

The GPS-coupled flight computer can provide the following information:

- Current position
- Previous position
- Speed and time to destination
- Distance to planned destination
- Height needed to glide to destination
- Current climb or descent rate
- The optimum airspeed to fly to the next thermal
- The optimum airspeed to fly to a location on the ground, such as the finish line in a race or the airport of intended landing at the end of a cross-country flight

The primary benefits of the flight computer divide into two areas: navigation assistance and performance (speed) enhancement.

Flight computers utilize the concept of a waypoint, which includes latitude, longitude, and altitude. Glider races and cross-country glider flights frequently involve flight around a series of waypoints called turnpoints. The course may be an out-and-return course, polygon shape, or just a series of waypoints. The glider pilot navigates from point to point, using available lift sources to climb periodically so that flight can continue to the intended goal. The GPS-enabled flight computer aids in navigation, summarizing flight progress, and logging completion of tasks or flight goals. When encountering strong lift, for example, the pilot can use the flight computer to mark the location of the thermal. If rounding a nearby turnpoint, the glider pilot might use the flight computer to return to the marked thermal for a rapid climb before continuing.

During the climb portion of the flight, the flight computer's variometer constantly updates the achieved rate of climb. During cruise, the GPS-coupled flight computer aids in navigating accurately to the next turnpoint. The flight computer also suggests the optimum cruise airspeed for the glider to fly, based on the expected rate of climb in the next thermal. During final glide to a goal, the flight computer can display glider altitude, altitude required to reach the goal, distance to the goal, the strength of the headwind or tailwind component, and optimum airspeed to fly.

When the flight computer detects rapid climbs, it suggests higher cruise airspeeds to enhance performance. When the computer detects a low rate of climb, it compensates for the weaker conditions by suggesting lower airspeeds. The flight computer frees the pilot to look for other air traffic, look for sources of lift, watch the weather ahead, and plot a strategy for the remaining portion of the flight.

As explained in the next chapter, water ballast increases the optimal speed to fly. The flight computer compensates for water ballast carried, adjusting speed-to-fly computations according to the weight and performance of the glider. Some flight computers require the pilot to enter data regarding the ballast load of the glider. Other flight computers automatically compensate for the effect of water ballast by constantly measuring the performance of the glider and deducing the operating weight of the glider from these measurements. If the wings of the glider become contaminated with bugs, glider performance declines. The flight computer can be adjusted to account for the resulting performance degradation.

Magnetic Compass

Most gliders do not come under regulation for powered aircraft as referenced in Title 14 of the Code of Federal Regulations (14 CFR) part 91, section 91.205, and only need to comply with regulations for "civil aircraft." For this reason, some gliders do not have a compass unless required per the aircraft's Type Certificate Data Sheets (TCDS).

Slip/Skid Indicators

Yaw String

A piece of yarn mounted in the free airstream and easily visible to the pilot, provides an effective slip/skid indicator. [*Figure 4-28*] During coordinated flight, the yarn points straight back. During a slipping turn, the tail (unattached end) of the yaw string offsets toward the outside of the turn. During a skidding turn, the tail of the yaw string offsets toward the inside of the turn. A pilot having difficulty sensing and correcting uncoordinated flight can apply pressure to the rudder pedal not aligned with the yaw string tail.

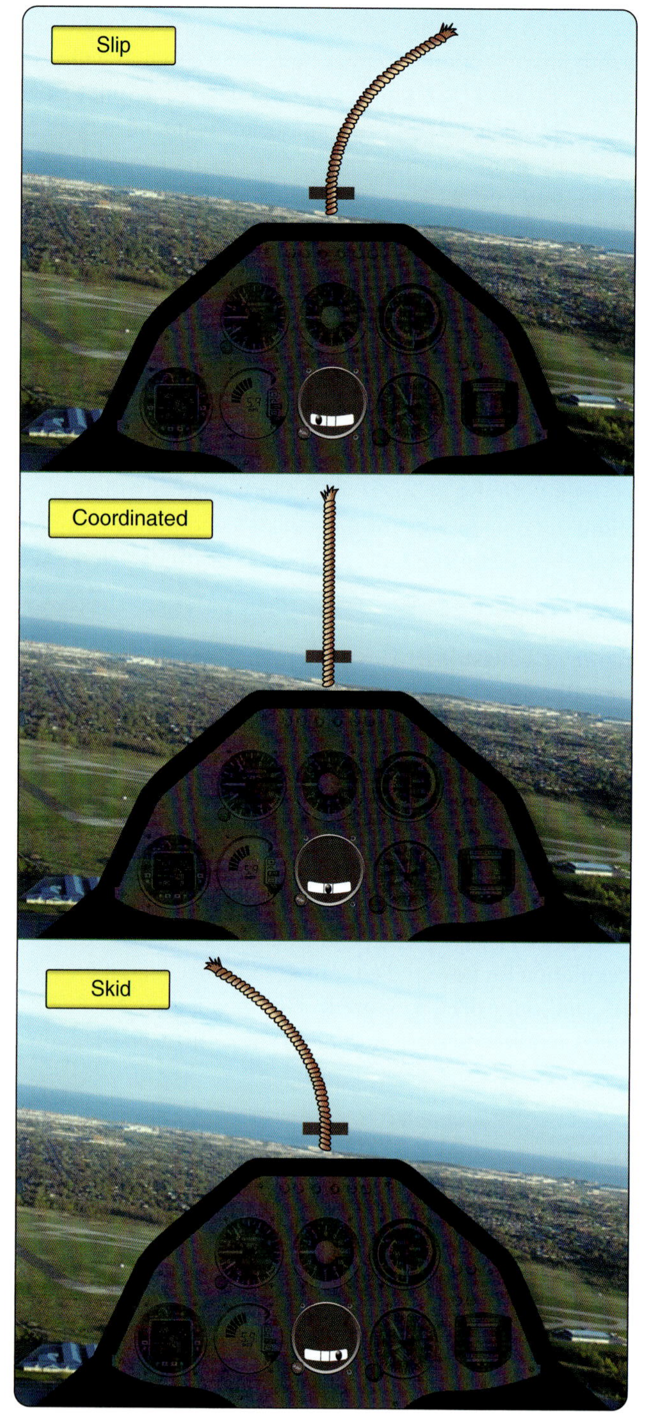

Figure 4-28. *Left turn condition indications of a yaw string and inclinometer.*

Inclinometer

An inclinometer also provides slip/skid indication. The inclinometer responds to centrifugal force and gravity. The inclinometer consists of a metal ball in an oil-filled, curved glass tube. When the glider flies in coordinated fashion, the ball remains centered at the bottom of the glass tube. The inclinometer differs from the yaw string during uncoordinated flight. The ball moves to the inside of the turn to indicate a slip and to the outside of the turn to indicate a skid. [*Figure 4-28*] The phrase, "step on the ball" explains how a pilot should respond to restore coordinated flight. During a spin, the inclinometer does not provide direction of rotation information, and pilots should not use the inclinometer for guidance.

Gyroscopic Instruments

Unpowered gliders do not usually have gyroscopic instruments, while self-launching gliders often have one or more gyroscopic instruments on the panel. Common gyroscopic instruments include the attitude indicator, heading indicator, and turn coordinator.

G-Meter

A panel-mounted G-meter registers positive G forces from climbs and turns, as well as negative G forces when diving down or pushing over from a climb. During straight, unaccelerated flight in calm air, the G-meter registers a load factor of 1 G (1.0 times the force of gravity). During flight in turbulent air, the glider and pilot experience G-loads greater than or less than 1 G when encountering updrafts or downdrafts.

Each glider type can withstand a specified maximum positive G-load and a specified maximum negative G-load. The operating limitations, as described in the GFM/POH or on placards, are the definitive source for this information. Exceeding the allowable limit loads may result in deformation of the glider structure. In extreme cases, exceeding permissible limit loads may cause structural failure of the glider. The G-meter allows the pilot to monitor G-loads from moment to moment during aerobatic flight and during flight in rough air. Most G-meters also record and display the maximum positive G-load and the maximum negative G-load encountered during flight. The recorded maximum positive and negative G-loads can be reset by adjusting the control knob of the G- meter. [*Figure 4-29*]

Figure 4-29. *The G-meter.*

FLARM Collision Avoidance System

A mid-air collision presents a risk to all pilots, and glider pilots have additional mid-air scenarios to consider when thermaling, cloud street flying, or ridge running. Periodically, mid-air collisions or near misses occur in club flying, during competitions, and between gliders and towplanes after release.

FLARM systems (the name being inspired from "flight alarm") can warn pilots of impending collisions with other FLARM-equipped gliders and give the location of non-threatening nearby FLARM-equipped gliders. *Figure 4-30* shows the interior of an ASW 19 glider with a FLARM unit on top of the instrument panel. FLARM transmits and receives information, models the unique flight characteristics of gliders, and stays quiet unless it detects a real threat. However, it can only provide information about other gliders with an operating FLARM system.

Figure 4-30. *FLARM unit on top of the instrument panel.*

FLARM obtains its position from an internal global positioning system (GPS) and a barometric sensor and then broadcasts this data with forecast data about the future 3D flight track. Its receiver receives signals from other FLARM devices typically within 3-5 kilometers and processes the information received. Motion-prediction algorithms predict potential conflicts for up to 50 other signals and warn the pilot using sound and visual means. FLARM can also store information about static aerial obstacles, such as cables, into a database.

Why use FLARM in a glider? Conventional Airborne Collision Avoidance Systems (ACAS) would provide continuous and unnecessary warnings about all aircraft in the vicinity. FLARM only gives selective alerts to aircraft posing a collision risk. It consumes much less power than a transponder or ADS-B and is relatively inexpensive to buy and install. While versions exist for use in light aircraft and helicopters, as well as gliders, the short range of the signal makes FLARM unsuitable for avoiding collisions with fast aircraft.

While gliders are exempt from carrying transponders and ADS-B out transmitters in most, but not all airspace, the Soaring Society of America strongly encourages pilots to install this equipment when operating near high density airspace. This increases their visibility to other users of the National Airspace System (NAS).

Transponder Code

The Federal Aviation Administration (FAA) assigned transponder code 1202 for use by gliders not in contact with air traffic control (ATC) as of March 7, 2012. Effective November 1, 2021 (JO 7110.66G), the FAA amended this practice to include gliders in contact with ATC. Glider pilots operating in areas with an agreement with local ATC to use a different code should contact the agreement sponsor for guidance.

Definitions

- SQUAWK CODE: The 4-digit code set in the transponder, such as 1202.
- IDENT or SQUAWK IDENT: A controller may direct a pilot to "ident" or "squawk ident" to verify the aircraft's location on the radar screen. When directed, the pilot pushes the button on the transponder marked IDENT. This

causes the target on the controller's radar screen to change for several seconds. The pilot should not push the ident button without direction from ATC.

- Tow planes normally squawk 1200 unless otherwise instructed by ATC.

Outside Air Temperature (OAT) Gauge

The outside air temperature gauge (OAT) mounts with the sensing element in contact with the outside air. OAT gauges display degrees Celsius, degrees Fahrenheit, or both, and provide the glider pilot with information about freezing temperatures which could affect water ballast or flight controls. [*Figure 4-31*]

Figure 4-31. *Outside air temperature (OAT) gauge.*

When flying a glider loaded with water ballast, knowledge of the height of the freezing level can affect safety of flight. Extended operation of a glider loaded with water ballast in below-freezing temperatures may result in frozen drain valves, ruptured ballast tanks, and structural damage to the glider.

Chapter Summary

This chapter introduced the pitot-static system and its associated instruments including the airspeed indicator, the altimeter, and the variometer. A glider has various V-speeds and some of these speeds correspond to specific markings on the airspeed indicator regardless of altitude. The chapter also explains what a pilot should know about setting an altimeter, nonstandard pressure and temperature effects on indicated altitude, and different kinds of altitude. Variometers normally indicate the sum of the descent rate of the glider added to the lift or sink of the surrounding air mass. A compensated variometer removes indications that result from changes in airspeed (stick thermals). A netto removes the sink rate of the glider and provides the vertical speed of the surrounding air mass. Flight computers can integrate information from different sources, including from memory, and provide a wealth of information to the glider pilot. Other instruments described in this chapter include a magnetic compass, yaw string, inclinometer, gyroscopes, G-meter, and outside temperature gauge (OAT). Pilots can use FLARM systems for short range collision avoidance, however the system can only detect other aircraft that use the FLARM system. A transponder-equipped glider automatically reports position and altitude to ATC.

Chapter 5: Glider Performance

Introduction

Glider performance depends on design, weather, wind, and other atmospheric phenomena. While the design of the glider affects performance to a great degree, design remains fixed and known to the pilot. Other factors change and affect launch, cruise, and landing.

Variable Performance Factors

For a specific glider flight, some factors that affect performance include density altitude, wind, and weight.

Density Altitude

In general, an increase in density altitude refers to thinner air, while a decrease in density altitude refers to thicker air. Conditions that result in high-density altitude include high elevation, low atmospheric pressure, high temperature, high humidity, or some combination of these factors. Lower elevation, high atmospheric pressure, low temperature, and low humidity reduce density altitude. Since high density altitudes may exist at lower elevations on hot days, pilots should consider density altitude based on actual conditions before a flight.

A chart provides one way to determine density altitude. [*Figure 5-1*] For example, knowing field elevation of 1,600 feet MSL with a current altimeter setting of 29.80 "Hg and temperature of 85 °F, what is the density altitude? The right side of the chart provides an adjustment for nonstandard pressure (29.80 "Hg) and suggests adding 112 feet to the field elevation. This step provides a current pressure altitude of 1,712 feet. The next step involves tracing a line vertically from the bottom of the chart from the temperature of 85 °F (29.4 °C) that intercepts the diagonal 1,712-foot pressure altitude line. The final step involves tracing a line horizontally to the left from the interception point and reading the density altitude of approximately 3,500 feet. Under these conditions, a self-launching glider or towplane will perform as if at 3,500 feet MSL on a standard day.

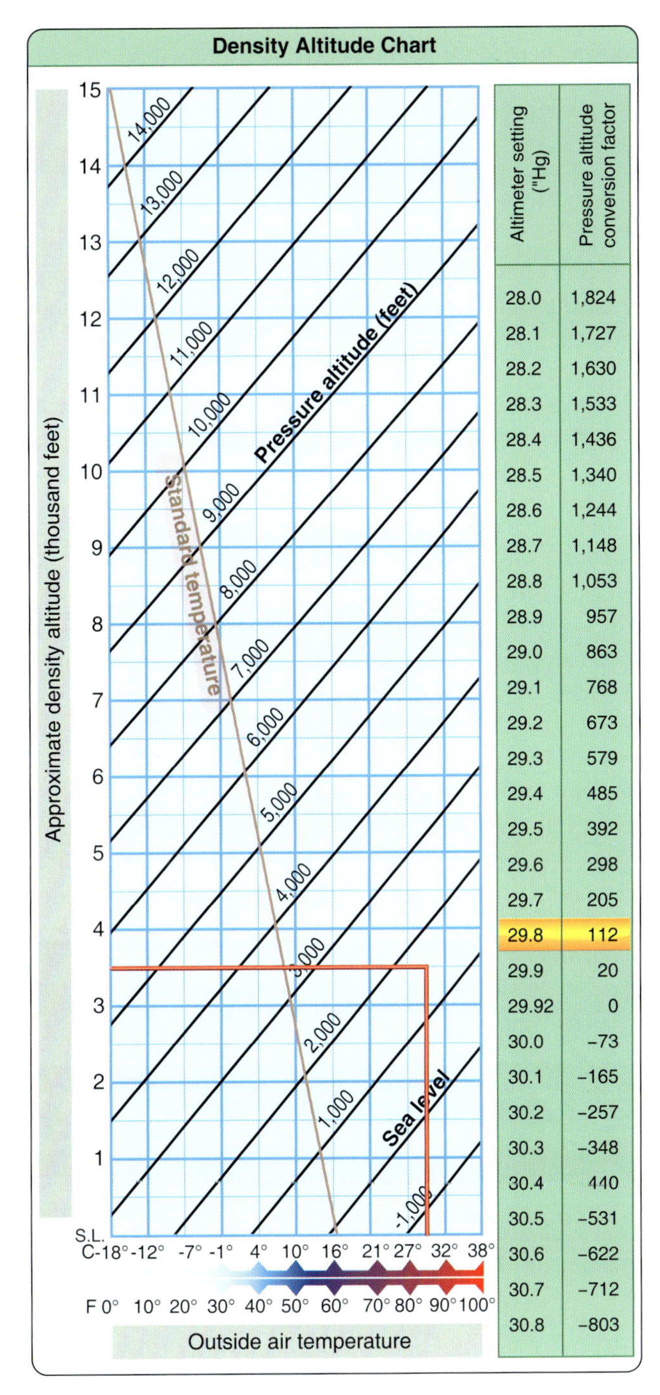

Altimeter setting ("Hg)	Pressure altitude conversion factor
28.0	1,824
28.1	1,727
28.2	1,630
28.3	1,533
28.4	1,436
28.5	1,340
28.6	1,244
28.7	1,148
28.8	1,053
28.9	957
29.0	863
29.1	768
29.2	673
29.3	579
29.4	485
29.5	392
29.6	298
29.7	205
29.8	112
29.9	20
29.92	0
30.0	−73
30.1	−165
30.2	−257
30.3	−348
30.4	440
30.5	−531
30.6	−622
30.7	−712
30.8	−803

Figure 5-1. *Density altitude chart.*

Many performance charts use pressure altitude and temperature inputs without requiring the pilot to calculate density altitude. However, if the pilot wants to know the density altitude, a density altitude chart or a flight computer can supply that information.

Atmospheric Pressure

Atmospheric pressure at a given location changes from day to day. The following sample Meteorological Aerodrome Report (METAR) indicates a local pressure of A2953, or altimeter setting of 29.53 inHg. When considering barometric pressure only, the lower than normal pressure reading results in a higher density altitude that decreases aircraft performance. This reduction affects takeoff and climb performance and increases the length of runway needed during landing for both the glider and towplane. On the other hand, if barometric pressure rises, the lower density altitude improves takeoff and climb performance, and the length of runway needed for landing decreases.

KDAL 232153Z 21006KT 7SM -RA BKN025 BKN060 OVC110 12/11 A2953 RMK AO2 PRESFR SLP995 P0005 T01220106

Temperature

Temperature changes have a significant effect on density altitude. Heated air expands—the molecules move farther apart, making the air less dense. Higher density altitude reduces glider and towplane takeoff and climb performance and increases the length of runway required for landing.

Consider the following METAR for two airports with same altimeter setting, temperature, and dewpoint. Love Field (KDAL) airport elevation of 487 feet versus Denver International (KDEN) at 5,431 feet.

KDAL 240453Z 21007KT 10SM CLR 25/15 A3010 RMK AO2...

KDEN 240453Z 24006KT 10SM FEW120 SCT200 25/15 A3010 RMK AO2...

The computed density altitude for Love Field is 1,774 feet; for Denver, 7,837 feet—Denver experiences almost twice the amount of increase compared to Love Field. The effects of altitude and temperature can surprise a pilot who does not consider the related performance issues.

Wind

Wind also affects glider performance. Headwind during launch or landing results in a shorter ground roll, while tailwind causes a longer ground roll. [*Figure 5-2*] Crosswinds during launch or landing require proper crosswind procedures or control input to track along the runway.

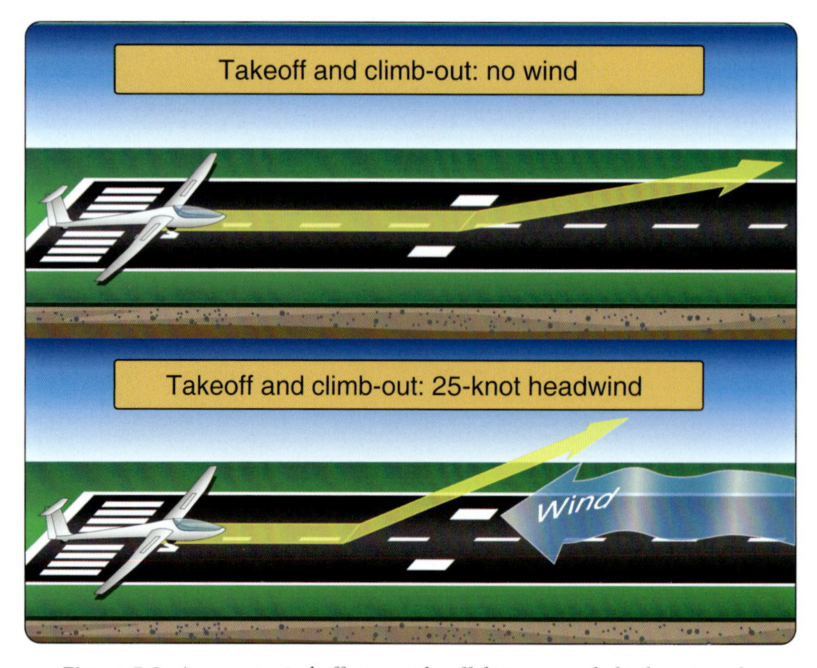

Figure 5-2. *Apparent wind effect on takeoff distance and climb-out angle.*

Due to diminishing ground friction between the wind and the ground, wind speed often increases with altitude. This "wind gradient" during a ground launch could result in an exceedance of the maximum launch speed. [*Figure 5-3*]

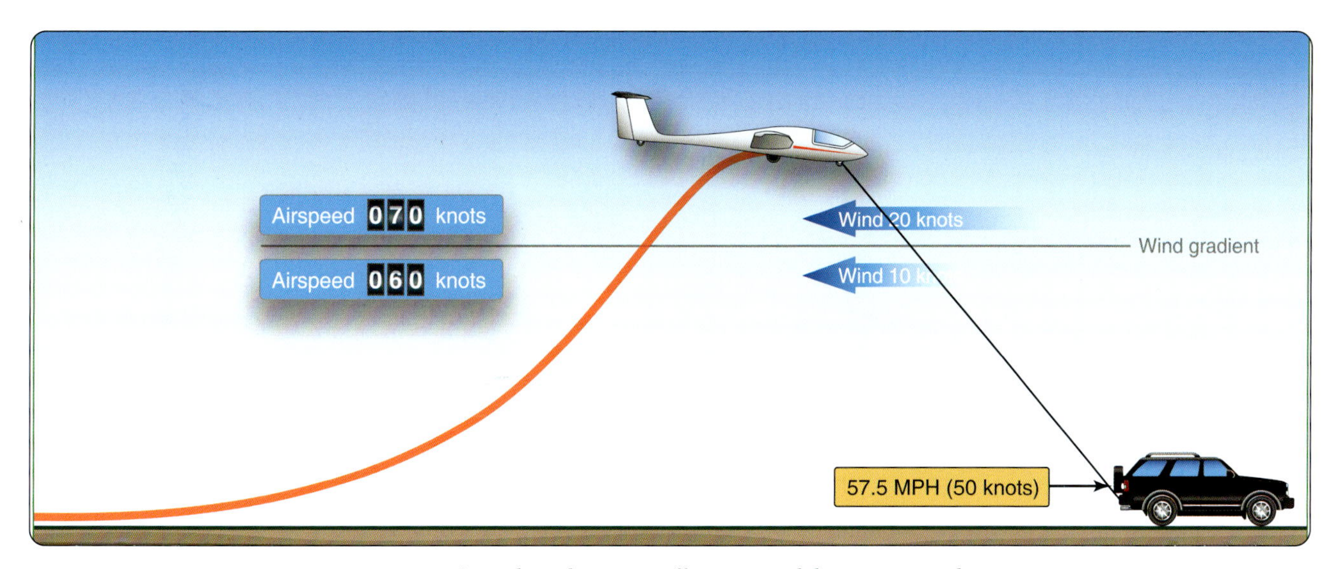

Figure 5-3. *A wind gradient may affect airspeed during a ground tow.*

During cruising flight, headwinds reduce the groundspeed of the glider. A glider flying at 60 knots true airspeed into a headwind of 25 knots has a groundspeed of only 35 knots. Tailwinds increase the groundspeed of the glider. A glider flying at 60 knots true airspeed with a tailwind of 25 knots has a groundspeed of 85 knots.

Some self-launching gliders can cruise for extended periods using the engine. The self-launching glider's fuel capacity normally limits maximum range and duration with the engine running. Wind has no effect on flight duration but does have a significant effect on range. During powered cruising flight, a headwind reduces range, and a tailwind increases range. The Glider Flight Manual/Pilot's Operating Handbook (GFM/POH) provides recommended airspeeds and power settings to maximize range when flying in no wind, headwind, or tailwind conditions.

When flying straight and level and following a selected ground track, the pilot can point the glider into the prevailing wind as the preferred method to correct for wind drift. The wind speed, the angle between the wind direction and the glider's longitudinal axis, and the airspeed of the glider determine the required wind correction angle. [*Figure 5-4*] Crosswinds may also have a head or tailwind component that results in a lower or higher groundspeed.

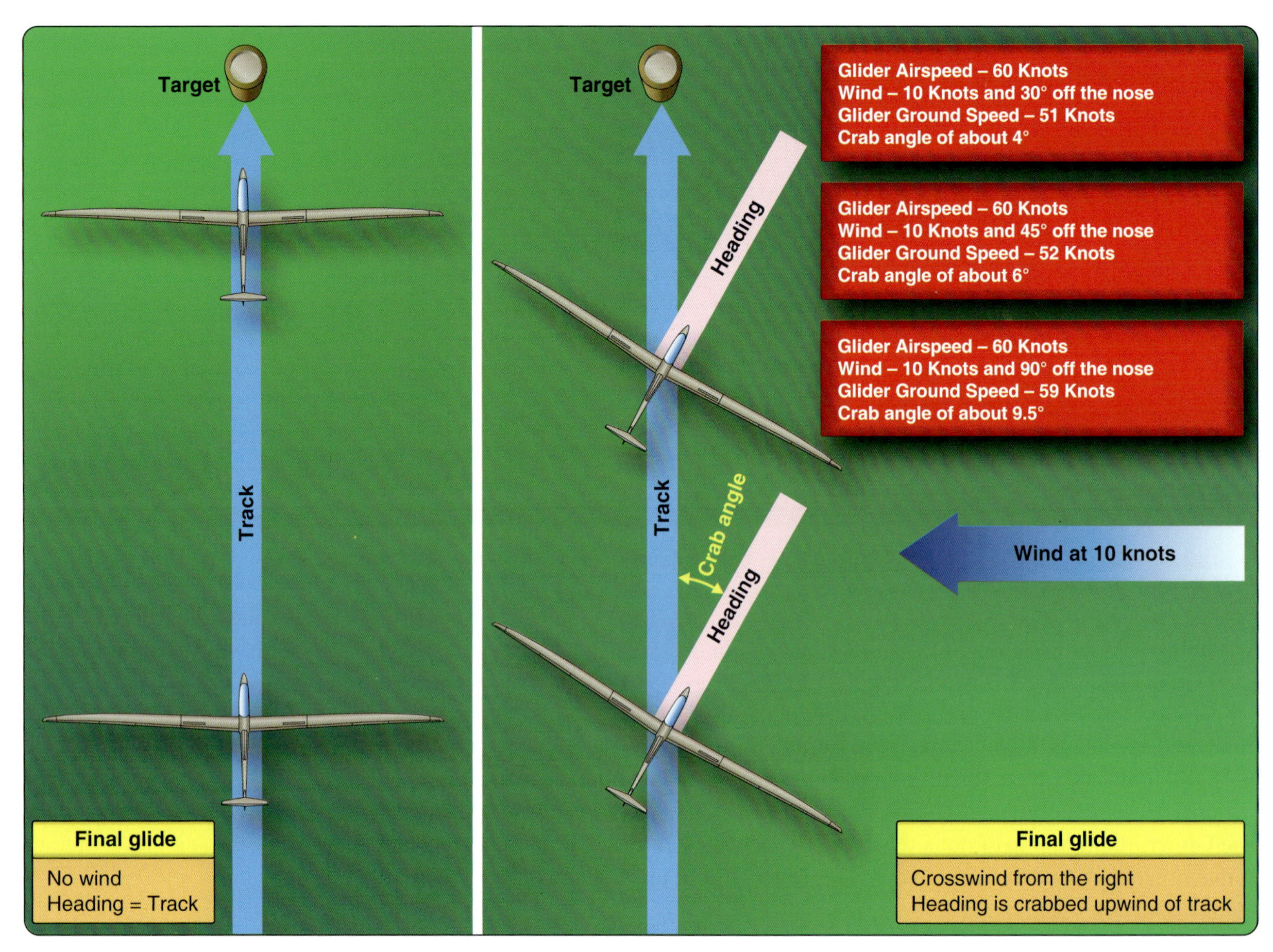

Figure 5-4. *Crosswind effect on final glide.*

A headwind during an approach results in greater altitude loss per distance traveled. The glider descends at a constant rate, but a lower groundspeed increases the observed approach angle. Pilots use various techniques to ensure safe touchdowns in different wind conditions. For example, if landing in a strong headwind, the glider pilot should plan for a base leg closer to the landing zone to allow for the steeper approach. Another technique uses delayed extension of spoilers or dive brakes with a faster airspeed to counter the headwind component. In the case of a tailwind and an apparent lower angle of approach, the glider pilot can use more spoiler or dive brake extension, slip, or use a combination of the two if allowed by the GFM. In any case, the pilot generally aims for a spot past the threshold of the runway to provide a safety factor that accounts for the effects of the wind gradient or other factors that may cause the approach to be shorter than expected. [*Figure 5-5*]

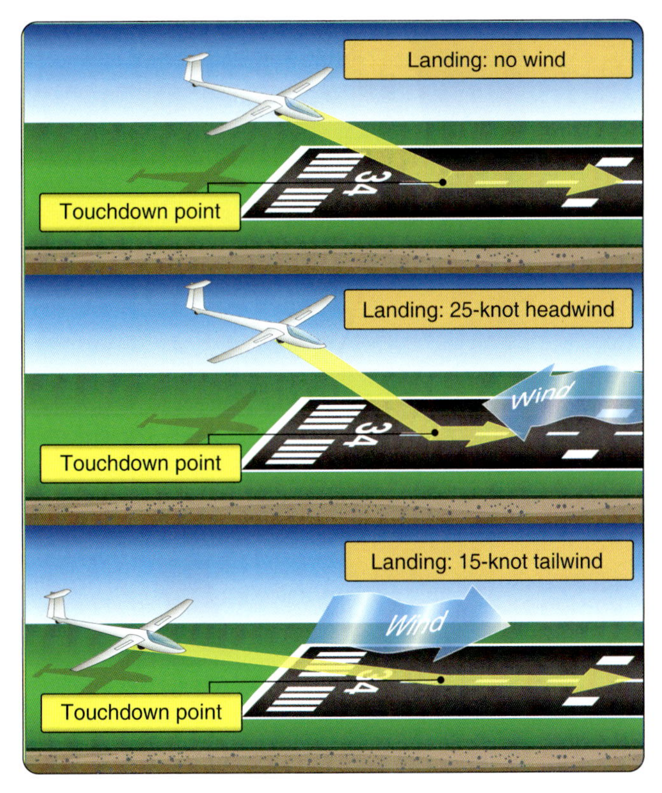

Figure 5-5. *The effect of wind on final approach and landing distance.*

When approaching to land during windy and gusty conditions, a pilot normally adds half of the difference between the steady wind and gusts to the approach speed to mitigate any variations in airspeed. Instead of holding the glider off the ground for a low kinetic energy landing during these conditions, the pilot can land a little faster than normal. Upon touchdown, extending the air brakes prevents the glider from becoming airborne from a gust during the landing roll.

The pilot usually expects any headwind to diminish during descent to a landing. This is called the wind gradient. If the wind gradient is abrupt, the pilot may experience a sudden airspeed loss. After the pilot lowers the nose to compensate, it takes a finite time interval to overcome the inertia of the glider and regain airspeed. A significant speed loss near the ground may preclude recovering any or all the lost speed. A wind gradient on final approach may cause the glider to land short of the point of intended touchdown therefore, a closer approach may be necessary.

The pilot landing with a tailwind has a higher groundspeed and should expect a longer landing roll. As the surface friction slows the winds, the pilot may observe an increase in airspeed.

A strong windshear gradient can affect a glider during a steep turn on a low altitude final approach at a low airspeed. The gradient can create different lift on the low wing and the raised wing. [*Figure 5-6*] The rolling force created by the gradient can overcome the ailerons, cause a loss of control, and explains why the pilot should limit bank angle while close to the ground and transitioning a strong wind gradient.

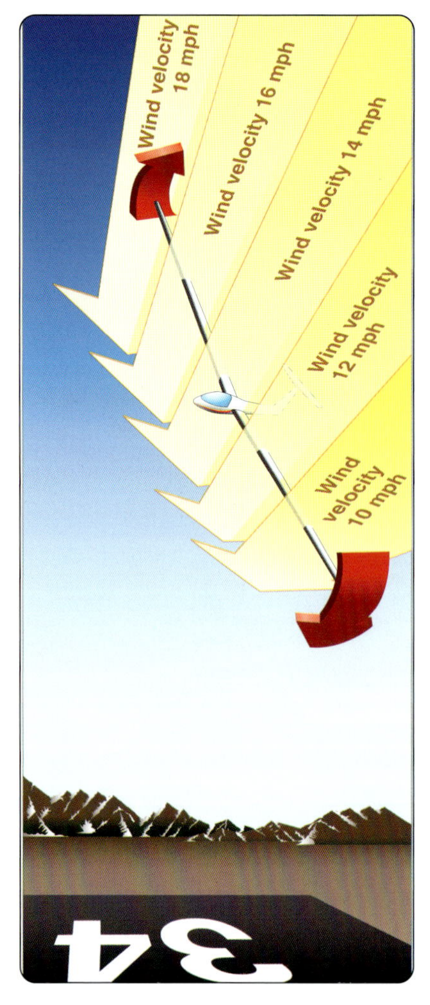

Figure 5-6. *Effect of wind velocity gradient on a glider. Stronger airflow over higher wing may cause bank to steepen.*

Weight

The weight of a glider affects acceleration during launch. A glider at maximum takeoff weight takes longer to attain flying speed. After takeoff, a tow plane with a heavy glider will climb more slowly. Increasing the weight of a powered glider causes it to accelerate more slowly and climb more slowly. [*Figure 5-7*]

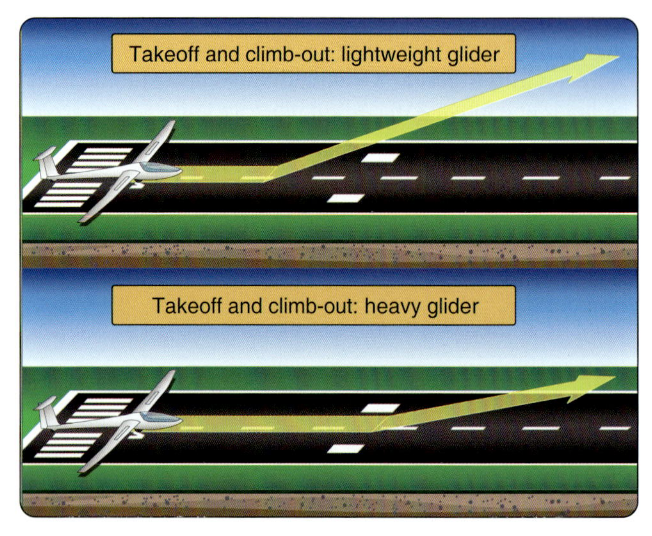

Figure 5-7. *Effect of weight on takeoff distance and climb-out rate and angle.*

The stall speed increases with the square root of any load on the glider. For example, if the weight or load factor on the glider doubles, the stall speed increases by the square root of 2 or 1.41. If a 540-pound glider has a stalling speed of 40 knots and the pilot adds 300 pounds of water ballast increasing the total weight to 840 pounds, the stall speed increases to approximately 50 knots (40 x √(840/540)).

The speed that results in the lowest altitude loss over time, the minimum sink airspeed, also increases with weight. At any given bank angle, a heavier glider uses higher airspeeds for efficiency, which result in larger diameter circles. The best lift in thermals often occurs in a narrow cylinder near the core, and large diameter circles generally reduce the glider's capability to exploit the strongest lift. [*Figure 5-8*]

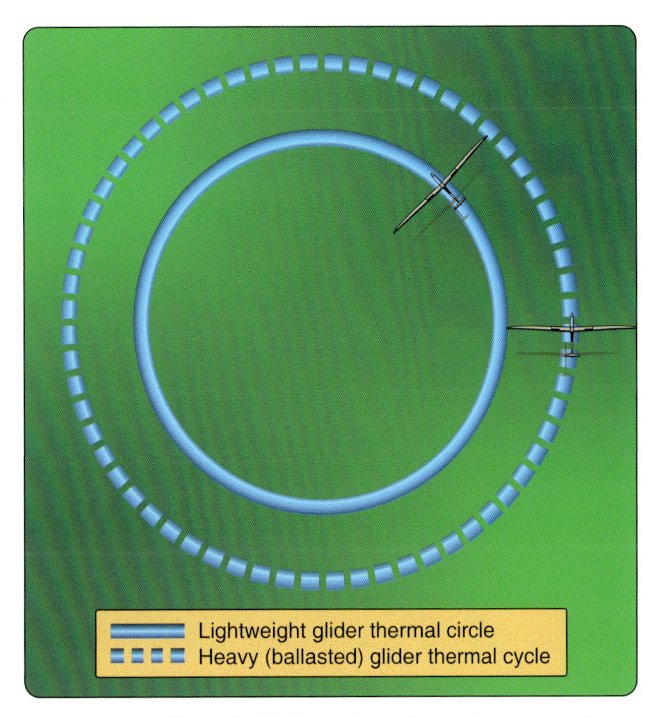

Figure 5-8. *Effect of added weight on thermaling turn radius.*

Increasing the operating weight of a given glider not only increases the stall airspeed and minimum sink airspeed, it also increases the best L/D airspeed [*Figure 5-9*]. Although a heavier glider sinks faster, it glides the same horizontal distance (at a higher speed) as a lighter glider with the same glide ratio and starting altitude.

Operating Weight	Stall Airspeed	Minimum Sink	Best L/D Airspeed
800 pounds	36 knots	48 knots	60 knots
1,200 pounds	44 knots	58 knots	73 knots
1,600 pounds	50 knots	68 knots	83 knots

Figure 5-9. *Effect of added weight on performance airspeeds.*

Figure 5-9 above shows that increasing weight from 800 to 1,200 pounds increases the best L/D airspeed from 60 knots to 73 knots. A glider with more weight can fly faster while maintaining the same lift-to-drag (L/D) ratio (glide ratio). The advantage of the heavier weight becomes apparent during faster flight between thermals in sink. In strong lift the heavy glider can climb reasonably well, and the advantage during the cruising portion of flight may outweigh the disadvantage during climbs.

To fly faster with efficiency, some gliders have water tanks that allow the pilot to add weight as ballast. The pilot normally jettisons the water ballast before entering the traffic pattern or earlier if conditions necessitate a higher climb rate. Reducing the weight of the glider before landing allows the pilot to make a normal approach, normal landing, and reduces load on the landing gear.

Rate of Climb

Rate of climb for a ground-launched glider depends on the power of the ground-launch equipment. With a powerful winch or tow vehicle, rate of climb can exceed 2,000 feet per minute (fpm). For an aerotow, the rate of climb depends on the power of the towplane. The towplane should have sufficient power to tow the glider safely, considering the existing conditions, which include glider weight.

The rate of climb of self-launching gliders may vary from as low as 200 fpm to as much as 800 fpm or more. The pilot should consult the GFM/POH to determine rate of climb under the existing conditions.

Flight Manuals & Placards

The GFM/POH provides the pilot with the necessary performance information to operate the glider safely. A GFM/POH may include the following information:

- Description of glider primary components
- Glider assembly instructions
- Weight and balance data
- Description of glider systems
- Glider performance data
- Operating limitations

Placards

Placards attached to the glider provide the pilot with essential information for safe operation. The GFM/POH lists all the required placards.

The amount of information that placards convey to the pilot increases as the complexity of the glider increases. High-performance gliders may have wing flaps, retractable landing gear, a water ballast system, drogue chute for use in the landing approach, and other features to enhance performance. [*Figure 5-10*]

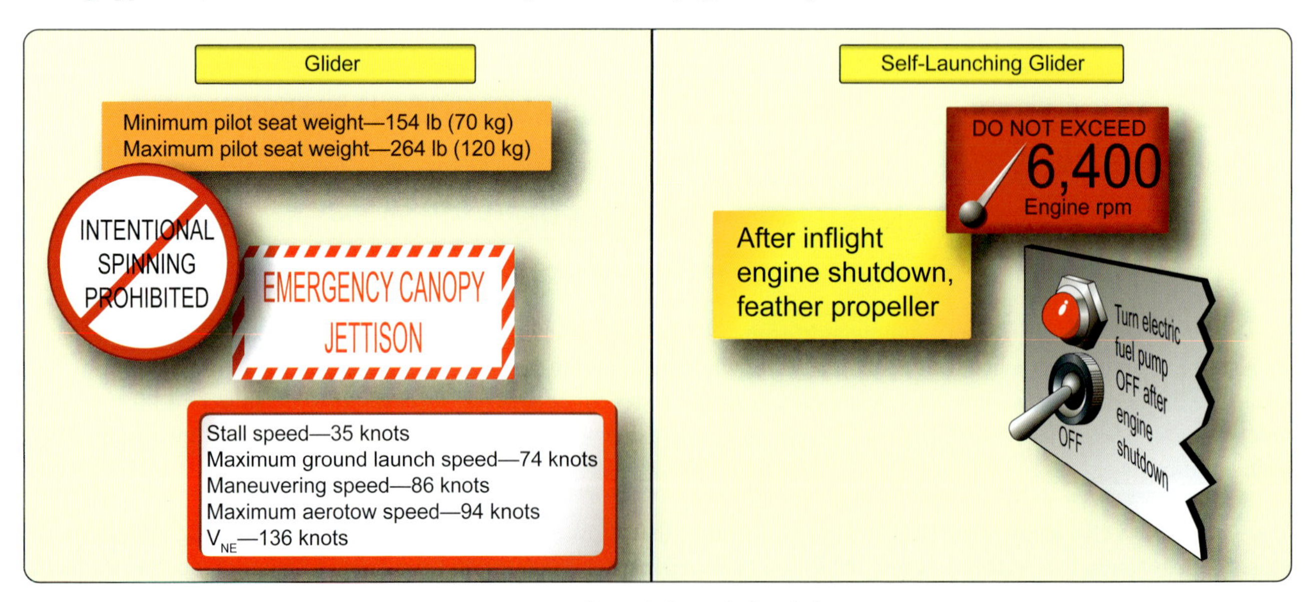

Figure 5-10. *Typical placards for gliders.*

Performance Information

The GFM/POH provided by the manufacturer contains glider performance information. The GFM/POH lists specific airspeeds such as stall speed, minimum sink airspeed, best L/D airspeed, maneuvering speed, rough airspeed, maximum aerotow speed, maximum ground launch speed, and the never exceed speed (V_{NE}). Some performance airspeeds apply only to gliders with certain equipment. For instance, gliders with wing flaps have a maximum permitted flap extended airspeed (V_{FE}).

Manuals for self-launching gliders include performance information about powered operations. These include rate of climb, engine and propeller limitations, fuel consumption, endurance, and cruise.

Glider Polars

The manufacturer provides information about the rate of sink in terms of airspeed summarized in a graph called a polar curve, or simply a polar.

The vertical axis of a polar shows the sink rate (increasing sink downwards), while the horizontal axis shows airspeed in the same units (knots). Every type of glider has a characteristic polar derived either from theoretical calculations or by actual inflight measurement of the sink rate at different speeds. The polar of each individual glider varies (even from other gliders of the same type) by a few percent depending on relative smoothness of the wing surface, the sealing around control surfaces, and even the number of bugs on the wing's leading edge. The polar forms the basis for speed to fly and strategies discussed in Chapter 11, Cross-Country Soaring.

The peak of the blue sink rate curve determines minimum sink rate. [*Figure 5-11*] In this example, a minimum sink of 1.9 knots occurs at 40 knots. Note that the sink rate increases between minimum sink speed and the stall speed (the left end point of the blue curve). A tangent from the origin to the polar indicates the best glide speed (best L/D). The best L/D speed is 50 knots with a sink speed of 2.1 knots. The glide ratio at best L/D speed is determined by dividing the best L/D speed by the sink rate at that speed, or 50/2.1, which is approximately 24 in this example. Thus, this glider has a best glide ratio in calm air (no lift or sink and no headwind or tailwind) of 24:1 at 50 knots.

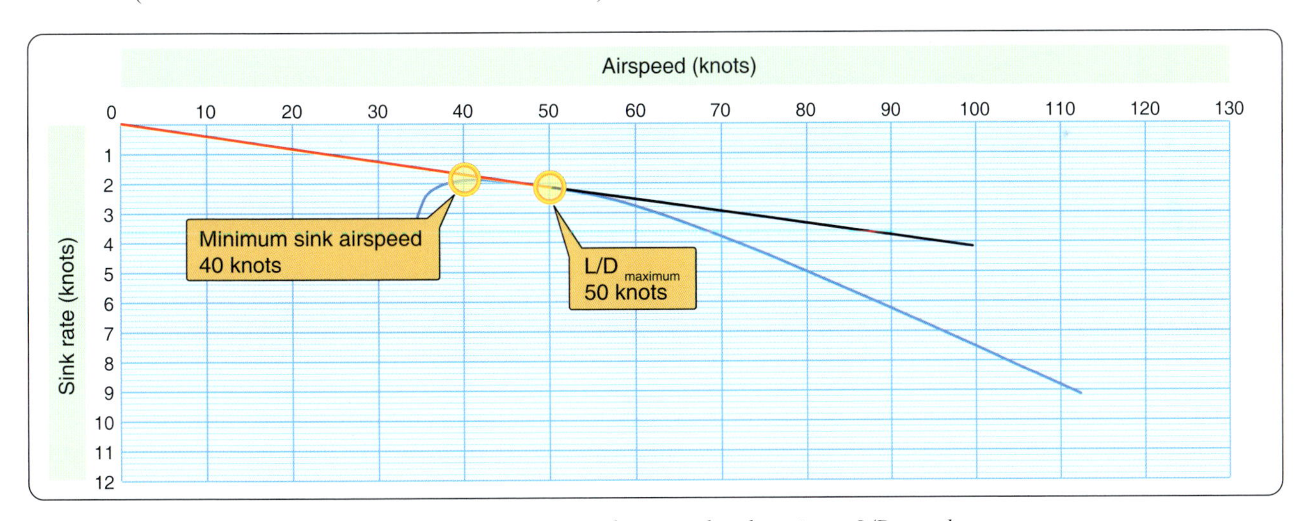

Figure 5-11. *Minimum sink airspeed and maximum L/D speed.*

To determine the best speed to fly for distance over the ground in a headwind, the pilot can shift the origin to the right along the horizontal axis by the speed of the headwind and draw a new tangent line to the polar. For tailwinds, the pilot shifts the origin to the left of the zero mark on the horizontal axis.

Figure 5-12 shows an example for a 20-knot headwind. The new tangent indicates 60 knots as the best glide speed. By repeating the procedure for different headwinds, the data show that flying faster as headwinds increase results in a greater distance traveled over the ground. Analysis of the data from many gliders leads to the following general rule: the pilot can add half the headwind component to the zero wind L/D to obtain maximum distance.

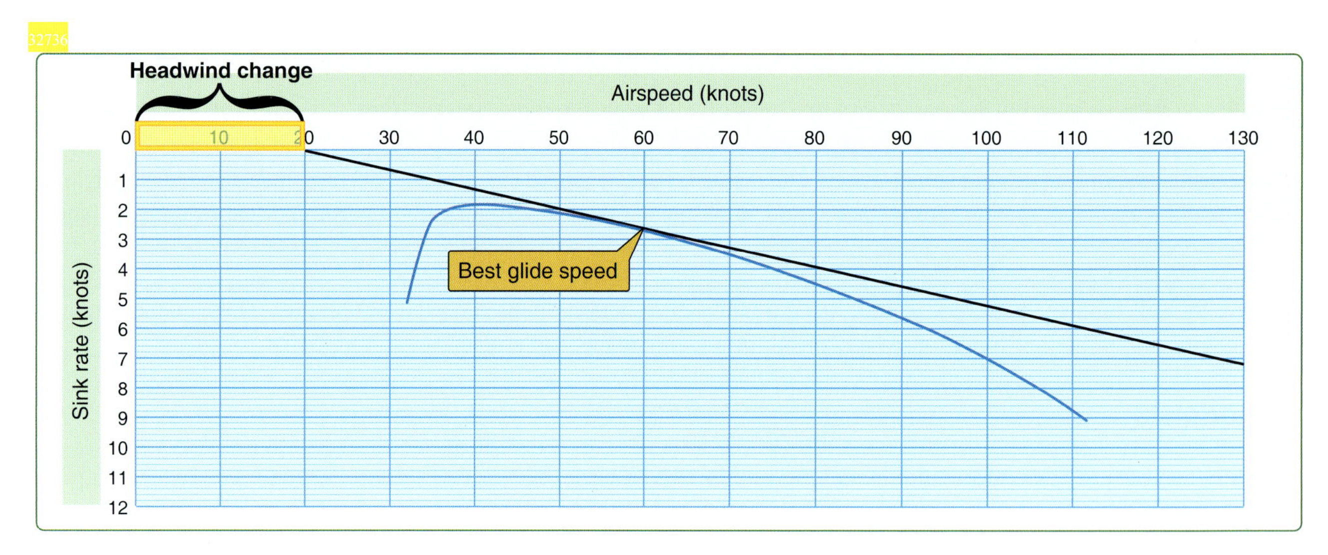

Figure 5-12. *Best speed to fly in a 20-knot headwind.*

The speed to fly in a tailwind lies between minimum sink and best L/D, but never lower than minimum sink speed.

Sinking air often exists between thermals and flying faster than best L/D can result in less time in sinking air and an increase in efficiency. The pilot can determine how much faster to fly using the glider polar, as illustrated in *Figure 5-13* for an airmass sinking at 3 knots between thermals. In this case, the pilot draws a tangent line to the polar that begins 3 knots above the origin to read the best speed to fly, 60 knots in this case. Note, that in this situation, the variometer would show a total sink of 5 knots (3 knots from sinking air and 2 knots for the constant aircraft descent) as highlighted in the figure.

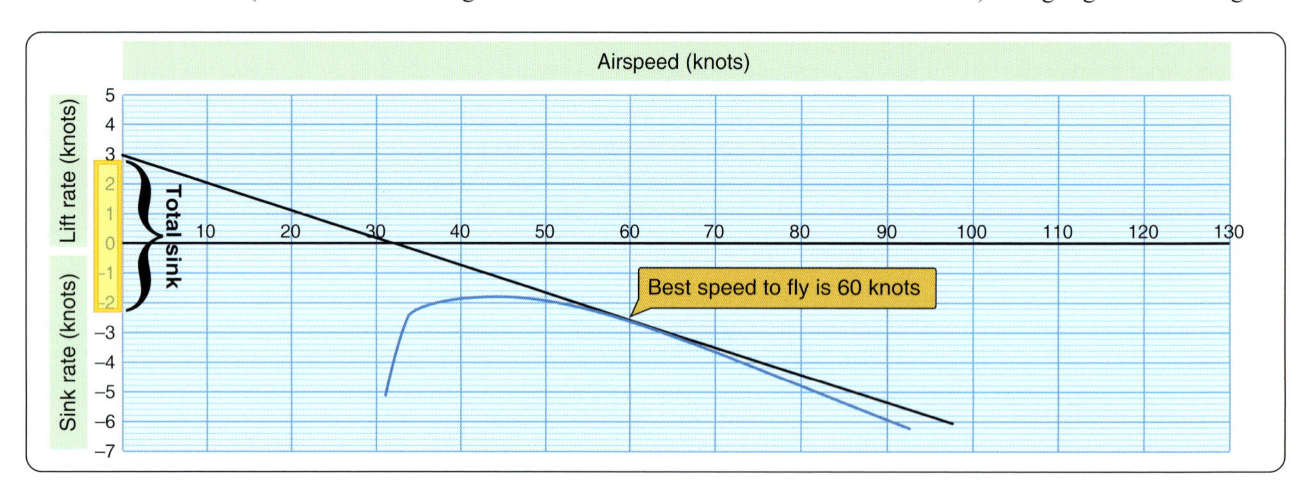

Figure 5-13. *Best speed to fly in sinking air.*

If the glider has water ballast, wing flaps, or wingtip extensions, the polar depicts the performance characteristics of the glider in those configurations. [*Figure 5-14*, *Figure 5-15*, and *Figure 5-16*] Comparing the polar with and without ballast shows that the minimum sink increases and occurs at a higher speed after adding weight. As a result, working weak thermals becomes more challenging with ballast. In addition, ballast lowers the sink rate at higher speeds. [*Figure 5-14*] The best glide ratio remains the same, but it occurs at a higher speed. Note that as expected, the stall speed increases with added ballast.

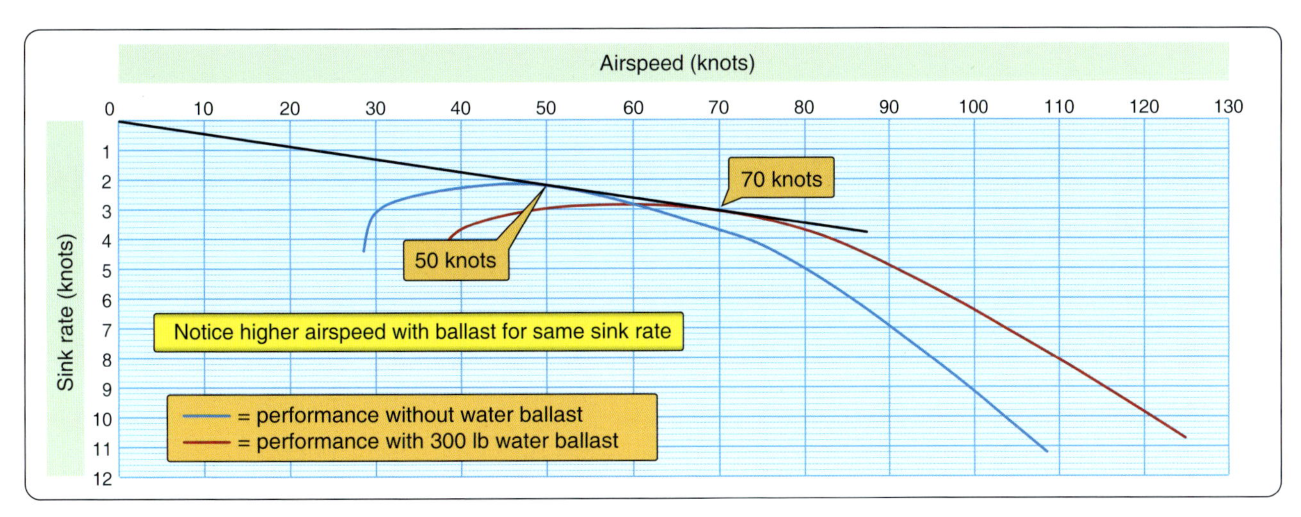

Figure 5-14. *Effect of water ballast on performance polar.*

Flaps with a negative setting as opposed to a 0-degree setting during cruise also reduce the sink rate at higher speeds, as shown in the polar. [*Figure 5-15*] Therefore, when cruising at or above 70 knots, setting the flaps to -8° would provide an advantage. The polar with flaps set at -8° does not extend to speeds lower than 70 knots since the negative flap setting loses its advantage there.

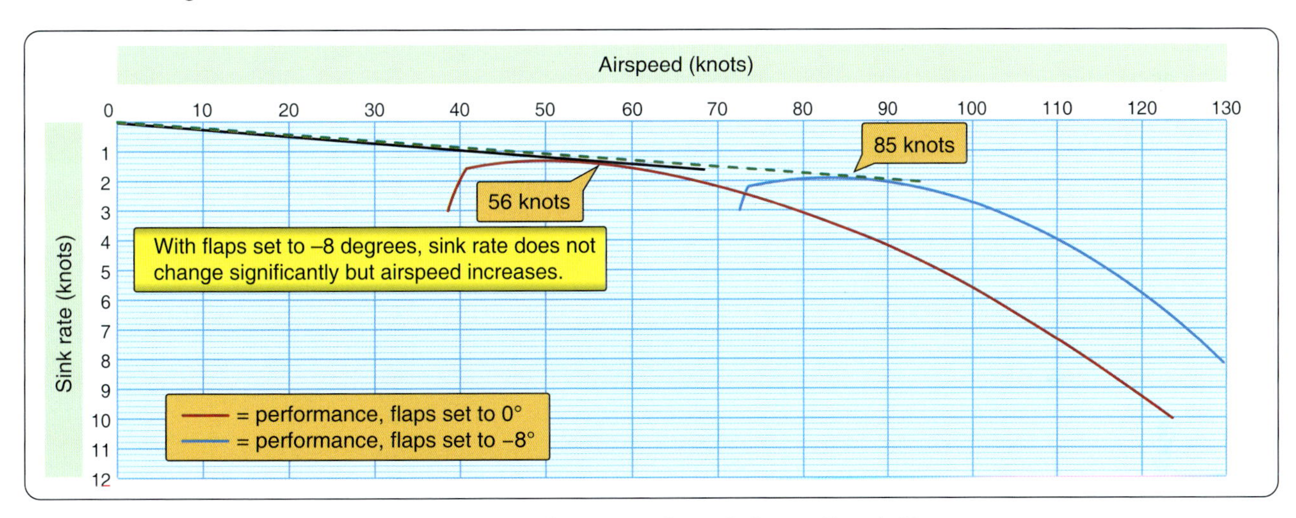

Figure 5-15. *Performance polar with flaps at 0° and –8°.*

Wingtip extensions also alter the polar data, as shown in *Figure 5-16*. The illustration shows that the additional 3 meters of wingspan creates an advantage at all speeds. In some gliders, the low-speed performance improves with the tip extensions, while high-speed performance diminishes slightly.

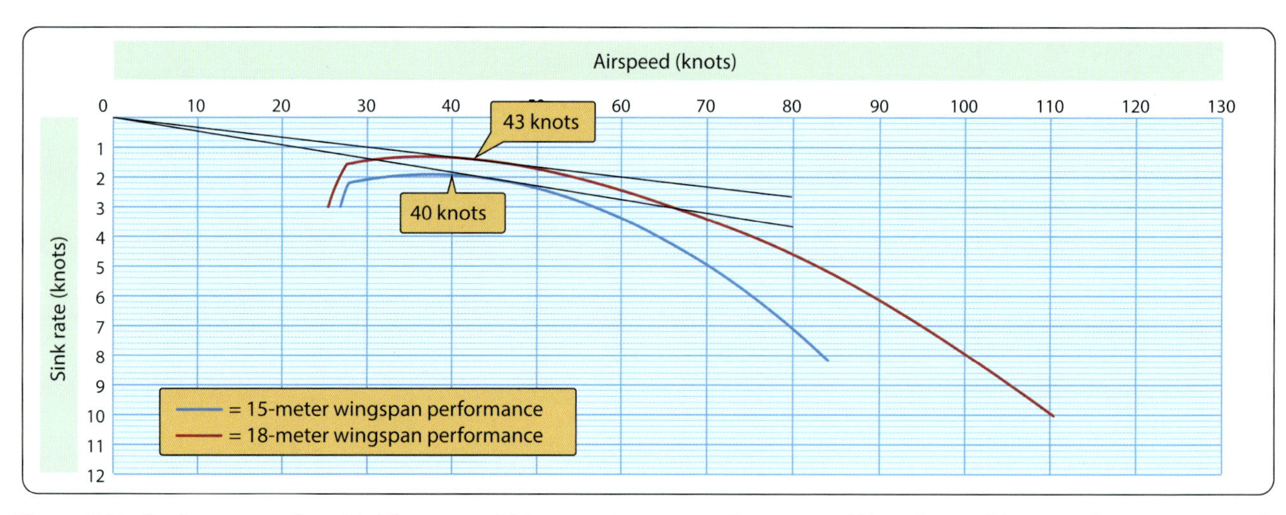

Figure 5-16. *Performance polar with 15-meter and 18-meter wingspan configurations. (Matt please add tangent lines to intersect the curves at the 43 and 40 knot annotated points.)*

Limitations

Regardless of glider complexity, designers and manufacturers provide operating limitations to ensure the safety of flight. The glider VG diagram provides the pilot with information on the design limitations, such as limiting airspeeds and load factors (L.F. in *Figure 5-17*). Pilots should become familiar with all the operating limitations of each glider flown and should not operate outside the limits. *Figure 5-17* shows different possible limiting conditions and the basic flight envelope for a high-performance glider.

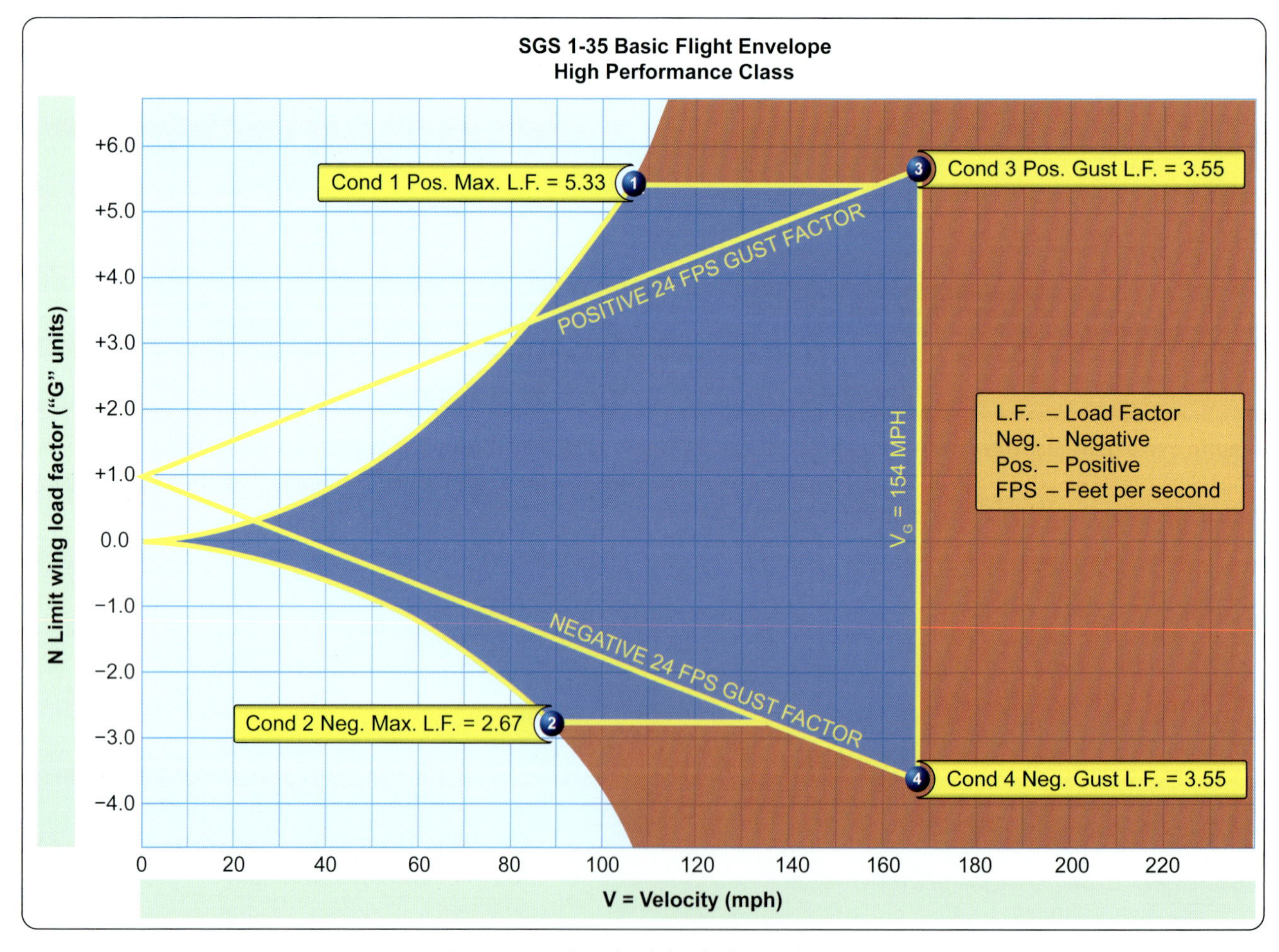

Figure 5-17. *Sample glider flight envelope.*

The curved yellow lines represent the maximum lift that the glider can generate at different airspeeds as Gs.

Condition 1 identifies the maximum speed at which the pilot can use full elevator up authority without damaging the glider. Up to this point along the curve and to the left at lower speeds, the glider would stall before the pilot exceeds the design load limit shown on the diagram. At higher speeds (to the right of condition 1), lift can exceed the maximum design load factor before a stall occurs. Such operation may impose damaging loads to the structure.

Condition 2 represents the speed at which the pilot can use full down elevator authority and not create a negative load that damages the glider. Above this speed the pilot can impose damaging loads to the structure.

Vertical Gusts During High-Speed Cruise

In high-speed cruise pilots should pay attention to load factor limitations. An encounter with an abrupt updraft during wings-level high-speed cruise increases the angle of attack, bends the wings upward, briefly increases the G-load, and stores elastic energy in the wing spars. As the wings release this energy, the wing spars spring downward and loft the fuselage higher. As the fuselage reaches the top of this motion, the wing spars, now bent downward, move upward again to release the stored energy. Since a negative G-load can occur as the fuselage drops downward, the seat belt and shoulder harness can prevent the pilot's head from banging against the top of the canopy.

During these excursions, the weight of the pilot's hand and arm may inadvertently move the control stick forward or aft. Positive G-loading and the increased apparent weight of the pilot's arm tend to move the control stick aft and further increase the angle of attack and G-load. Negative G-loading and the decreased apparent weight of the pilot's arm tend to move the control stick forward and further decrease the angle of attack and G-load.

To minimize the intensification of vertical gusts and avoid high-speed pilot induced oscillations (PIOs), the pilot should reduce speed when cruising through turbulent air. The pilot may also brace both arms and use both hands on the control stick to prevent unwanted input. Some glider designs incorporate a parallelogram control stick linkage to reduce the likelihood of PIOs during high-speed cruise.

Weight & Balance

The pilot should understand proper weight and balance management and the consequences of overloading or improperly loading the glider.

Weight & Balance Information

The GFM/POH provided by the manufacturer gives information about the weight and balance of the glider. Since addition or removal of equipment, such as radios, batteries, flight instruments, or airframe repairs affect the CG position, aviation maintenance technicians (AMTs) record changes to the weight and balance data in the GFM/POH and glider airframe logbook. They also update weight and balance placards.

Center of Gravity

Longitudinal balance affects stability around the lateral axis of a glider. To achieve satisfactory pitch attitude handling, manufacturers position the center of gravity (CG) of a properly loaded glider forward of the center of lift (CL) and publish the glider CG limits in the GFM/POH.

On most gliders, the horizontal stabilizer and elevator provide a down force to balance the CG and center of lift arrangement. As the airspeed changes the pilot adjusts the trim, and the tail-down force exactly balances the forward CG. A glider in this configuration tends to resume its previous pitch attitude after an upset about the lateral axis. Should an upset occur that pitches the nose upward, the resultant slower airspeed and decrease in tail-down force lowers the nose and allows the airspeed to return toward its pre-upset value. Conversely, if the upset places the aircraft in a nose-down attitude, the increase in airspeed increases tail-down force and raises the nose toward the pre-upset condition. This arrangement creates positive stability. However, if the tail stalls, this stabilizing action will not begin until the tail begins producing down force.

Problems Associated with CG forward of the Published Limit

Loading the glider with the CG forward of the limit makes it difficult to raise the nose on takeoff and requires considerable back pressure on the controls to regulate pitch attitude. At low airspeeds the tail may stall or not provide sufficient down force. Any tail stall results in a sudden nose-down pitch change and potential for a slow recovery. The pilot may not have sufficient elevator authority to perform the landing flare due to nose heaviness. Inability to flare could result in a nose-first hard landing.

A CG forward of the limit might occur for these reasons:

- The pilot weight exceeds the maximum permitted.
- Installed ballast weights added to the weight of the pilot exceed the maximum permitted.

Problems Associated with CG forward of the Published Limit

Loading a glider with the CG location behind the aft limit creates a tail-heavy condition. Tail heaviness can make pitch control of the glider difficult or impossible.

A CG aft of the limit might occur for these reasons:

- The pilot weighs less than the specified minimum pilot seat weight without necessary ballast installed in the glider.
- Tailwheel dolly not removed prior to flight.
- A heavy, non-approved tailwheel or tail skid installed on the aft tail boom of the glider.
- Foreign matter or debris (water, ice, mud, sand, or nests) accumulation in the aft fuselage.

Sample Weight & Balance Problems

Some glider manufacturers provide weight and balance information in a graphic presentation. A well-designed graph provides a convenient way to determine whether the glider is within weight and balance limits.

Sample *figure 5-18* indicates that the minimum weight for the front seat pilot is 125 pounds (the lowest number on the x-axis) to a maximum of 250 pounds (the highest number on the x-axis). It also indicates a maximum rear seat pilot weight of 225 pounds (the highest number on the y-axis). If each pilot weighs 150 pounds, the intersection of pilot weights falls within the envelope. Therefore, the glider load falls within the envelope for safe flight. If each pilot weighs 225 pounds, the intersecting lines intersect in the yellow portion of the graph and indicate a load outside of weight and balance limits.

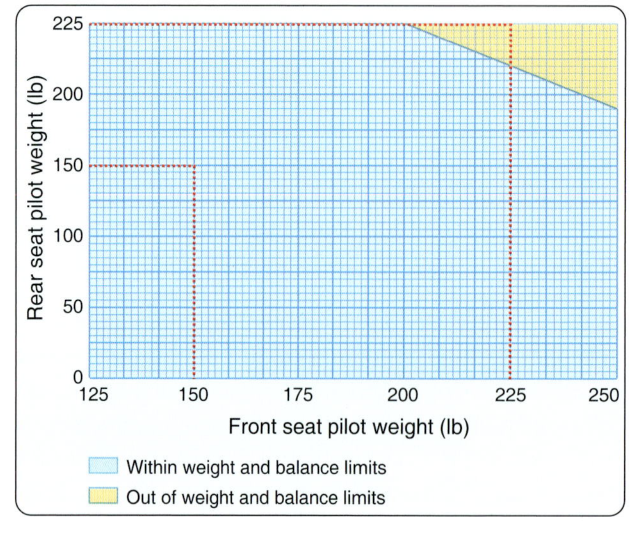

Figure 5-18. *Sample weight and balance envelope.*

Weight along the longitudinal axis of the glider affects the CG location. Pilots calculate the CG using of the arm or distance of known weights from a specific point (datum) on the longitudinal axis. The GFM/POH supplies the arm from the datum for the empty glider, each occupant seat, and for any cargo storage.

The pilot can determine the CG position using the following formulas:

- Weight × Arm = Moment.
- Total Moment ÷ Total Weight = CG Position (in relation to the datum).

The computational method involves the application of basic math functions as follows:

Given:

Maximum gross weight: 1,040 lb

Empty weight: 669 lb

CG range: 14.8–18.6 in

Front seat occupant: 180 lb

Rear seat occupant: 200 lb

To determine the loaded weight and CG, follow these steps:

1. List the empty weight of the glider and the weight of the occupants.
2. Enter the moment for each item listed. Remember, weight × arm = moment.
3. Total the weight and moments.
4. To determine the CG, divide the total moments by the total weight. [*Figure 5-19*]

Note: *The weight and balance records for a particular glider provide the empty weight and moment, as well as the information on the arm distance. [Figure 5-19]*

Item	Weight (pounds)	Arm (inches)	Moment (inch·pounds)
Empty weight	669	+93.7	+62,685
Front seat pilot	180	+43.8	+7,884
Rear seat pilot	190	+74.7	+14,193
	1,039 total weight	+81.58	+84,762 total moment

Figure 5-19. *Sample weight and balance: front and rear seat pilot weights and moments.*

In *Figure 5-19* above, the weight of each pilot appears in the appropriate block in the table. For the front seat pilot, multiplying 180 pounds by 43.8 inches yields a moment of 7,884 inch-pounds. For the rear seat pilot, multiplying 190 pounds by 74.7 inches yields a moment of 14,193 inch-pounds. The next step is to find the sum of all weights (980 pounds) including the empty weight of the glider. Then, find the sum of all moments (+84,762 inch-pounds). To determine the CG position of the loaded glider, divide the total moment by the total weight to in inches from the datum: 84,762 inch-pounds ÷ 1039 pounds = 81.58 aft of the datum.

For the final step the pilot determines whether total weight and CG location values are within acceptable limits. The GFM/POH lists the maximum gross weight as 1,040 pounds. The operating weight of 1039 pounds does not exceed the 1,040 pounds maximum gross weight. The GFM/POH lists the approved CG range as between 78.2 inches and 86.1 inches from the datum. The operating CG of 81.58 inches from the datum falls within these limits. Therefore, the calculation shows the glider within operating limits if loaded as planned.

Ballast

Ballast includes nonstructural weight added to a glider. In soaring, ballast weight serves two purposes. Trim ballast adjusts the location of the CG of the glider to remain within acceptable limits. Performance ballast improves high-speed cruise performance.

Trim Ballast

Removable trim ballast weights, often made of metal, attach to a ballast receptacle incorporated in the glider structure. These weights compensate for a front seat pilot who weighs less than needed to maintain the CG within acceptable operating limits. The ballast weight mounted well forward in the glider cabin can move the CG within permissible limits with the minimum addition of weight.

Whenever an approved POH or Glider Flight Manual limitation section includes specific instructions on the use of trim ballast, pilots must follow the approved method stated in the GFM/POH for placement of that ballast. For gliders without limitations or placards regarding placement of trim ballast, pilots may consider using a seat cushion with sand or lead shot sewn into the unit to provide additional weight. Since this type of ballast may shift position during maneuvering, pilots should not rely on seat cushion ballast during acrobatic or inverted flight. The pilot should develop a means to verify the presence, absence, weight, and appropriateness of any trim ballast before a flight.

Trim ballast may also include water in a tail tank in the vertical fin. Water weighs 8.35 pounds per gallon. Because of its far-aft location, the pilot can use a small amount of water in the tail tank to offset the moment of any main wing tank performance ballast. Even though a tail tank generally holds less than two gallons of water, a calculation error leading to excess water in the tail could result in flight with a CG aft of the limit.

Performance Ballast

Adding weight enhances high-speed performance in gliders. Increasing the operating weight of the glider increases the optimum speed to fly during wings-level cruising flight. The resulting higher ground speed provides an advantage in cross-country soaring and in glider racing.

Manufacturers commonly install water tanks in the main wing panels. That water acts as performance ballast. Personnel add clean water through fill ports in the top of each wing. The amount of water introduced depends on the pilot's choice of operating weight. After adding water, replacement of the filler caps prevents water from sloshing out of the filler holes. Vents in the filler caps allow air to enter the tanks to replace the volume of water drained from the tanks. [*Figure 5-20*] Pilots should ensure that the vents work properly to prevent wing damage when draining water ballast.

Figure 5-20. *Water ballast tank vented filler cap.*

Drain valves fit to the bottom of each tank, and the pilot controls the valves from inside the glider. [*Figure 5-21*] The pilot can fully or partially drain the tanks with the glider on the ground to reduce weight prior to launch. The pilot can also manipulate the valves to drain the ballast tanks partially or completely in flight—a process called dumping ballast, which normally occurs prior to landing. The long streaks of white spray behind an airborne glider indicate water draining in the air.

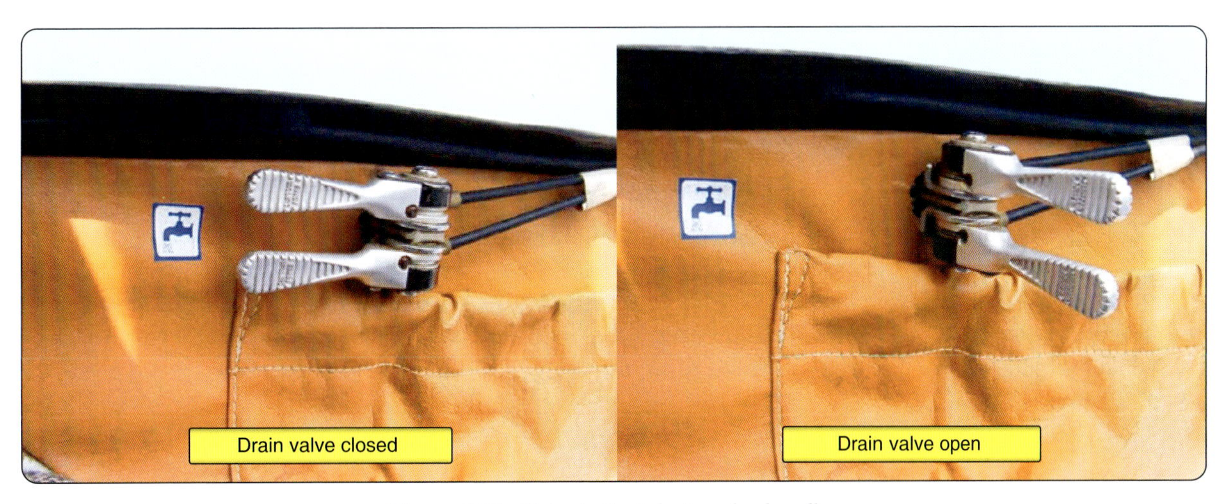

Figure 5-21. *Water ballast drain valve handles.*

Pilots should check the drain valves for correct operation prior to flight. Water ballast should drain from each wing tank at the same rate. Unequal draining leads to a wing-heavy condition that makes inflight handling, as well as landings, more difficult. If the wing-heavy condition becomes extreme, the pilot may lose control of the glider.

Water ballast should drain into the air outside the glider rather than leak into the fuselage. Water trapped in the fuselage may flow through or over bulkheads, causing unmanaged changes to the glider CG. Sufficient CG movement could lead to control difficulty or even total loss of control.

The flight manual provides guidance regarding the length of time it takes for the ballast tanks to drain completely. When preparing for landing, the pilot should dump ballast early enough to give the ballast drains sufficient time to empty the tanks.

Use of water ballast in low ambient temperatures can result in water freezing the drain valve, making dumping ballast difficult or impossible. If only one valve freezes, uneven dumping may occur as discussed above. If water in the wings freezes, serious wing damage may occur because water expands while freezing. The resulting increased volume can deform ribs and other wing structures or delaminate glued bonds. In cold weather or when expecting cold flight conditions, pilots should not use water ballast unless adding antifreeze to the water. The GFM has information on antifreeze compounds approved for use in the glider.

A glider carrying large amounts of water ballast has noticeably different handling characteristics than the same glider without water ballast. Water ballast:

- Reduces the rate of acceleration of the glider at the beginning of the launch due to the increased glider weight.
- Increases the length of ground roll prior to glider liftoff.
- Increases stall speed.
- Reduces aileron control during the takeoff roll, increasing the chance of uncontrolled wing drop and resultant ground loop.
- Reduces rate of climb during climb-out.
- Reduces aileron response during free flight. The addition of large amounts of water increases lateral stability substantially. This makes quick banking maneuvers difficult or impossible to perform.

The pilot routinely dumps water ballast before landing to reduce the weight of the glider. Dumping ballast:

- Decreases stall speed.
- Decreases the optimum airspeed for the landing approach.
- Shortens landing roll.
- Reduces the load that glider structures must support during landing and rollout.

While performance advantages from ballast occur during strong soaring conditions, pilots should consider that ballast degrades takeoff performance, climb rate, and low-speed handling. Before committing to a launch with water ballast aboard, the pilot should review operating limitations to ensure safety of flight.

Chapter Summary

Factors that affect all glider flights include temperature, atmospheric pressure, humidity, wind, and operating weight. Pilots should consider the design of the glider and its operating characteristics and know the expected performance before flight. Pilots should only fly when weight and balance conditions remain within limits since these conditions affect stability and control. Glider polars indicate the performance speeds that pilots can expect at different weights and under different wind conditions and can assist in maximizing performance. Glider pilots flying models that use water ballast should understand how to verify proper system operation before flight and know when to drain any water when necessary or before landing.

Chapter 6: Preflight & Ground Operations

Introduction

Safe operation of a glider depends on careful assembly and preflight. Proper assembly techniques, followed by a close inspection of the glider using checklists contained in the Glider Flight Manual/Pilot's Operating Handbook (GFM/POH), provide for that safety. Pilots should conduct launch procedures systematically and consistently for each flight using appropriate methods and teamwork. Students and pilots unfamiliar with glider assembly should seek instruction from a glider flight instructor.

Assembly & Storage

Personnel preparing to assemble a glider should consider the following elements: location, number of helpers, tools, parts, and checklists that detail the appropriate assembly procedures. The GFM/POH should contain checklists for assembling and preflighting a glider. If not, personnel should develop one and follow it. Glider pilots should avoid distractions that may occur during assembly and check for required documents on board the glider.

During assembly of the glider, pilots normally check for documents carried on board, which include:

1. Airworthiness certificate displayed in the aircraft
2. Registration
3. Required placards
4. GFM/POH

Proper tools and accessories support glider assembly. [*Figure 6-1*] Some individuals place all the necessary hardware on a shelf or canvas with a spot for each item. An inventory of the hardware after assembly reduces the risk of a misplaced tool in the glider structure that could restrict or jam a flight control. Since detached tape could also cause control issues in flight, any person applying tape should use care when taping the wing roots, turtle deck (cover where the flight controls are attached), elevator, or other areas.

Figure 6-1. *Wing stands and wing dollies may assist assembling the glider.*

Depending on the type of glider, assembly may involve two or more people. After assembly, a thorough inspection of all attach points ensures proper installation and security of bolts and pins.

On many gliders, the wings routinely detach from the fuselage for storage. On some motor gliders, the wings fold. After folding or detaching the wings, persons can move the glider in and around the hangar area. [*Figure 6-2*] This allows assembly, disassembly, and storage in a normal size hangar. Hangars usually range between 35 and 45 feet in width. Because most glider wings span a minimum of 40 feet (12 meters) total length, a large hangar may save time and effort.

Figure 6-2. *Folding the wings of the glider reduces storage space.*

Once a glider has been completely assembled and before flight, the pilot inspects all critical areas to ensure completion of all flight control attachments. Many manufacturers provide a critical assembly checklist (CAC) for use after assembly. A positive control check (PCC) provides an additional means of verification that should occur even if appropriate personnel complete a CAC. The pilot should refer to a written checklist provided by the manufacturer or a commercial source that accurately reproduces glider checklists.

Trailering

Glider transport, storage, or retrieval uses specially designed trailers [*Figure 6-3*], and the glider components should fit snugly in the space provided, without being forced, and in a manner that prevents chafing. Glider trailers may also have storage areas to secure parts.

Figure 6-3. *Components of a glider stored and transported in a trailer.*

When towing a trailer, standard motor vehicle towing precautions and experience apply.

Tiedown & Securing

When tying down at airports shared by powered aircraft, propeller blast may cause damage to an improperly secured glider. Personnel should tie down a glider with the canopy closed and latched before leaving it unattended. If possible, the tie-down spot should face into the wind. Permanent tie-downs often use straps, ropes, or chains for the wings and tail and do not adjust for variations in wind. Pilots should carry a proper tie down kit for a cross-country trip if no permanent tie-down exists at the destination. Pilots should consult the GFH/POH for specific tie-down procedures.

If expecting strong winds, pilots can tie the spoilers open with seat belts, or place a padded stand under the tail to reduce the angle of attack of the wings. When securing the glider for an extended period, gust locks on the control surfaces prevent them from banging against their stops. Pitot tube and total energy probe covers prevent entry of insects or debris. [*Figure 6-4*]

Figure 6-4. *Protecting the pitot tube and total energy probe.*

Glider canopy covers provide additional protection from damage while shielding the interior of the flight deck from ultraviolet (UV) rays. [*Figure 6-5*]

Figure 6-5. *Protecting the canopy.*

Water Ballast

If transporting a glider or parking a glider in cold weather, personnel should check for and drain any remaining water ballast. If unable to drain water ballast, adding antifreeze prevents freezing.

Ground Handling

Moving a glider on the ground requires special handling procedures, especially during high winds. When moving a glider, all personnel should receive a briefing on procedures and signals.

When using a vehicle to move a glider, personnel should use a tow line more than half the wingspan of the glider. If one wingtip stops moving for any reason, this tow line length prevents the glider from pivoting and striking the tow vehicle with the opposite wingtip. Since any wind or air currents from another aircraft can cause the glider canopy to close abruptly, a closed and latched canopy can prevent damage to the canopy frame or glass.

When starting to move the tow vehicle, the driver should slowly take up slack in the line to prevent jerking the glider. The towing speed should not exceed that of a brisk walk. The driver should use at least one wing walker. The wing walker(s) and the driver of the tow vehicle function as a team, alert for obstacles, wind, and any other factor that may affect the safety of the glider. The driver should stay alert for any signals from the wing walker(s). [*Figure 6-6*]

Figure 6-6. *Positioning the glider for the tow vehicle.*

Strong winds and gusts can damage the glider during ground handling. During ground movement in high winds, two or more crewmembers normally hold the wingtips and a pilot sits at the controls to deploy spoilers and hold the controls appropriately.

Another method of towing uses specially designed towing gear like a trailer tow bar that attaches directly to the vehicle towing a trailer. The tow bar and other equipment make guiding the glider much easier and allow the wing walkers to focus on wingtip clearances. [*Figure 6-7*]

Figure 6-7. *A wing dolly (left) ready for attachment and a tail dolly (right) being used for glider ground movement.*

Launch Equipment Inspection

While a glider pilot may inspect the tow rope, 14 CFR part 91, section 91.309(a) gives the person operating a civil aircraft towing a glider responsibility for the tow rope strength. [*Figure 6-8*] Since a weak spot could develop at any time, tow ropes should undergo inspection prior to every flight. The tow line should be free from excess wear; all strands should be intact, and the line should be free from knots. A knot in the tow line reduces its strength by up to 50 percent and creates a thick spot in the tow line more susceptible to wear. The ring to which the glider attaches may create a high-wear area. Operators may replace the tow rope periodically due to ultraviolet (UV) exposure from the sun or after a specific number of tows.

Figure 6-8. *Inspecting the tow line.*

Tow ropes and cables are made of materials such as nylon, step-index fiber from Red Polyester, Hollow Braid POLYPRO, Dacron, steel, and Polyethylene. [*Figure 6-9*]

Figure 6-9. *Tow rope materials.*

Figure 6-10 shows the strength of some ropes typically used. The rope manufacturer can provide exact specifications. 14 CFR part 91, section 91.309(a)(3), requires tow line strength within a range of 80 to 200 percent of the maximum certificated weight of the glider. However, the tow line may have a breaking strength more than twice the maximum certificated operating weight of the glider if using safety links as specified in 14 CFR part 91, section 91.309(a)(3). Note that a specific model of glider may have a more stringent limitation on the maximum tow line strength. The POH would list this limitation.

Diameter	Nylon	Dacron	Polyethylene	Polypropylene	Polypropylene
			Hollow braid	Monofilament	Multifilament
3/16"	960	720	700	800	870
1/4"	1,500	1,150	1,200	1,300	1,200
5/16"	2,400	1,750	1,750	1,900	2,050

Figure 6-10. *Typical rope strengths.*

If using a tow line with a breaking strength more than twice the maximum certificated operating weight of the glider, 14 CFR part 91, section 91.309(a)(3) requires installation of two safety/weak links, either of which can break before excess tension on the tow line causes structural damage to either the glider or the tow plane. The weak link near the glider attachment point [*Figure 6-11*] requires the same breaking strength range as the tow rope, 80 to 200 percent of the maximum certificated operating weight of the glider. The weak link installed near the tow plane attachment point must have a breaking strength greater, but not more than 25 percent greater, than that of the weak link at the glider end of the tow line and not more than twice the maximum certificated operating weight of the glider. See *figure 6-12* for a sample weak link strength analysis.

Figure 6-11. *A safety link. The foam added to the tow line prevents tow line abrasion.*

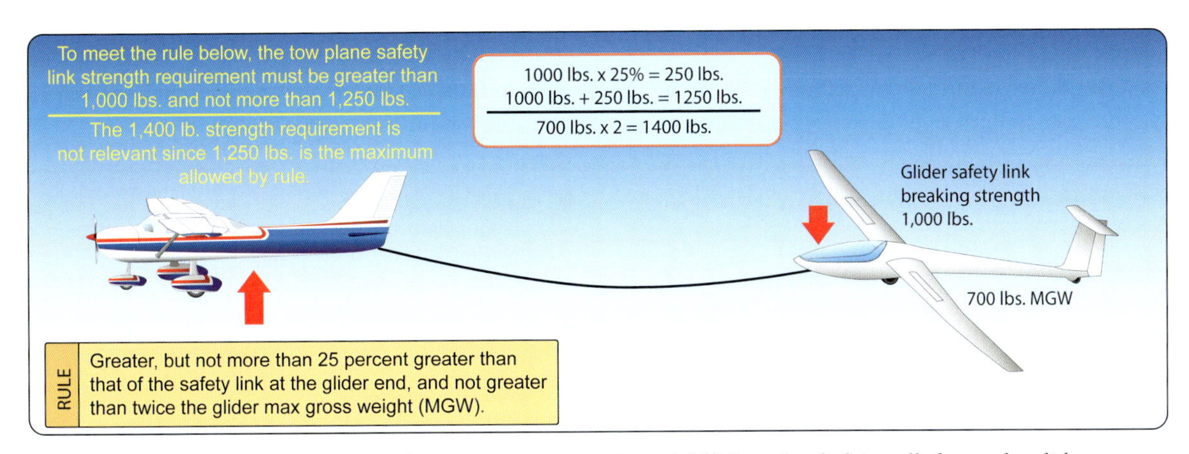

Figure 6-12. *Safety link strength requirements assuming a 1,000 lb. safety link installed near the glider.*

The two most common types of tow hook are an over-the-top design, such as a Schweizer hook, or a grasping style, such as a Tost hook. Pilots should inspect and verify that any tow hook is free from damage and operates freely. [*Figure 6-13* and *Figure 6-14*]

Figure 6-13. *Schweizer-type tow hook.*

Figure 6-14. *Tost hook.*

Glider Preflight Inspection

Pilots should conduct a thorough inspection of the glider before launch. If the glider GFM/POH does not contain a preflight checklist, the pilot should develop a checklist using the guidelines contained in *Figure 6-15*.

- Begin by assessing the overall condition of the fiberglass or fabric.
- Be alert for signs of damage or excessive wear.
- Ensure that the canopy is clean and free from damage.
- Verify the interior wing and control connections are safe and secure.
- If a battery is used, ensure that it is charged and safely fastened in the proper spot.
- Ensure that seat harnesses are free from excessive wear.
- Buckle and tighten any harness that will not be used to prevent it from inadvertently interfering with controls.
- Test the tow hook to ensure it is operating correctly.
- Inspect top, bottom, and leading edge of wings, ensuring they are free from excess dirt, bugs, and damage.
- Inspect spoilers/dive brakes for mechanical damage. They should be clear of obstructions. Perform control continuity check.
- Inspect the wingtip and wingtip skid or wheel for general condition.
- Inspect ailerons for freedom of movement, the condition of hinges and connections, and the condition of the gap seal. Perform control continuity check.
- Check the condition of flaps for freedom from damage and for appropriate range of motion. Perform control continuity check.
- Inspect the general condition of the empennage.
- Check static ports, pitot tube, and total energy probe to ensure they are free from obstruction.
- Check top, bottom, and leading edge of tailplane for bugs, dirt, and damage.
- Check the landing gear for signs of damage or excessive wear. The brake pads should be checked if they are visible; otherwise, the brakes can be checked by pulling the glider forward and applying the brakes. Note that the landing gear is frequently a problem area for gliders used in training.
- Check elevator and trim tab for condition of connections, freedom of movement, and condition of gap seal. Perform control continuity check.
- Check rudder freedom of movement and condition of connections. Perform control continuity check.

Figure 6-15. *A glider preflight inspection checklist.*

Prelaunch Checklist

The pilot should conduct a positive control check with the help of an additional crewmember, especially if the glider was just assembled. The pilot moves the control stick and rudder pedals while the crewmember provides resistance, holding each aileron, the elevator, spoilers, flaps, and the rudder to ensure correct and secure control connections. [*Figure 6-16*] A stick or pedal that moves freely with the corresponding control surfaces restricted indicates the connections are not secure, and the glider is not airworthy. Pilots should adjust the pilot or passenger seats, as well as adjustable controls, such as rudder pedals, prior to buckling the seat belt and shoulder harness. Caution should be exercised to avoid crimping or clamping any onboard oxygen supply.

Figure 6-16. *Positive control check of spoilers.*

If the GFM/POH does not provide a specific prelaunch checklist, then some good generic checklists are CB SWIFT CBE and ABBCCCDDE, which are explained in *Figure 6-17*. Pilots should follow each step carefully regardless of the checklist used.

Before Takeoff Checklist

Phase I:
- Controls (plural)
- Ballast

Phase II:
- Straps (plural)
- Wind
- Instruments (plural)
- Flaps (singular)
- Trim

Phase III:
- Canopy
- Brakes (singular for most gliders, plural for a Blanik, since the wheel brake is not at the end of the air brake travel)
- Emergency plan (the three phases plus the return speed that can and should be prebriefed prior to closing the canopy)

A	Altimeter set to correct elevation
B	Ballast
B	Seat belts and shoulder harnesses fastened and tightened
C	Controls checked for full and free movement
C	Canopy closed, locked, and checked
C	Cable or towline properly connected to the correct hook
D	Dive brakes closed and locked
D	Direction of wind checked
E	Emergency Plan reviewed

Figure 6-17. *Generic prelaunch checklists.*

Glider Care

Depending on the type of glider, different cleaning methods should be employed. After all flights, the pilot or ground personnel can wipe down the glider with a soft cloth or a wet chamois. This removes any debris or bugs. Delaying this process allows bugs to dry and makes removal difficult.

Gliders made of fabric should be cleaned with spray-on and wash-off products. Anyone cleaning a glider should avoid using large amounts of water as the water may penetrate cracks and holes, especially on earlier wood or fabric gliders, which damages and reduces the life of the wood or fabric. Persistent and prolonged moisture exposure can damage any glider over time. On metal gliders, personnel can use a low-pressure hose and mild detergent for cleaning. High-performance fiberglass gliders use a special spray or paste for cleaning. Care must be taken when using a powered buffer, as the buffer may burn the fiberglass if not used in a proper manner. Anyone cleaning a canopy should use only recommended cleaners for that purpose. The GFH/POH or a glider supply store or organization can suggest appropriate cleaning materials based on the type of glider.

After cleaning, a coat of wax or a sealer may be applied to fabric, metal, and fiberglass gliders. However, wax containing silicon adheres to the pores of fiberglass and complicates future fiberglass repair. After applying any wax, personnel may use a clean, soft cloth to wipe off excess wax and buff the area by hand.

An owner or pilot may perform minor repairs on a fiberglass glider such as fixing a scratch or a chip. Individuals should consult a fiberglass expert prior to making any minor repair and use approved parts and materials when conducting these repairs. All gliders, including those certified as standard or experimental, should be repaired in accordance with the manufacturer's recommended procedures and 14 CFR part 43.

Preventive Maintenance

Pilots may conduct preventive maintenance on their glider. Preventive maintenance does not involve complex assembly operations. For the list of permissible pilot-performed maintenance activities and for more information on preventive maintenance, refer to Appendix A of 14 CFR Part 43, Major Alterations, Major Repairs, and Preventive Maintenance found at www.ecfr.gov.

Chapter Summary

This chapter introduces recommended safety practices for assembly, storage, tie-down, securing, and ground handling for a glider. The final authority for any glider is the specific Flight Manual or Pilot's Operating Handbook (POH) developed by the aircraft manufacturer and approved by the FAA.

Chapter 7: Launch, Flight Maneuvers, Landing, & Recovery Procedures

Introduction

In the early days of soaring, a crew might launch a glider from the top of a hill using a bungee cord. With the tail of the glider tied down, the ground crew would attach the center of a bungee cord to a hook on the nose of the glider. Members of the ground crew stretched the separate ends of the bungee cord ahead, into the wind, and offset from the glider. With sufficient tension on the bungee cord and upon release of the tail tie-down, the glider would accelerate as if launched from a slingshot.

As gliders got larger, pilots looked for better ways to launch. Enthusiasts began using cars to pull gliders. Over time, powered winches and airplane towing became preferred launching methods. This chapter discusses glider launch techniques and procedures as well as takeoff procedures, traffic patterns, flight maneuvers, and landing and recovery procedures.

Glider pilots should understand risks associated with the large wingspan of a glider and ground operations. The wings can strike runway lights and other obstructions near the runway during reposition, takeoff, or landing. Impact with a wingtip could lead to ground loops during takeoff or landing. A cartwheel could occur if a wingtip strikes the ground before the glider touches down, leading to extensive damage and serious injury. Training, pilot proficiency, hazard consideration, and risk mitigation before and during flight reduce the likelihood of these undesirable events.

Aerotow Takeoff Procedures

Signals

Visual signals enhance communication and coordination between the glider pilot, the pilot of the towing aircraft, and the ground crew.

Prelaunch Signals

Aerotow prelaunch signals facilitate communication between pilots and launch crewmembers/wing runners when preparing for the launch. *Figure 7-1* illustrates these hand signals. The raise wingtip to level position is given by the pilot.

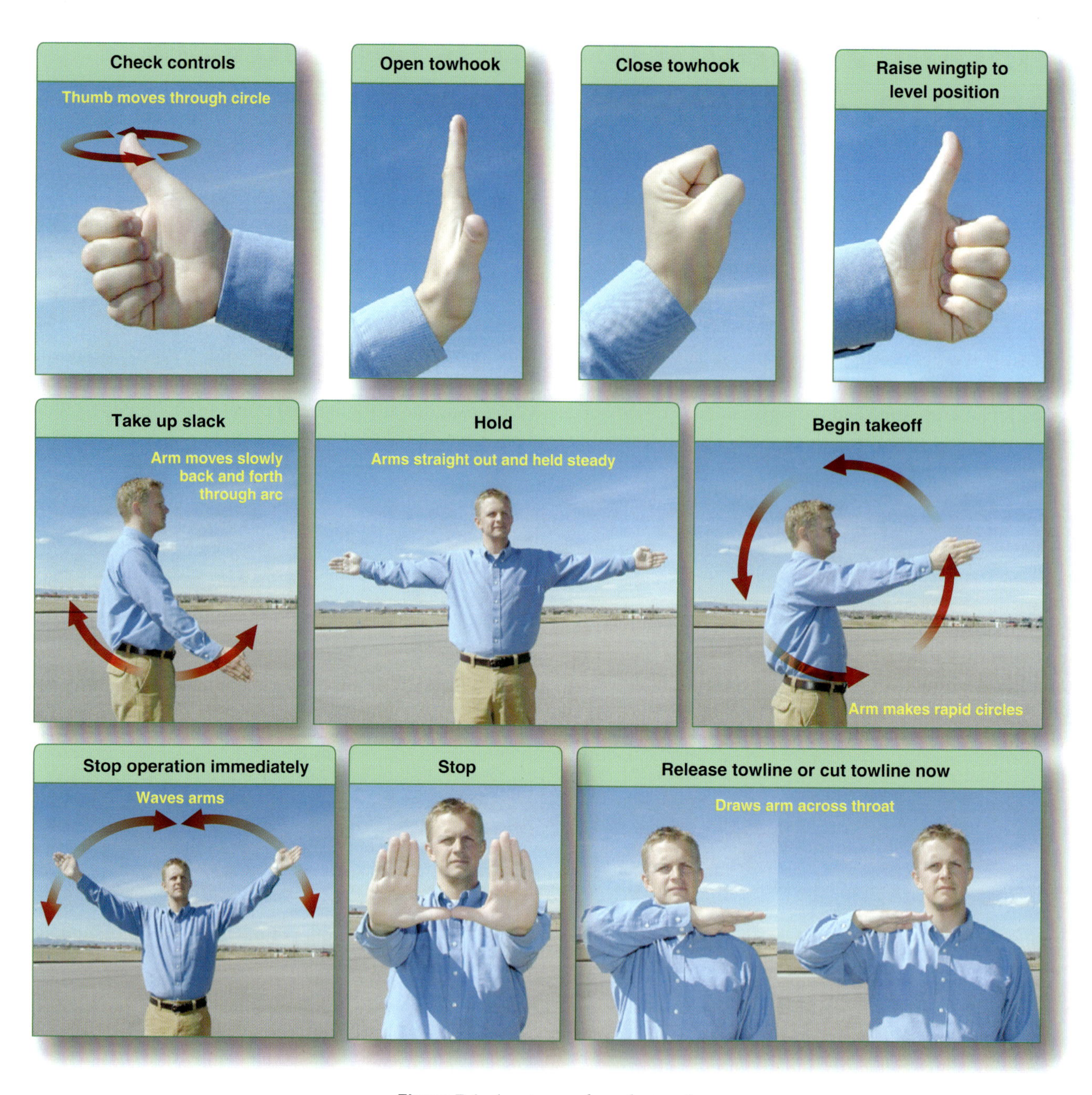

Figure 7-1. *Aerotow prelaunch signals.*

Inflight Signals

When airborne, pilots use the flight controls to create visual signals that allow the tow pilot and the glider pilot to communicate. The signals divide into two types: those from the tow pilot to the glider pilot and signals from the glider pilot to the tow pilot. *Figure 7-2* depicts these signals.

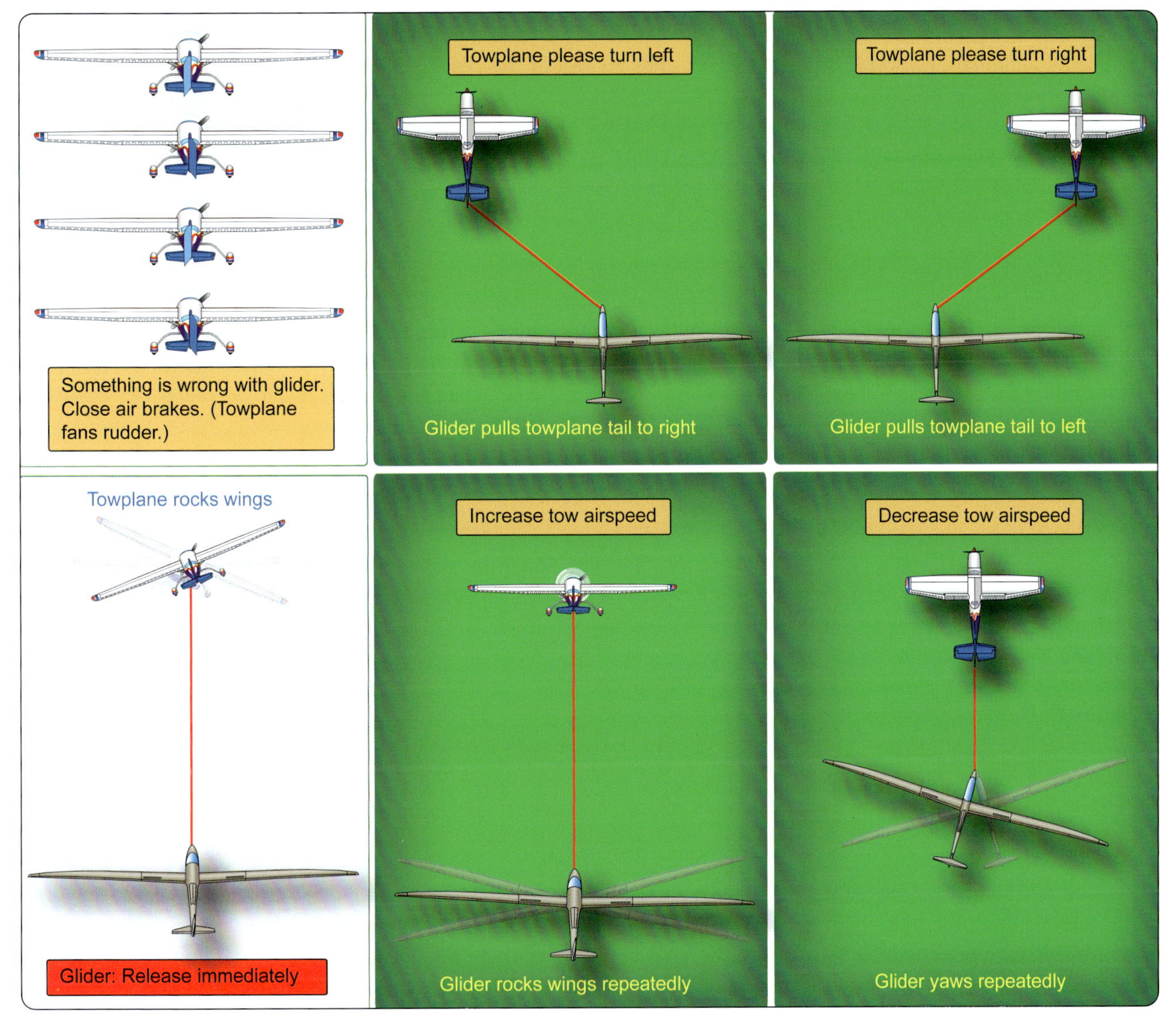

Figure 7-2. *Inflight aerotow visual signals.*

The tow pilot could use the aerotow signals shown in the two panels on the left side of *figure 7-2* when close to the ground. The glider pilot should know how to differentiate between these signals to avoid an unnecessary release close to the ground. Even with two-way radio in both aircraft, radio communications could distract either pilot when operating near the ground and could increase risk of loss of control.

The two green panels of the top row of *figure 7-2* illustrate how a glider pilot requests a turn. Since large or abrupt lateral offset has the potential to interfere with tow-plane control, glider pilots should only use lateral offset signals as depicted in at or above 1,000 feet AGL.

The two green panels of the bottom row of *figure 7-2* illustrate how a glider signals for a change in speed when at or above 1,000 feet AGL. The glider pilot yaws repeatedly or rocks the wings as depicted, and the force on tow line oscillates and causes a series of small accelerations and decelerations. Signaling in this manner should get the attention of the tow pilot who then can look back and interpret the signal.

Takeoff Procedures & Techniques

Takeoffs benefit from a crewmember on the ground who can scan for traffic and provide general assistance during the takeoff. An assisted takeoff includes a crewmember or wing runner who maintains the glider wing in a level position as the glider begins its takeoff roll. An unassisted takeoff does not include a wing runner or other ground crew. Glider and tow pilots should only perform an unassisted launch if trained on the procedure and if conditions allow for a safe unassisted takeoff. An unfamiliar glider or lack of proficiency adds the risk of this type of takeoff.

Prior to takeoff, the tow pilot and glider pilot should agree on a plan for the aerotow. The glider pilot should also ensure the launch crewmember has sufficient knowledge of the plan. Some items to consider include the intended ground path, pattern clearing procedures, and glider configuration checks (spoilers closed, tailwheel dolly removed, canopy secured). Takeoffs normally occur into the wind.

Connecting the tow line to a glider in preparation for takeoff should only occur with the glider pilot aboard and ready for flight. The launch crewmember presents the tow rope end to the pilot so the pilot can ensure it is in good conditions and with the correct ring and weak leak, if required. When the required checklists have been completed with both the glider and towplane ready for takeoff, the launch crewmember/wing runner starts to hook the towline to the glider. If the pilot exits the glider for any reason, the pilot or launch crewmember should disconnect the towline to prevent accidental tow of an unoccupied glider.

Normal Assisted Takeoff

A deliberate tow rope hookup should occur, which includes a check of the release mechanism for proper operation. The launch crewmember should apply tension to the tow line and signal the glider pilot to activate the release. The launch crewmember should verify that the release works properly and communicate that information to the glider pilot. With the tow line again hooked up to the glider, the launch crewmember moves to the wingtip on the ground and clears both the takeoff and landing areas.

When the glider pilot signals the launch crewmember at the wingtip to lift the wing, that crewmember picks up and holds the wing in a level position and signals the tow pilot to "take up slack" in the tow line. With the slack out of the tow line, the glider pilot signals ready for takeoff by wagging the rudder, and the crewmember simultaneously signals the tow pilot for takeoff. If using a radio, the glider pilot could indicate the takeoff signal to the tow pilot by stating, "Canopy locked and ready for takeoff."

As the aerotow begins and the glider accelerates, the launch crewmember runs alongside the glider, holding the wing in a level attitude until the glider pilot gains roll control or the speed of the glider exceeds the crewmember's safe running speed. An increase in resistance to aileron movement indicates aileron effectiveness. Holding the wings level with the ailerons may require full deflection of the flight controls until sufficient airspeed increases effectiveness of the controls.

Risk of collision with runway lights, signage, and other obstructions alongside the runway during the takeoff roll increases due to the combination of long wings and short landing gear when compared to airplanes. The pilot can mitigate this risk by steering the glider solely with the rudder so as not to have one wing low. [*Figure 7-3*]

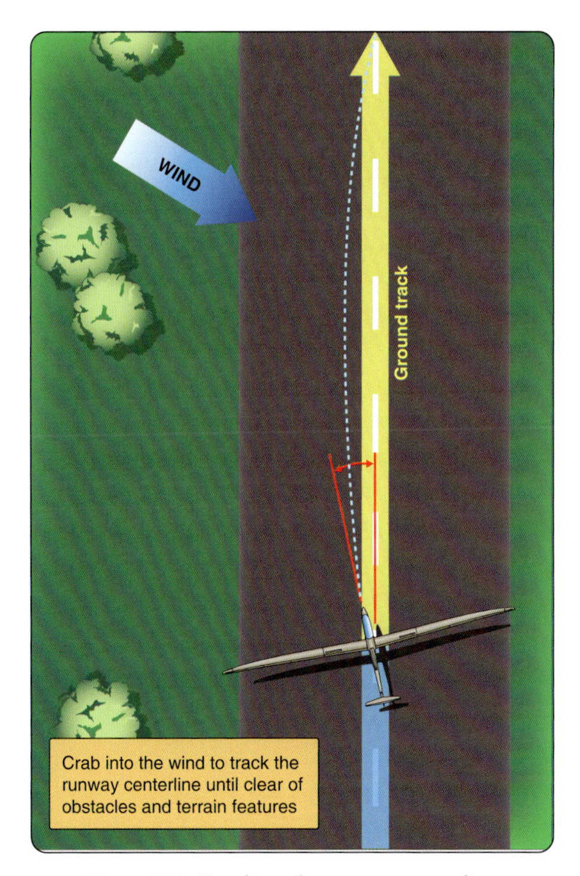

Figure 7-3. *Tracking the runway centerline.*

When the glider achieves lift-off speed, the glider pilot should maintain the glider at a low altitude of 2 to 4 feet—the exact altitude depends on the specific glider. As the glider and tow plane accelerate, the glider pilot should maintain altitude by applying stick pressure, as necessary. If the glider climbs above the towplane's tail
during the takeoff, tension on the towline pulls up on the towplane tail and could force the towplane's propeller into the runway surface. Once lift-off occurs and since lateral deviation can force the towplane off the runway, the glider pilot can use coordinated aileron and rudder to remain directly behind the tail of the towplane.

During most takeoffs, the glider achieves flying airspeed before the towplane. In this case and once the towplane lifts off, it accelerates in ground effect to the desired climb airspeed, and the climb begins for both the towplane and glider. However, a glider loaded with ballast might not achieve liftoff airspeed before the towplane. In this situation, the towplane should remain in ground effect until the glider becomes airborne.

Unassisted Takeoff

The unassisted takeoff begins with the glider positioned slightly off the runway heading (runway centerline) by approximately 10–20° with one wing on the ground. If the glider is canted to the right, then the left wing should rest on the ground. If canted to the left, the right wing should rest on the ground. When ready for takeoff, the glider pilot advises the tow pilot either by radio or by signaling the tow pilot with the "ready for takeoff" rudder waggle signal. As the towplane accelerates, the wing on the ground accelerates at a slower rate due to the increased drag due to the ground contact. This imparts a yawing motion that will help straighten out the glider. The pilot should use rudder to raise the lower wing until sufficient speed is obtained to allow aileron control of the bank angle. If the glider begins the takeoff roll aligned with the towplane during the takeoff, the wing on the ground tends to drag and severe swerving or a ground loop becomes more likely.

Crosswind Takeoff

Most gliders have a crosswind limit up to approximately 10–12 knots. Pilots should refer to the Glider Flight Manual/ Pilot's Operating Handbook (GFM/POH) for model specific information.

Crosswind takeoff procedures compensate for the following:

1. The glider tends to weathervane into the crosswind with the weight on the main wheel.

2. After lift-off, the glider tends to drift off the runway centerline with the crosswind.

Assisted

Prior to takeoff, the glider pilot should direct the launch crewmember to hold the upwind wing slightly low during the initial takeoff roll. In a crosswind, the pilot should hold full aileron into the wind as the takeoff roll begins. The pilot maintains this control position while the glider accelerates until the ailerons become effective. At the same time, the pilot uses downwind rudder to maintain a straight takeoff path and offset any tendency to weathervane while on the ground. [*Figure 7-4*] Note that takeoff using a CG hook makes the glider more sensitive to crosswind forces as there is no force from the tow line acting to keep the nose of the glider aligned with the direction of motion.

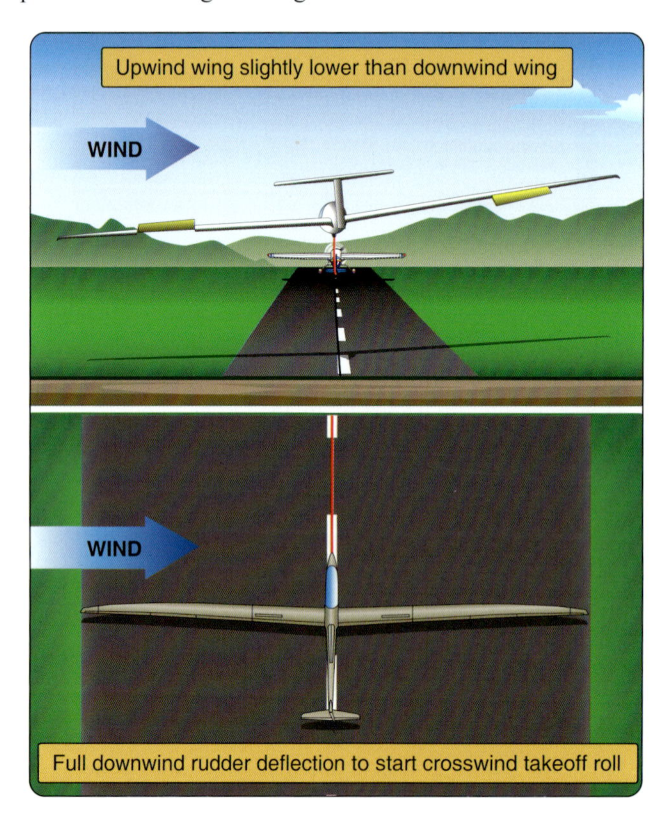

Figure 7-4. *Crosswind correction for takeoff.*

As the glider's forward speed increases, the crosswind becomes more of a relative headwind, and the pilot reduces the application of aileron into the wind. However, the pilot maintains sufficient aileron pressure throughout the takeoff roll to prevent the crosswind from raising the upwind wing.

If the upwind wing rises and exposes more wing surface to the crosswind, a skipping action or series of small bounces may result as the glider begins to fly and then settles back onto the runway. This side skipping imposes side loads on the landing gear. If the downwind wingtip touches the ground, the resulting friction may cause the glider to yaw in the direction of the dragging wingtip, which could lead to a loss of directional control and runway departure.

While on the runway during takeoff, the glider pilot uses rudder to control direction and alignment behind the towing aircraft. The pilot should avoid yawing back and forth behind the towplane, as this affects the ability of the tow pilot to maintain control. If glider controllability becomes a problem, the glider pilot should release and stop the glider on the remaining runway. In this case as the glider slows, the crosswind may cause the glider to weathervane.

After becoming airborne, but before the towplane lifts off, the glider pilot should maintain the crosswind correction to remain behind the towplane. Once the towplane becomes airborne and is clear of obstacles, the glider pilot repositions as needed to align behind the towplane.

Unassisted

Experienced pilots may consider using an unassisted crosswind launch procedure. An unassisted crosswind takeoff uses different wing positioning and glider alignment. The crosswind strikes the fuselage of the glider, tending to push it downwind, making it necessary to position the glider on the upwind side of the runway. If unable to offset, the towplane pilot may angle into the wind to reduce the crosswind component for the glider.

The glider should rest offset on the runway with the downwind wing on the ground and the glider angled approximately 20–30° into the wind. [*Figure 7-5*] As in a normal unassisted takeoff, the drag on the downwind wing imparts a yawing moment that swings the upwind wing forward at a faster rate than the downwind wing, aiding the pilot in leveling the wings. If the pilot begins the takeoff run with the downwind wing on the ground, a ground loop may result since the downwind wing will drag along the ground. The pilot should execute crosswind takeoff procedures as described above once the upwind wing rises and maintain a normal position directly behind the towplane.

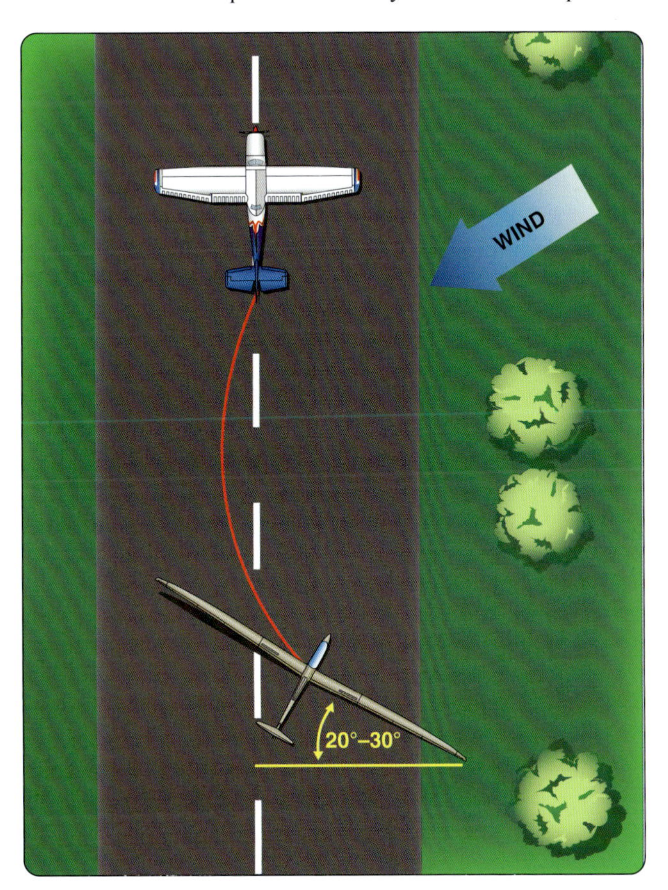

Figure 7-5. *When setting up for a crosswind takeoff, the glider should start on the upwind side of the runway.*

Pilot Induced Oscillations (PIOs) During Launch

During the first moments of the takeoff roll, as airflow begins to impact the control surfaces, it takes considerable displacement of the flight controls to affect the glider's flightpath. The pilot also experiences a higher control lag time due to reduced control effectiveness at low speed. As the glider accelerates, aerodynamic response improves, lag time decreases, and PIOs become less likely.

Several pilot techniques reduce the likelihood and severity of PIOs during aerotow launch. A pilot should not attempt to lift off until the glider responds sufficiently to aerodynamic control. Just after the moment of lift-off, the pilot should bring the glider to two to four feet above the runway to prevent ground contact from any minor excursion in pitch attitude. [*Figure 7-6*]

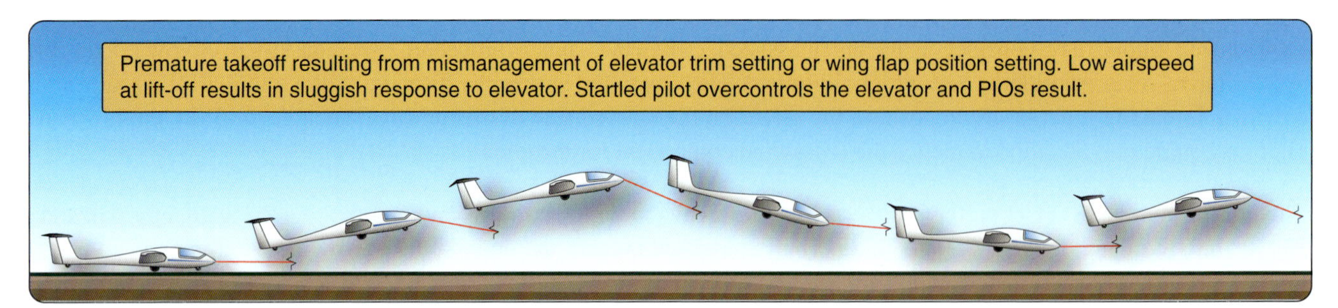

Figure 7-6. *Premature takeoffs and PIOs.*

Improper Elevator Trim Setting and PIOs

Gliders with an aerodynamic elevator trim tab or an anti-servo tab on the elevator may experience more challenging control issues when improperly trimmed. Pilots find that a simple spring-system elevator trim tends to help prevent PIOs, but they can still occur.

Improper Wing Flap Setting and PIOs

With an incorrect positive flap setting, the glider may lift off the runway prematurely. In response, the pilot exerts forward pressure on the controls and exerts an increasing nose-down force on the glider. When the glider eventually pitches down, the pilot may exert considerable back pressure on the stick to arrest the descent. A cycle of PIOs could result, which could lead to hard contact with the runway surface, glider damage, and personal injury.

An incorrect negative flap setting decreases wing camber and wing lift, and the glider may remain on the runway even after the tow plane lifts off and begins to climb out. The pilot may exert significant back pressure on the control stick to lift off, and ballooning may occur as the elevator becomes more effective. A series of PIOs may result, which could require termination of the tow to prevent ground contact or tow plane loss of control.

Gust Induced Oscillations

Gusty headwinds may induce pitch oscillations due to changes in the speed of the airflow over the elevator while crosswind gusts can induce yaw and roll oscillations. In gusty crosswinds, the effects on glider control change rapidly depending on the speed and angle of the crosswind component.

Nearby obstacles, such as hangars, trees, or hills and ridges can affect low altitude winds, particularly on the downwind side of the obstruction. In general, an upwind obstacle induces additional turbulence and gustiness in the wind. Pilots may encounter these conditions from the surface to an altitude of 300 feet or more. If flying in these conditions, the pilot should use a faster-than-normal speed prior to lift-off.

The additional speed increases the responsiveness of the controls, simplifies correcting for turbulence and gusts, and provides a measure of protection against PIOs. The added speed also provides a safety margin above the stall speed since variations in the headwind component affect airspeed.

Caution: The pilot should not exceed the glider's tow speed limitations when adding speed for takeoff in windy conditions.

Pilot Induced Roll Oscillations During Launch

As the tow pilot applies power, the glider moves forward, balanced laterally on its main wheel by the wing runner. After the wing runner lets go and before the glider achieves significant speed, a wingtip may drop toward the ground. In response, the pilot may apply considerable control displacement leading to a series oscillations and potential wingtip ground contact. [*Figure 7-7*]

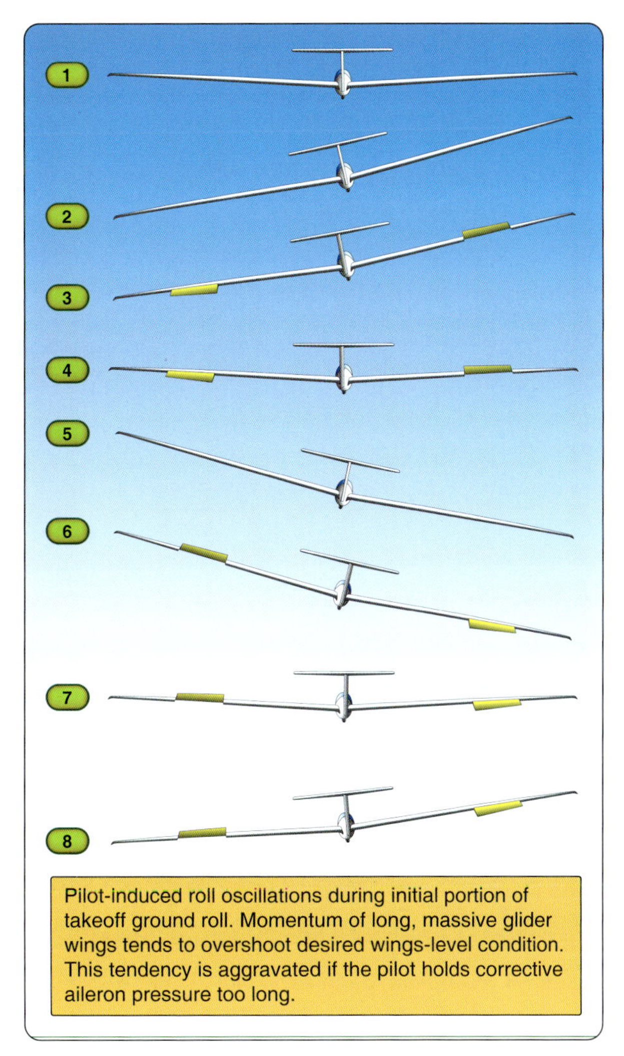

Figure 7-7. *Pilot-induced roll oscillations during takeoff roll.*

If a glider's wingtip contacts the ground during takeoff roll, the drag of the wingtip on the ground induces a yaw in the direction of the grounded wingtip. Mild drag and yaw result with a wingtip on smooth pavement, but strong drag and yaw develop from a wingtip dragging through tall grass. If appropriate aileron pressure fails to raise the wingtip off the ground quickly, the pilot should release the tow line before losing control of the glider.

Pilot Induced Yaw Oscillations During Launch

If the glider veers away from the tow plane while on the ground, rudder application in the appropriate direction corrects the situation. As the glider continues to accelerate, the effect of the rudder increases, and the lag time decreases. The pilot should anticipate the momentum of the glider about the vertical axis and reduce pressure on the rudder pedal when the nose of the glider begins to yaw in the desired direction. [*Figure 7-8*] If the pilot holds rudder pressure too long, momentum of the glider results in an overshoot of the desired yaw position. In extreme cases, and after a series of PIOs, the glider may veer off the runway or force the tow plane off the runway.

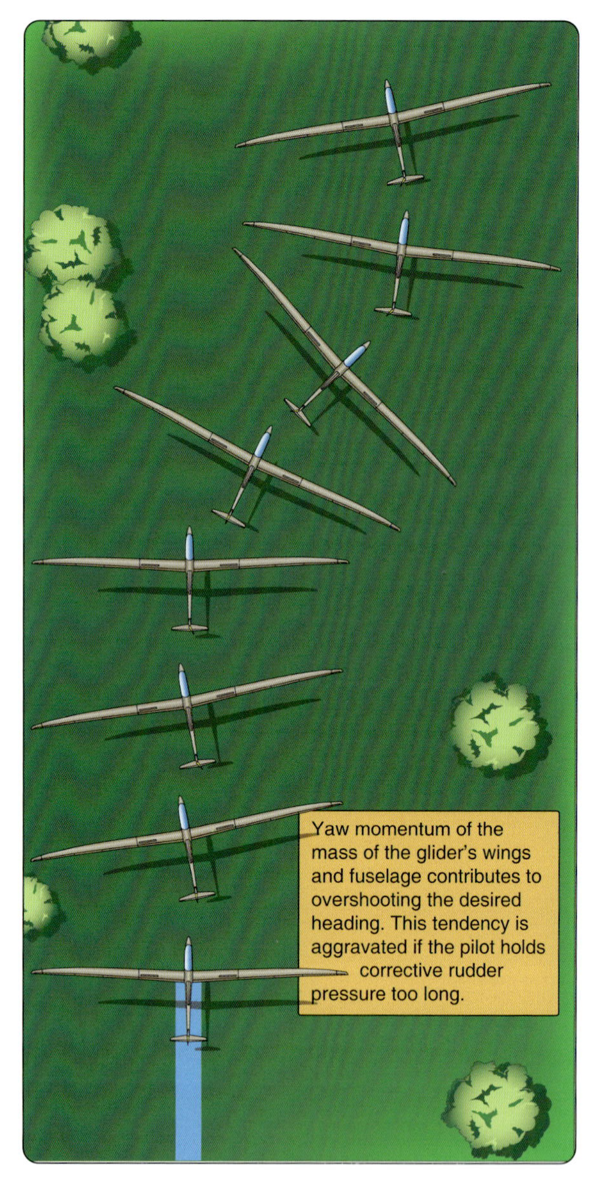

Figure 7-8. *Pilot-induced yaw oscillations during takeoff roll.*

Common Errors

Common errors in aerotow takeoffs include:

- Improper glider configuration for takeoff.
- Improper initial positioning of flight controls.
- Improper alignment of the glider (unassisted takeoff).
- Improper use or interpretation of visual launch signals.
- Failure to maintain alignment behind towplane before towplane becomes airborne.
- Improper alignment with the towplane after becoming airborne.

- Climbing too high after lift-off and causing a towplane upset.

Aerotow Climb-Out

The towplane's wake drifts down behind the towplane, and the glider can climb either above or below that wake. During high-tow, the glider pilot maintains a position slightly above the wake of the towplane. During low-tow flight, the pilot positions the glider just below the wake of the towplane. [*Figure 7-9*] Pilots should use both positions when learning coordinated towing procedures and aerotow dynamics. For gliders with retractable gear, the pilot normally leaves the undercarriage down until after release.

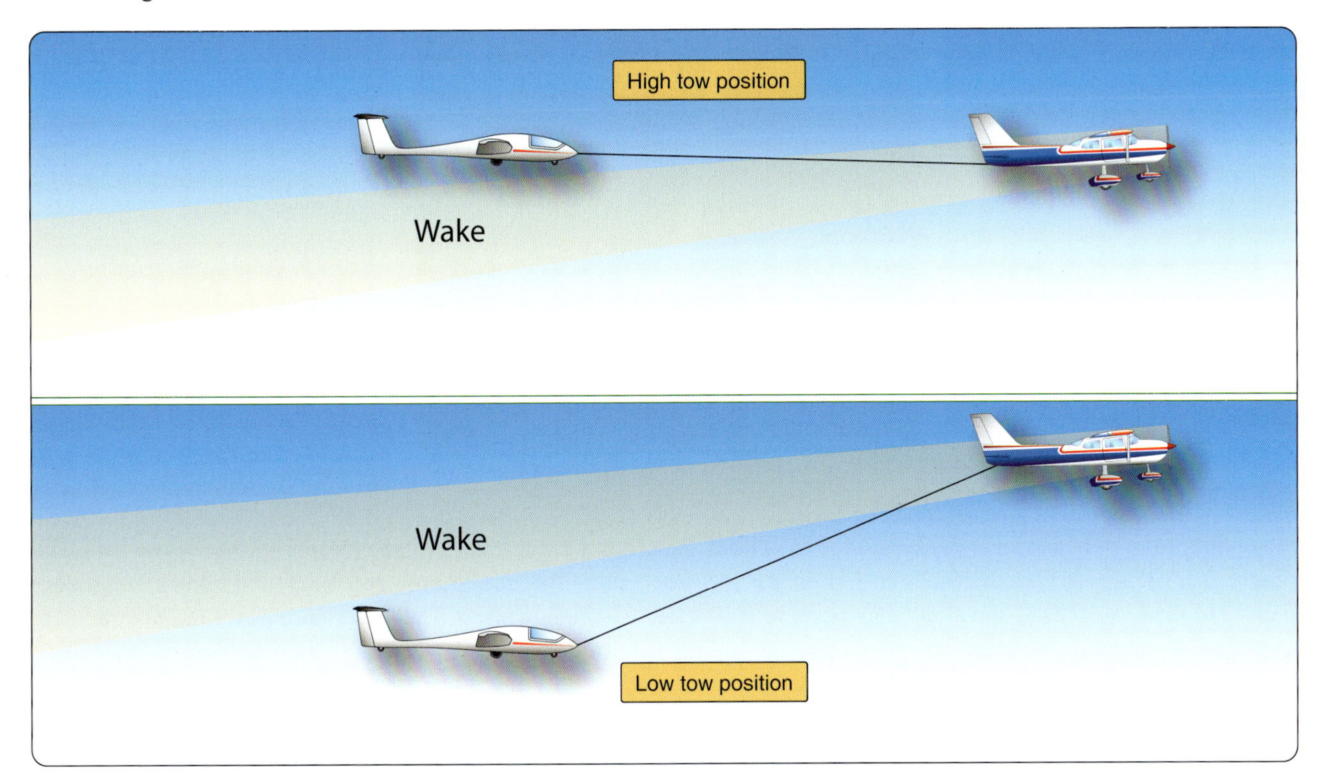

Figure 7-9. *Aerotow climb-out positions.*

The tow pilot strives to maintain a steady pitch attitude and a constant power setting for the desired climb airspeed. Any excessive deviation from the low- or high-tow position by the glider causes the tow pilot to use abnormal control inputs, which generate more drag and degrade climb performance during the tow. The glider pilot uses visual references on the towplane to maintain a proper lateral and vertical position. The glider pilot may use different sight pictures, including adjusting relative to the image of the towplane's wings on the horizon or maintaining the towplane's rudder centered over a point on the fuselage of the towplane.

Low-tow offers the glider pilot a better view of the towplane and results in a more aerodynamically efficient tow, especially during climb, as the towplane requires less upward elevator deflection due to the downward pull of the glider. However, low tow increases the risk of towline fouling from a broken towline or release by the towplane during a climb. Low tow works well for a level-flight tow during a cross-country flight.

The tow pilot normally makes climbing turns using shallow bank angles. During turns, the glider pilot observes the towplane, matches the bank angle in a coordinated turn, and aims the nose of the glider at the outside wingtip of the towplane. [*Figure 7-10*]

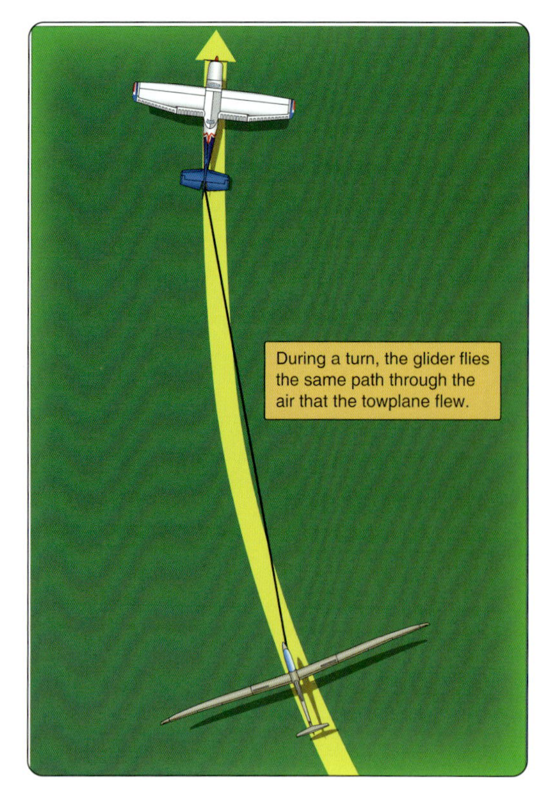

Figure 7-10. *Aerotow climbing turns.*

If the glider pilot uses a steeper bank than the towplane, the glider's turn radius becomes less than that of the towplane. [*Figure 7-11*] If this occurs, reduced tension on the tow line causes the line to bow and slack and allows the glider's airspeed to slow. As a result, the glider may begin to sink relative to the towplane. The glider pilot can correct by reducing the glider's bank angle, so the glider flies the same radius of turn as the towplane. A following section in this chapter describes how using the spoilers or performing a slip can correct for slack line.

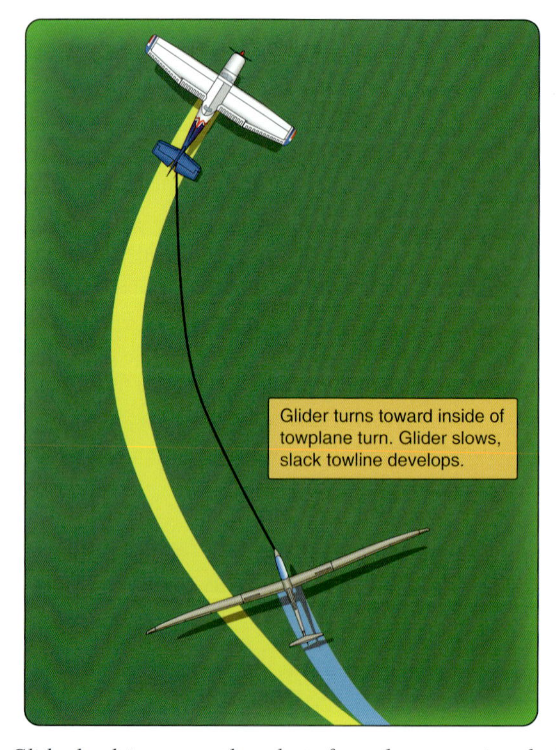

Figure 7-11. *Glider bank is steeper than that of towplane, causing slack in tow line.*

If the glider pilot uses a shallower bank than the tow plane, the glider's turn radius exceeds than that of the tow plane. [*Figure 7-12*] If this occurs, the increased tension on the tow line causes the glider to accelerate and climb. The glider pilot can correct by increasing the glider's bank angle, so the glider flies the same radius of turn as the tow plane. Without timely corrective action and if the glider climbs too high above the tow plane, the tow plane may lose rudder and elevator control. If this occurs, the glider pilot should release the tow line and turn to avoid the tow plane.

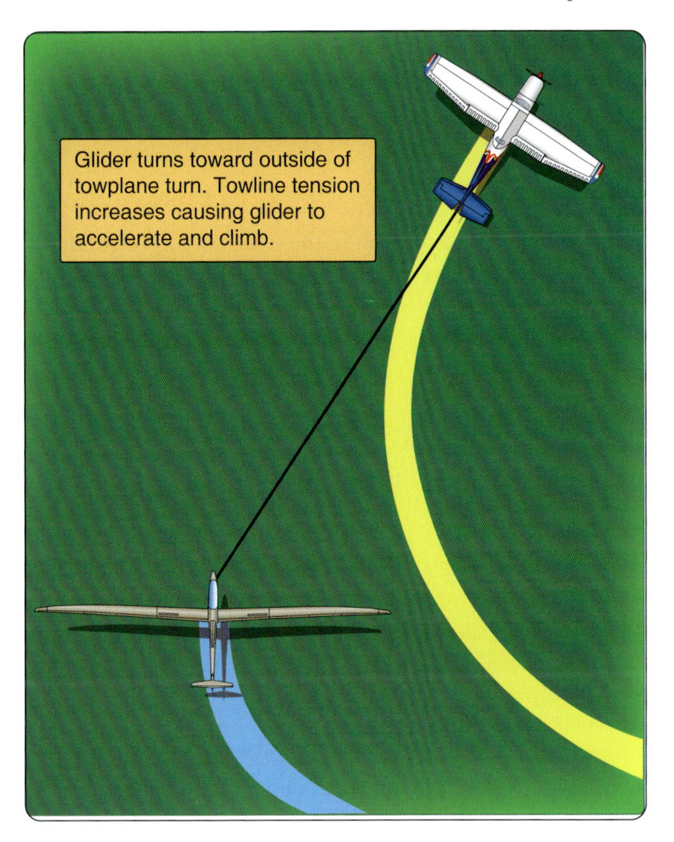

Figure 7-12. *Glider bank too shallow, causing turn outside towplane turn.*

Towline/Tow Hook Characteristics and PIOs

A short tow line keeps the glider close to the towplane and its turbulent wake and complicates glider control. Using a tow line of adequate length—200 feet minimum for normal towing operations—minimizes the influence of the towplane's wake and reduces the likelihood of PIOs.

The characteristics of the tow hook/tow line combination may cause changes in pitch attitude during the tow. On many gliders, the tow hook resides below the pilot enclosure or just forward of the landing gear. An increase in tension on the tow line causes an uncommanded pitch-up of the glider nose as shown in *Figure 7-13*. Decrease in tow line tension results in an uncommanded pitch-down. Even if rapid changes in tow line tension during a turbulent aerotow of a bellyhook-equipped glider leads to these pitch changes, the pilot should make control inputs that avoid overcontrol and PIOs.

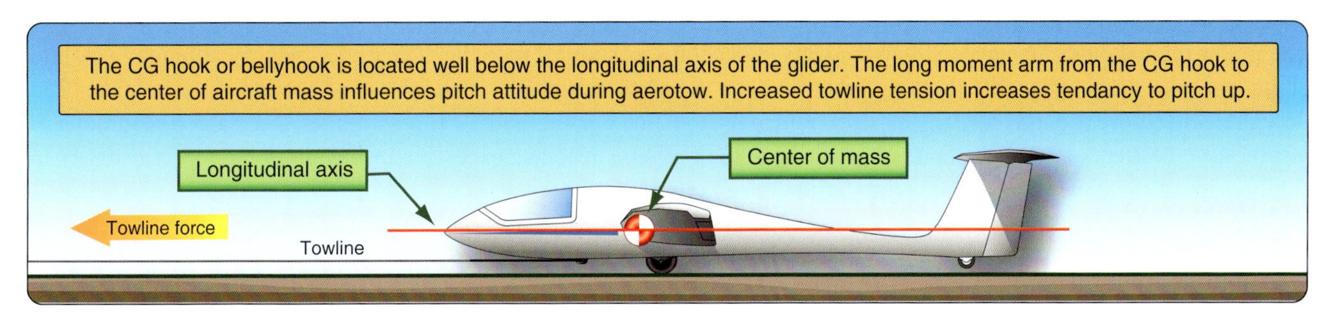

Figure 7-13. *Effects of increased tow line tension on pitch altitude of bellyhook-equipped gliders during aerotow.*

Common Errors

Common errors in aerotow climb-out include:

- Not maintaining proper vertical and lateral position during high- or low-tow.
- Inadvertent entry into towplane wake.
- Failure to maintain glider alignment during turns on aerotow.

Slack Line

Most cases of slack line are minor, require no corrective action on the part of the glider pilot, and resolve using a stabilized flight path. If severe enough, slack line or reduced tension in the tow line might entangle the glider and result in damage to the glider or towplane. Therefore, if the pilot loses sight of either the rope or the towplane, an immediate release should be accomplished.The following situations may result in a slack line:

- Abrupt power reduction by the towplane
- Aerotow descents
- Glider turns inside the towplane turn radius [Figure 7-11]
- Updrafts and downdrafts
- Abrupt recovery from a wake corner

If the towplane precedes the glider into an updraft, the glider pilot first perceives the towplane climbing faster and higher than the glider. Then, as the glider enters the updraft, it climbs more efficiently than the towplane. As a result, the glider pilot pitches the glider over to regain the proper tow position. A resulting increase in airspeed creates the slack tow line. [*Figure 7-14*] To avoid slack, the glider pilot should control the descent and closure rate to the towplane.

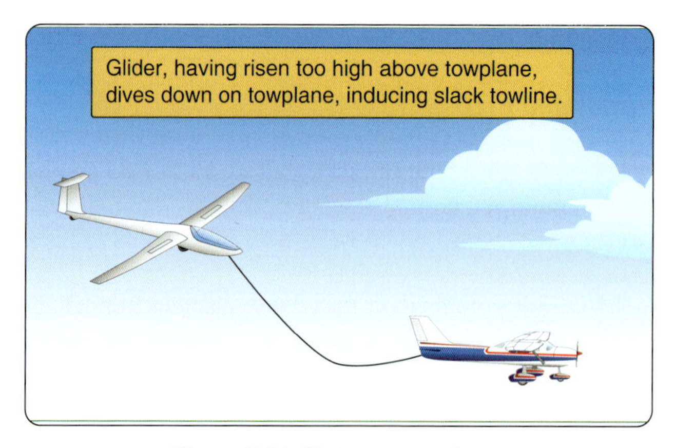

Figure 7-14. *Diving on towplane.*

The glider pilot should initiate slack line recovery procedures as soon as possible. The glider may try slipping back into alignment with the towplane. If slipping fails to reduce the slack sufficiently, careful use of spoilers or dive brakes can decelerate the glider to gently take up the slack. When the tow line tightens and the tow stabilizes, the glider pilot gradually resumes the desired aerotow position. When slack in the tow line becomes excessive or beyond the pilot's capability to safely recover, the glider pilot should release from the aerotow.

Common errors regarding a slack line recovery include:

- Failure to take corrective action at the first indication of a slack line.

- Improper procedure to correct slack line causing excessive stress on the tow line, towplane, and the glider.

Boxing the Wake

The towplane generates two types of wake turbulence. Propwash generates a light chop while wingtip vortices induce a strong rolling motion. Boxing the wake demonstrates a pilot's ability to maneuver the glider around the towplane's wake accurately and safely during aerotow. [*Figure 7-15*] A pilot can maneuver either clockwise or counterclockwise around the wake. The example below uses a clockwise example.

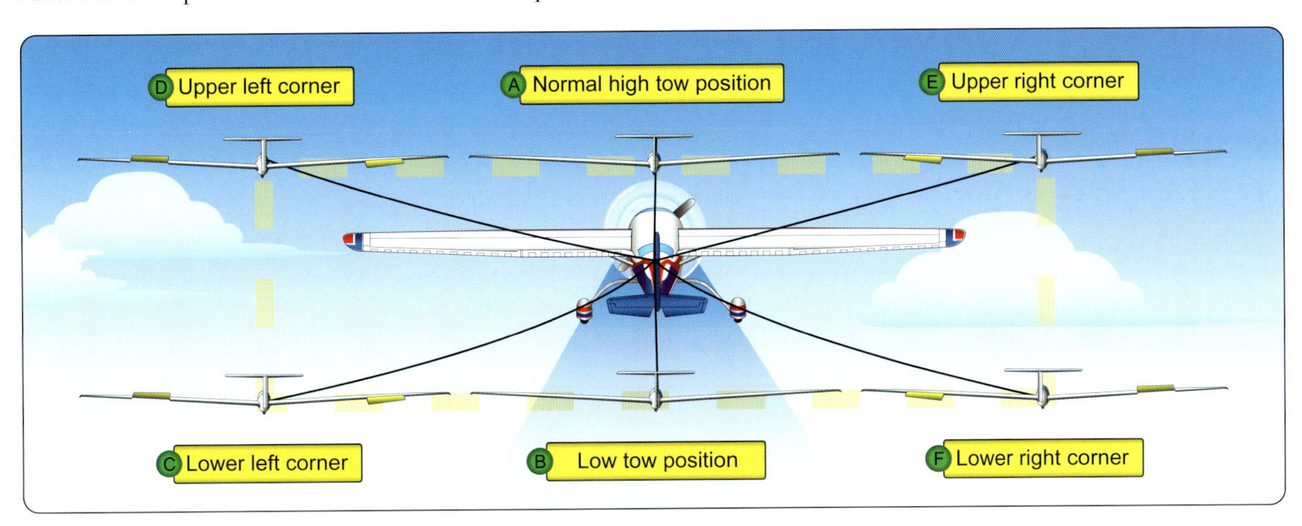

Figure 7-15. *Boxing the wake.*

Boxing the wake involves flying a rectangular pattern around the wake of the towplane. Prior to takeoff, the glider pilot should advise the tow pilot of the intention to box the wake. Boxing the wake should commence outside the traffic pattern area and no lower than 1,000 feet AGL.

Before starting the maneuver, the glider should move to the high tow position [*Figure 7-15* A] and descend from the high tow position through the wake to the center low-tow position [*Figure 7-15* B] as a signal to the tow pilot that the maneuver will begin. The pilot uses coordinated control inputs to move the glider over to the left side of the wake and holds that lower corner of the rectangle [*Figure 7-15* C] momentarily with sufficient rudder and aileron pressure.

The pilot applies sufficient control stick back pressure using the elevator to start a vertical ascent. During the ascent, the pilot uses aileron and rudder pressure to maintain constant lateral distance from the wake. The pilot holds the wings near level with the ailerons. When the glider reaches high left corner position [*Figure 7-15* D], the pilot momentarily maintains this position with sufficient rudder and aileron pressure.

As the maneuver continues, the pilot reduces the rudder pressure and uses coordinated flight control inputs to fly along the top side of the rectangle. The glider proceeds to the top right corner [*Figure 7-15* E] using aileron and rudder pressure, as appropriate. The pilot maintains this position momentarily with aileron and rudder pressure.

The pilot applies sufficient control stick forward pressure using the elevator to start a vertical descent. During the descent, the pilot uses aileron and rudder pressure to maintain constant lateral distance from the wake. The pilot holds the wings near level with the ailerons. When the glider reaches the low right corner position [*Figure 7-15* F], the pilot momentarily maintains this position with sufficient rudder and aileron pressure.

As the maneuver continues, the pilot reduces the rudder pressure and uses coordinated flight control inputs to fly along the bottom side of the box until reaching the original center low tow position [*Figure 7-15* B]. From center low tow position, the pilot maneuvers the glider through the wake to the center high tow position [*Figure 7-15* A], completing the maneuver.

Common Errors

Common errors when boxing the wake include:

- Performing an excessively large rectangle around the wake or too small a rectangle that enters the wake. Note that Figure 7-16 has exaggerated positions and is not to scale.
- Improper control coordination and procedure.
- Abrupt or rapid changes of position.
- Allowing or developing unnecessary slack during position changes.

Aerotow Release

When the aerotow reaches release position, the glider pilot should clear the entire area for other aircraft. Scanning should include to the right, which is the direction the glider will turn after release. Scanning should also include to the left, the direction the towplane turns after release. A normal release occurs with the tow line under tension since hook-type towing attachments may need that force to make the hook swing open. When ready to release, … but the glider pilot should visually confirm the tow line release prior to beginning a 90° right turn. [*Figure 7-16A*]

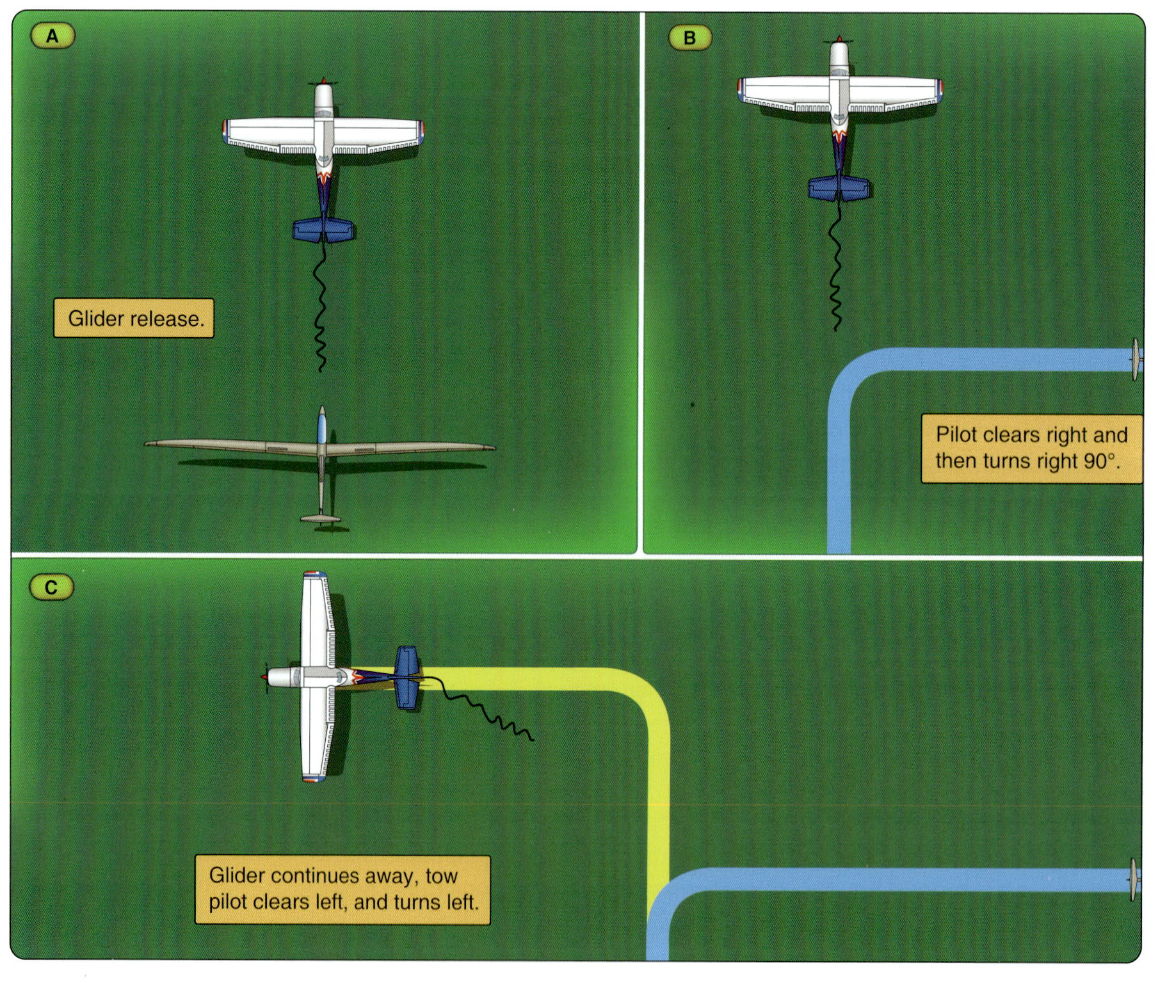

Figure 7-16. *Aerotow release.*

Figure 7-16B shows how this 90° change of heading achieves safe separation between towplane and glider in minimum time. After confirming glider release and seeing the glider turn right 90° away from the towplane, the tow pilot turns left, descending away from the release point.

Shown in *Figure 7-16C*, once clear of the glider and other aircraft, the tow pilot continues a descent toward the airport for landing. The tow pilot should continue to observe the glider as the glider pilot may start thermaling procedures and lose sight of the towing aircraft.

Common Errors

Common errors in aerotow release include:

- Lack of normal tension on tow line or slack in tow line.
- Failure to clear the area prior to release.
- Failure to visually confirm seeing the released tow line falling away prior to turning away from the towplane.
- Failure to make a 90° right turn after release.
- Release near other aircraft.
- Glider pilot or tow pilot losing sight of the other's aircraft.

Proper release procedures ensure proper aircraft separation in case the pilots lose sight of each other's aircraft. In any case, the tow pilot should exit the immediate area of the glider release. If the glider releases in a thermal or other lift, the glider normally remains in that lift to gain altitude, whereas the tow pilot can use power to return to the airport. However, both the tow pilot and glider pilot should maintain awareness of other gliders near areas of lift.

Ground Launch Takeoff Procedures

Ground launch uses a center of gravity (CG) tow hook that has an automatic back release feature. This protects the glider if the pilot cannot release the tow line during the launch. Since the failure of the tow release could cause the glider to be pulled toward the ground as it flies over the launching vehicle or winch, the pilot and ground crew should test the tow release feature prior to every flight. [*Figure 7-17*]

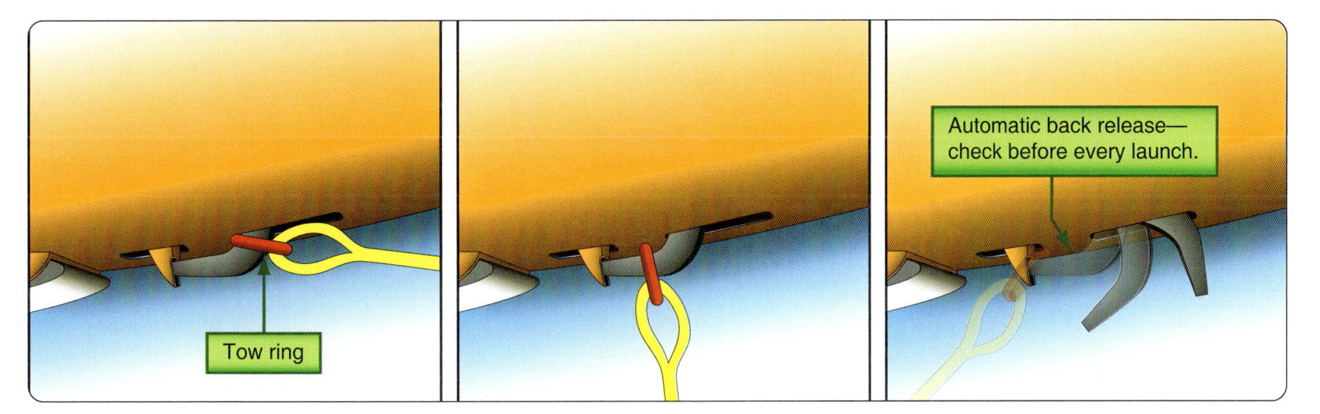

Figure 7-17. *Testing the tow hook.*

CG Hooks

Since attachment of a ground-launch tow line to a nose hook would pull the glider into the ground and overload the horizontal stabilizer and elevator, some training and high-performance gliders use a tow hook located near the CG. A CG hook gives the glider pilot control, free from any undue down moment during a ground tow. A CG hook uses a location either ahead of the landing gear, in the landing gear well, or on a bracket that attaches outside on the fuselage. If a glider only has a CG hook, and the CG hook does not have sufficient movement to fully release from an aerotow line under pressure, no aerotow should occur.

If the CG hook sits in the landing gear well, retracting the gear on tow may interfere with the tow line. For any type of tow, retracting the gear should wait until the glider achieved the desired tow altitude and releases the tow cable. Leaving the gear down also allows the glider pilot more time to assess the best landing options.

A CG hook makes a crosswind takeoff more difficult since the glider can weathervane into the wind more easily. In addition, a CG hook makes the glider more susceptible to kiting or rising too rapidly on an aerotow takeoff, especially with the CG near the aft limit.

Signals

Prelaunch Signals (Winch/Automobile)

Prelaunch visual signals for a ground launch operation allow the glider pilot, the wing runner, the safety officer, and the launch crew to communicate over considerable distances. When launching with an automobile, the glider and launch automobile may be 1,000 feet or more apart. When launching with a winch, the glider may start the launch 4,000 feet or more from the winch. Because of the distances involved, members of the ground crew use colored flags or large paddles to enhance visibility, as shown in *Figure 7-18.* When relaying information over large distances, direct voice communication between crewmember stations augments visual prelaunch signals, adds protection against premature launch, and facilitates an aborted launch if an unsafe condition arises.

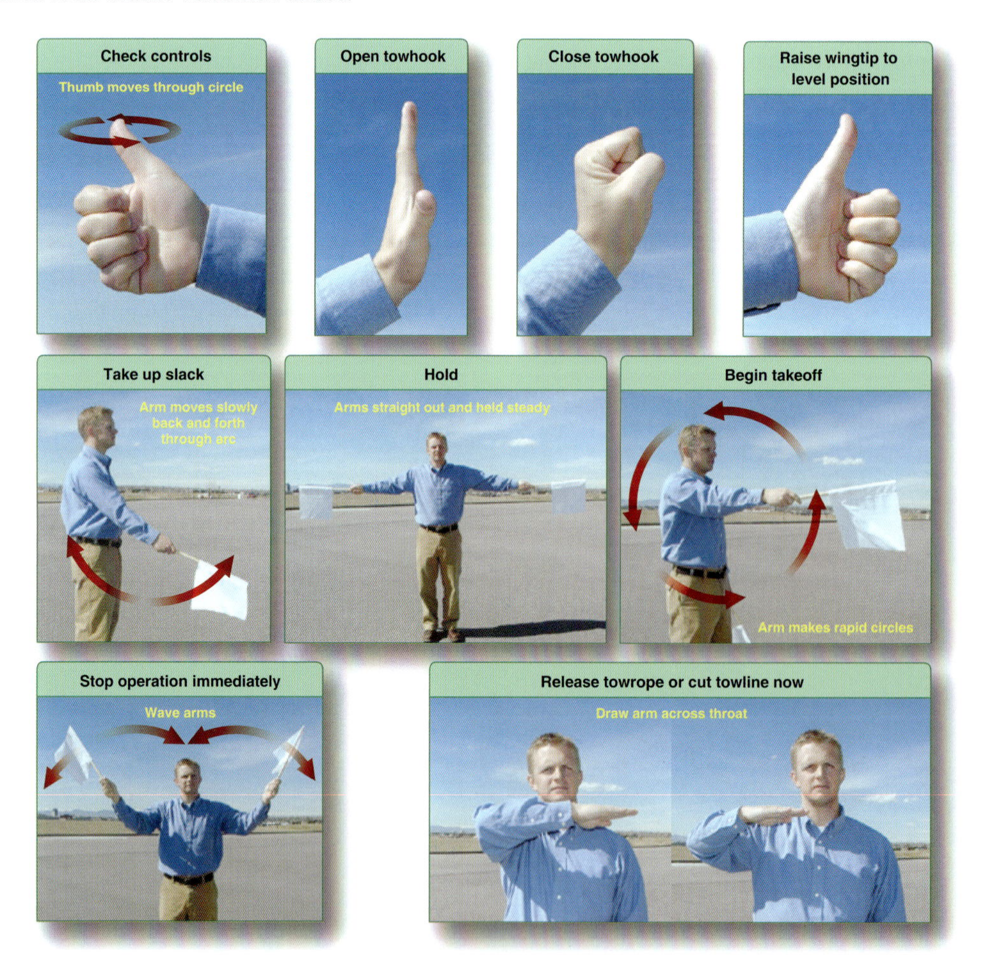

Figure 7-18. *Winch and aerotow prelaunch signals. The raise wingtip to level position is given by the pilot.*

Inflight Signals

Since ground launches occur quickly once the tow begins, inflight signals from the glider pilot to ground personnel only inform the winch operator or ground vehicle driver to increase or decrease speed. [*Figure 7-19*]

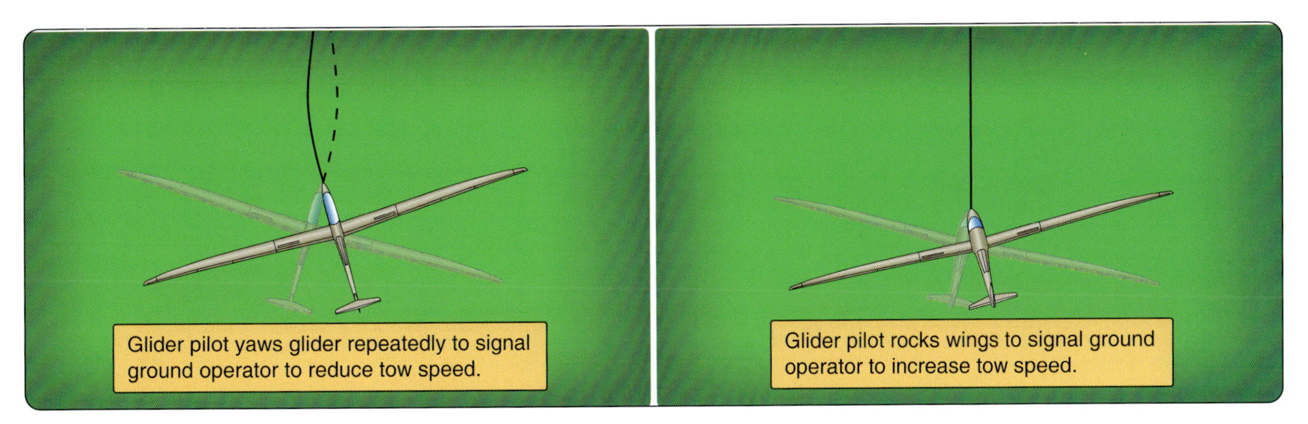

Figure 7-19. *Inflight signals for ground launch.*

Tow Speeds

The pitch attitude to airspeed relationship for a tethered glider during a ground tow presents a unique flight experience. Provided the tow mechanism has enough power to meet the energy demands of launch, a greater diversion of tow force from horizontal to vertical results in an acceleration to higher glider airspeed if the pilot pitches up. During the launch, pulling back on the stick tends to increase airspeed, and pushing forward tends to reduce airspeed.

Proper ground launch tow speeds ensure a safe launch. *Figure 7-20* compares various takeoff profiles that result when tow speeds vary above or below the correct speed.

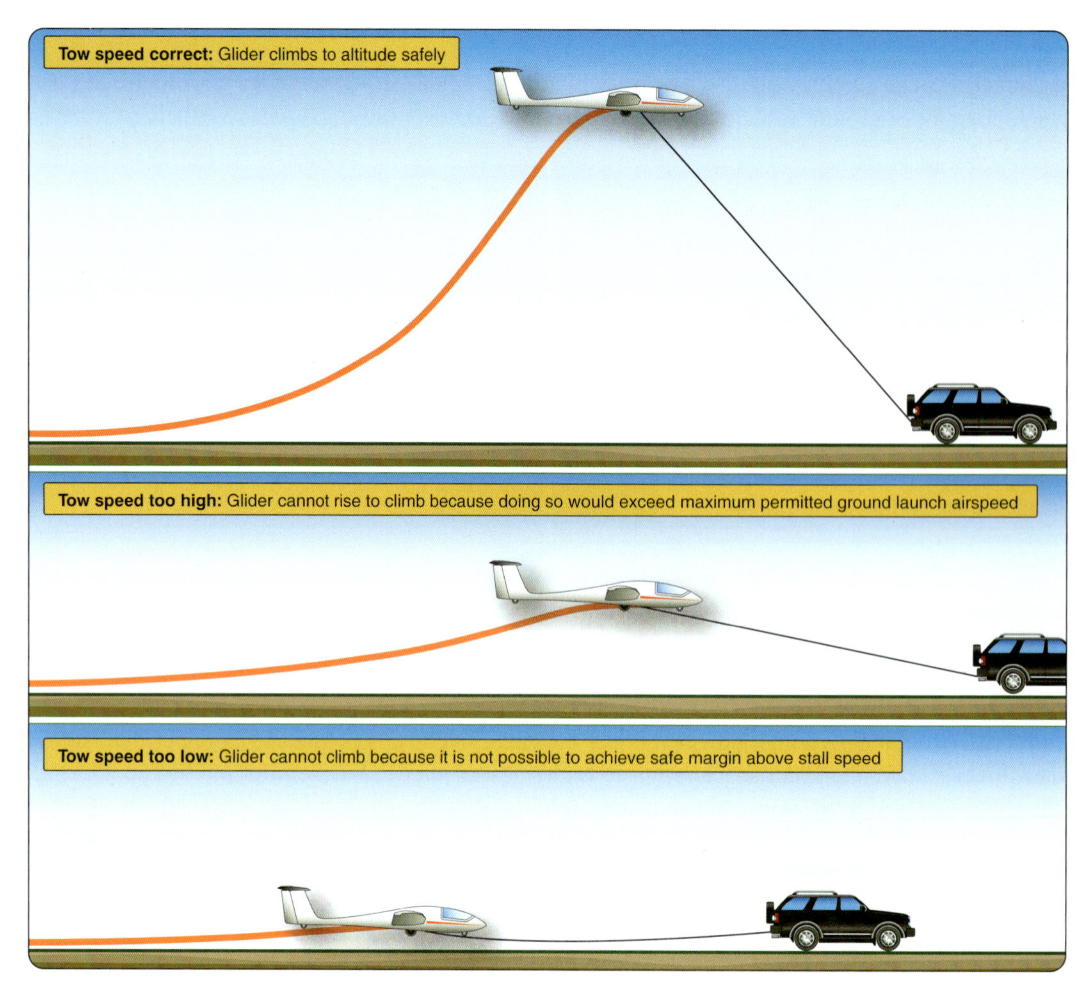

Figure 7-20. *Ground launch tow speeds.*

Each glider certified for ground launch operations has a placarded maximum ground launch tow speed. This speed usually applies to both automobile and winch launches. The glider pilot should stay at or below this speed during the ground tow to prevent structural damage to the glider.

Automobile Launch

Before the first launch, the pilot and vehicle driver should determine the appropriate vehicle ground tow speeds by considering the surface wind velocity, the expected wind gradient for the climb, and the glider speed increase during launch. The pilot and driver should agree on a maximum vehicle ground tow speed as a safety factor.

If a crosswind condition exists, the crew should position the glider slightly downwind of the takeoff heading and angled into the wind to help eliminate glider control issues during the initial portion of the tow. Due to the slow acceleration of the glider during an automobile ground launch, the tow line lay out should allow time for the glider to obtain sufficient speed while still in a headwind. [*Figure 7-21*]

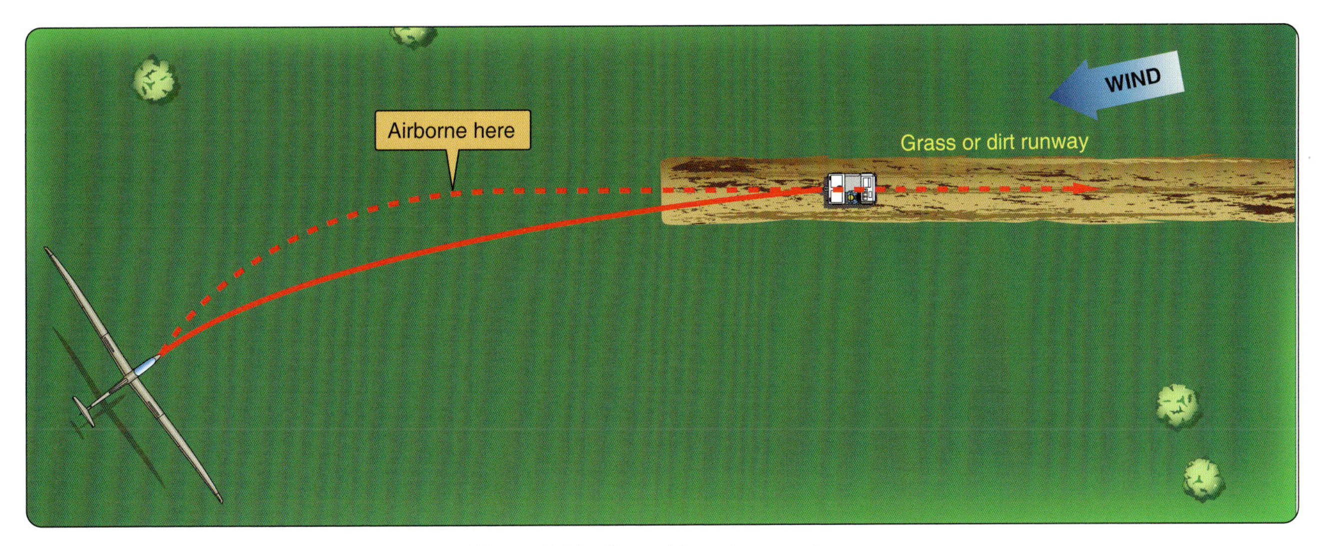

Figure 7-21. *Ground launch procedures.*

The tow speed calculation works as follows:

1. Subtract the surface winds from the maximum placarded ground launch tow speed for the glider.
2. Subtract an additional five miles per hour (mph) for the airspeed increase during the climb.
3. Subtract the estimated wind gradient increase encountered during the climb.
4. Subtract a 5 mph safety factor.

- As an example for a glider with a placarded maximum ground launch tow speed of 75 mph and with the following conditions:

  Surface winds 10 mph .... –10 mph

  Airspeed increase during climb 5 mph.......... –5 mph

  Estimated climb wind gradient 5 mph .... –5 mph

  Safety factor of 5 mph .... –5 mph

- The calculated automobile tow speed works out to.....................50 mph

Winch Launch

During winch launches, the winch operator applies full power smoothly and rapidly until the glider reaches an angle of 30° above the horizon. At this point, the operator begins to reduce the power, reaching approximately 20% power as the glider reaches a point about 60° above the horizon. As the glider reaches the 70° point above the horizon, the operator reduces power to idle. [*Figure 7-22*] The winch operator monitors the glider continuously during the climb for any signal to increase or decrease power, which would result in a change in the glider's speed.

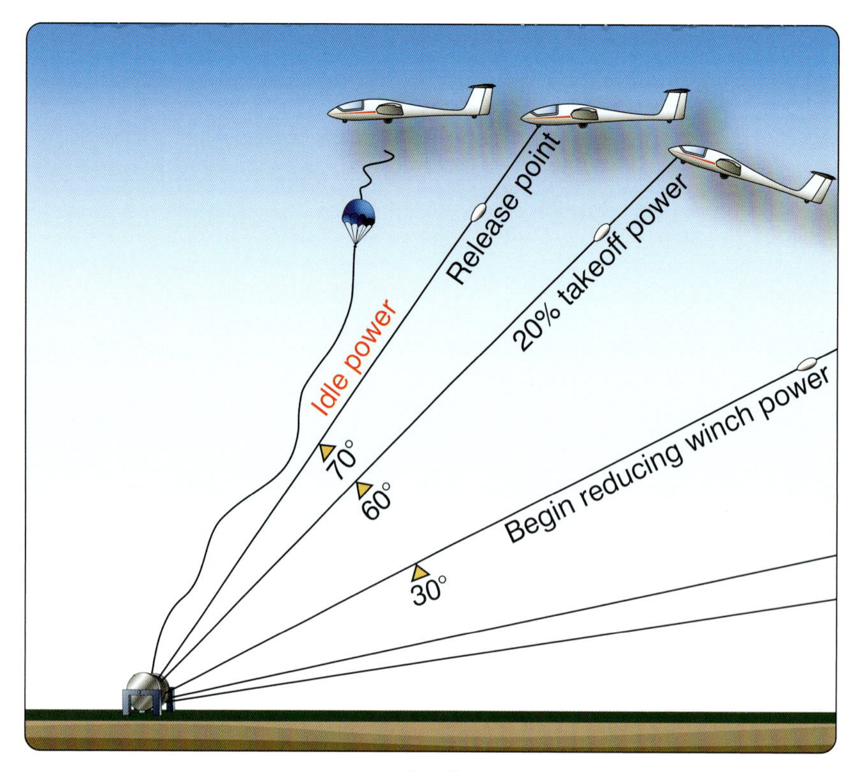

Figure 7-22. *Winch procedures.*

Crosswind Takeoff & Climb

The following are the main differences between crosswind takeoffs and climb procedures and normal takeoff and climb procedures:

- During the takeoff roll, the glider tends to weathervane into the wind.
- After liftoff, the glider drifts toward the downwind side of the runway.
- Strong crosswinds create a greater tendency for the glider to drift downwind.
- If space is available in the takeoff area, the tow line or cable should be laid out in a manner that the initial takeoff roll is slightly into the wind to reduce the crosswind component of the glider. [*Figure 7-23*]

Figure 7-23. *Wind correction angle for winch procedures.*

After lift-off, the glider pilot should establish a wind correction angle and fly toward the upwind side of the runway as shown in *Figure 7-24*. After release, the tow line tends to drift back toward the centerline of the launch runway. This helps keep the tow line from fouling or damaging any wires, poles, fences, aircraft, and other obstacles off to the side of the runway.

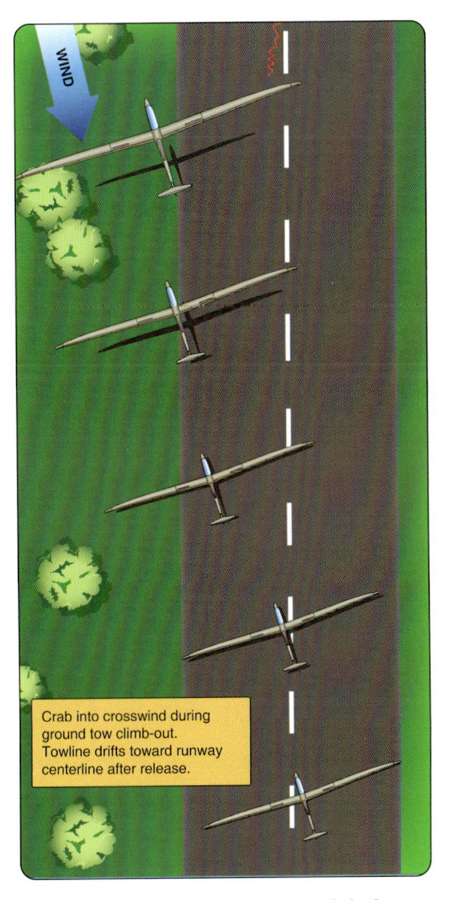

Figure 7-24. *Ground launch crosswind drift correction.*

Normal Into-the-Wind Launch

Prior to launch, the glider pilot, ground crew, and launch equipment operator brief launch signals and procedures. After completing checklists for the glider and ground launch equipment, the glider pilot should signal the ground crewmember to hook the tow line to the glider. For the release mechanism check, the ground crewmember applies tension to the tow line and signals the glider pilot to activate the release. The ground crewmember verifies that the release works properly and signals the glider pilot. After reconnecting the tow line, the ground crewmember takes a position at the wingtip of the down wing. When the glider pilot signals "ready for takeoff," the ground crewmember clears both takeoff and landing areas. After the ground crewmember ensures a clear traffic pattern, the ground crewmember then signals the launch equipment operator to "take up slack" in the tow line. With the slack removed from the tow line, the ground crewmember again verifies that the glider pilot is ready for takeoff. Then, the ground crewmember raises the wings to a level position, does a final traffic pattern check, and signals to the launch equipment operator to begin the takeoff.

The ground crewmember should never connect a glider to a tow line without the pilot onboard and ready for flight. If the pilot exits the glider for any reason, the pilot or ground crewmember should disconnect the tow line. Glider pilots should expect a takeoff anytime the tow line connects a glider to the source of the tow. If the launch begins before the pilot gives the launch signal, the glider pilot should promptly pull the tow line release handle.

The length, elasticity, and mass of the tow line used for a ground launch have several effects. First, a taut tow line often causes the glider to move forward. For this reason, the tow line should display a small amount of slack prior to beginning the launch. As the launch begins and for a few seconds, the glider pilot should hold the stick forward to avoid kiting. During the launch, the glider pilot should track the runway centerline and monitor the airspeed. [*Figure 7-25, position A*]

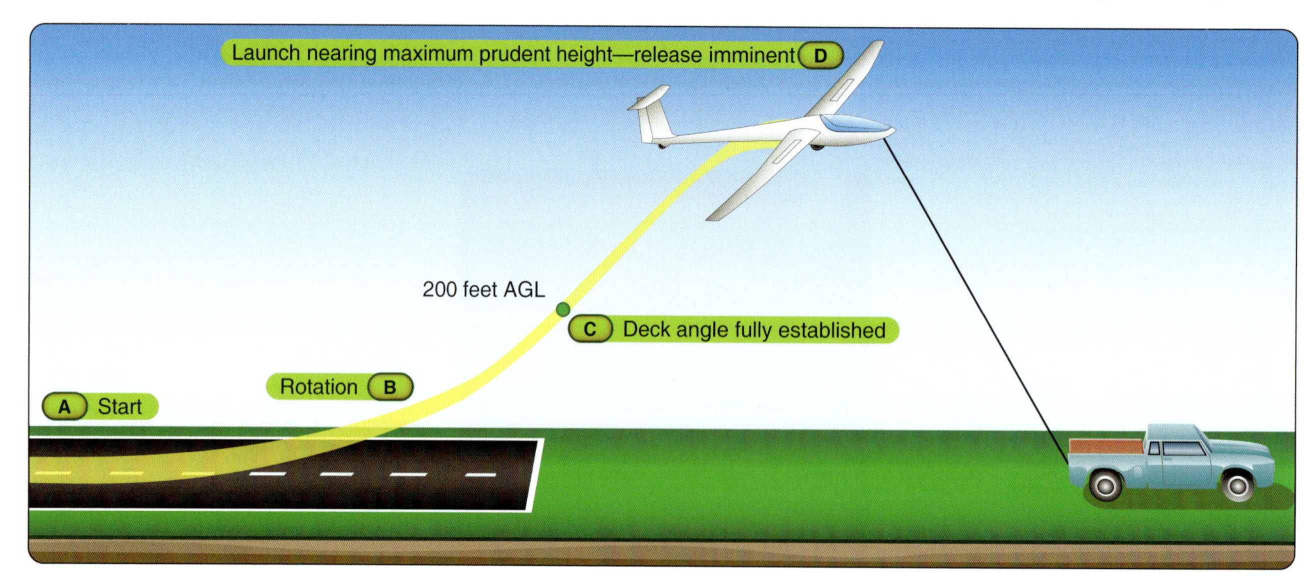

Figure 7-25. *Ground launch takeoff profile.*

When the glider accelerates and attains lift-off speed, the glider pilot eases the glider off the ground. The time interval from standing start to lift-off may be as short as 3 to 5 seconds. After the initial lift-off, the pilot should smoothly raise the nose to the proper pitch attitude, watching for an increase in airspeed. If the pilot raises the nose too soon or too steeply, the pitch attitude could become excessive while at low altitude. If the tow line breaks or the launching mechanism loses power, the pilot may find recovery from such a high pitch attitude difficult or impossible. Conversely, if the nose comes up too slowly, the glider may exceed the maximum ground launch tow speed. In addition, a shallow climb may result in the glider not reaching the planned release altitude. If this situation occurs, the pilot should pull the release and land straight ahead, avoiding any obstacles or equipment.

As the launch progresses, the pilot should ease the nose up gradually [*Figure 7-25*, position B] while monitoring the airspeed. The optimum pitch attitude for climb occurs, [*Figure 7-25*, position C] with the glider approximately 200 feet AGL. The pilot should monitor the airspeed during this phase of the climb-out to ensure an airspeed sufficient to provide a safe margin above stall speed but below the maximum ground launch airspeed. If the tow line breaks, or if the launching mechanism loses power at or above this altitude, the pilot should have sufficient altitude to release the tow line, lower the nose from the climb attitude to an approach attitude, and land straight ahead.

As the glider nears its maximum altitude [*Figure 7-25*, position D], the pilot should begin to level off above the launch winch or tow vehicle and reduce the rate of climb. In this final phase of the ground launch, the tow line pulls down on the glider. The pilot should gently lower the nose of the glider to reduce tension on the tow line and then pull the release handle two to three times to ensure tow line release. The pilot should feel the release of the tow line as it departs the glider and enter a turn to visually confirm the fall of the tow line. A broken tow line with a portion still attached to the glider may explain seeing only a portion of the tow line fall to the ground.

If pulling the release handle fails to release the tow line, the back-release mechanism of the tow hook should automatically release the tow line as the glider overtakes and passes the launch vehicle or winch.

Common Errors

Common errors in ground launching include:

- Improper glider configuration for takeoff.
- Improper initial positioning of flight controls.
- Improper use of visual launch signals.
- Improper crosswind procedure.
- Improper climb profile.
- Faulty corrective action for adjustment of airspeed and pitch.
- Exceeding maximum launch airspeed.
- Improper tow line release procedure.

Self-Launch Procedures

Preparation & Engine Start

A self-launching glider [*Figure 7-26*] has more systems than a nonmotorized glider, and manufacturers supply a more extensive preflight inspection checklist. Additional systems may include the fuel system, electrical system, engine, propeller, cooling system, and mechanisms that extend or retract the engine or propulsion system.
Whenever the engine runs, the pilot should consider the noise level and the need for hearing protection.

Figure 7-26. *Types of self-launching gliders.*

After preflighting a self-launching glider and clearing the area, the pilot starts the engine in accordance with the manufacturer's instructions. Typical items on a self-launching glider engine-start checklist include fuel mixture control, fuel tank selection, fuel pump switch, engine priming, propeller pitch setting, throttle setting, magneto or ignition switch setting, and electric starter activation. After starting and running through an after-start checklist and if the engine and propulsion systems appear within normal limits, the pilot may begin taxi operations.

Common Errors

Common errors in preparation and engine start include:

- Failure to use or improper use of checklist.
- Improper or unsafe starting procedures.
- Excessively high revolutions per minutes (rpm) after starting.
- Failure to ensure proper clearance of propeller.

Taxiing

Self-launching glider designs use different landing gear systems. Some designs use tricycle or tailwheel landing gear configurations commonly found on airplanes. Other types of self-launching gliders rest on a main landing gear wheel in the center of the fuselage and use outrigger wheels or skids on the wings to prevent the wingtips from contacting the ground.

Due to the long wingspan and low wingtip ground clearance of gliders, the self-launching glider pilot should consider airport layout and runway configuration. Some taxiways and airport ramps may not accommodate the long wingspan of the glider or limit maneuvering. Additionally, the pilot should consider the glider's crosswind capability during taxi operations. The pilot should manipulate the flight controls to prevent any crosswind from lifting a wing or causing the tail to rise. A general rule with a quartering headwind is to position the controls so as to climb into the wind while a quartering tailwind requires positioning the controls so as to dive away from the wind. Taxiing on soft ground requires additional power. Self-launching gliders with outrigger wingtip wheels may lose directional control if a wingtip wheel bogs down, and wing walkers can hold the wings level during low-speed taxi operations on soft ground.

Common Errors

Common errors in taxiing a self-launching glider include:

- Improper use of brakes.
- Failure to comply with airport markings, signals, and clearances.
- Taxiing too fast for conditions.
- Improper control positioning for wind conditions.
- Failure to consider wingspan and space required to maneuver during taxiing.

Pretakeoff Check

The manufacturer provides a before takeoff checklist. As shown in *Figure 7-27*, the complexity of many self- launching gliders makes a written takeoff checklist an essential safety item. Pretakeoff items on a self-launching glider may include checking fuel quantity and pressure, oil temperature and pressure, and other aircraft systems as applicable, conducting an engine runup, and setting throttle/rpm, propeller pitch, and cowl flaps. The pilot should also ensure seat belts and shoulder harnesses secure, doors and windows closed and locked, canopies closed and locked, air brakes closed and locked, altimeter set, communication radio set to the proper frequency for traffic advisory, and flight instruments adjusted for takeoff.

Figure 7-27. *Self-launching glider instrument panels.*

Common Errors

Common errors in the before takeoff check include:

- Improper positioning of the self-launching glider for runup.
- Failure to use or improper use of checklist.
- Improper check of flight controls.
- Failure to review takeoff emergency procedures.
- Improper radio and communications procedures.

Normal Takeoff

After completing the pretakeoff checklist, the pilot should check for traffic and prepare for takeoff. The pilot should make a final check for conflicting traffic, then taxi out onto the active runway and align the glider with the centerline. If operating from an airport with an operating control tower, the pilot must request and receive an air traffic control (ATC) clearance prior to taxi and before using any runway for takeoff.

The pilot should apply full throttle smoothly, begin the takeoff roll while tracking the centerline of the runway, fly the self-launching glider off the runway at the recommended lift-off airspeed, and allow the glider to accelerate in ground effect (IGE) until reaching the appropriate climb airspeed. If the runway has an obstacle ahead, the pilot should climb at the best angle of climb airspeed (V_X) until the obstacle is cleared. If no obstacle is present, the pilot should use either best rate of climb airspeed (V_Y) or the airspeed for best engine cooling during climb. The pilot should monitor the engine and instrument systems during climb-out. If the self-launching glider has a time limitation on full throttle operation, the pilot should adjust the throttle as necessary during the climb.

PIOs in Self-Launching Gliders

Power changes affect the glider's pitch attitude in some self-launching gliders equipped with an engine above the CG. Power changes can also cause variations in elevator effectiveness. [*Figure 7-28*] In most self-launching gliders, the effect becomes more noticeable when flying at or near minimum controllable airspeed (V_{MCA}). For this reason, self-launching glider pilots should avoid slow flight when flying at low altitude under power.

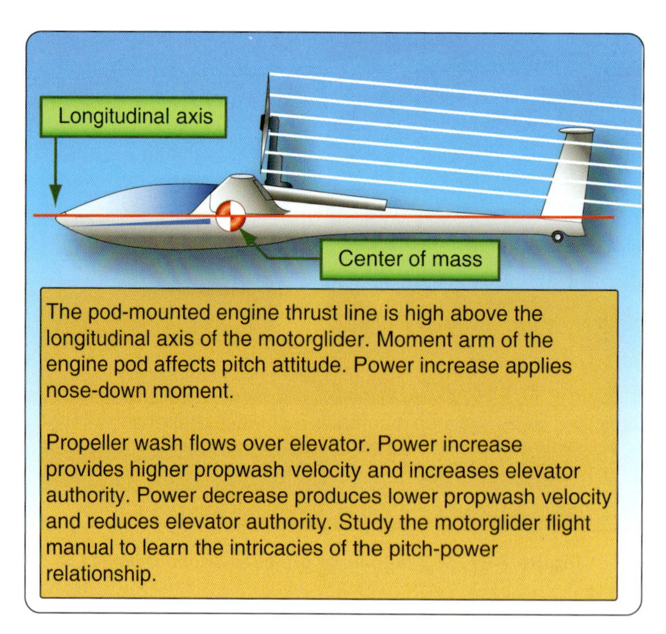

Figure 7-28. *Pitch attitude power setting relationships for self-launching glider with engine pod.*

The likelihood of PIOs around the lateral axis of self-launching gliders increases during the takeoff roll or landing with power because of power changes. The GFM/POH may contain information describing how to deal with these effects. In general, good pilot technique involves moving the throttle control smoothly, gradually, and in coordination with pitch control input.

Crosswind Takeoff

The long wingspan and low wingtip clearance of the typical self-launching glider make it vulnerable to striking a wingtip on runway signs or runway lights. In a glider with a single main wheel and no wing runner, the takeoff roll should start with the upwind wing on the ground with the aileron and rudder controls set for the current wind situation. For example, with a crosswind from the right, the right wing should be down, the control stick should be held to the right, and the rudder should be held to the left. The aileron input keeps the crosswind from lifting the upwind wing, and the downwind rudder minimizes the tendency of the self-launching glider to weathervane in a crosswind. As airspeed increases, control effectiveness improves, and the pilot can gradually decrease the crosswind control setting while maintaining the upwind wing slightly low. The self-launching glider should lift off at the appropriate lift-off airspeed and accelerate to climb airspeed. During the climb, the pilot should establish a wind correction angle and level the wings so that the self-launching glider tracks the extended centerline of the takeoff runway. [*Figure 7-29*]

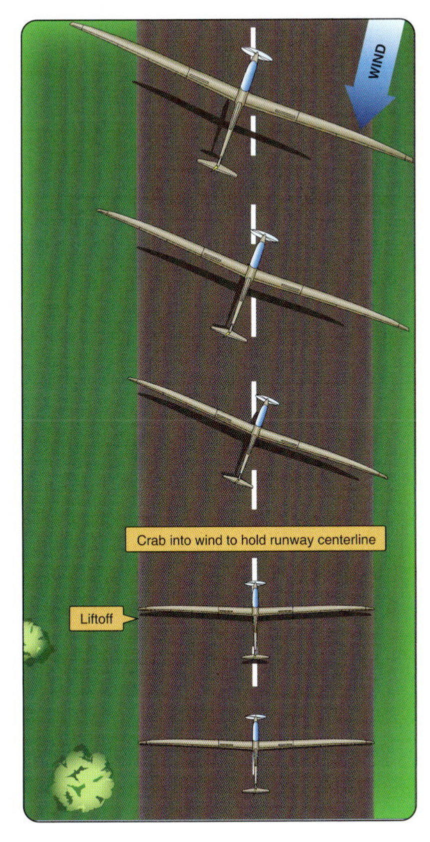

Figure 7-29. *Self-launching gliders—crosswind takeoff.*

Common Errors

Common errors in crosswind takeoff include:

- Improper initial positioning of flight controls.
- Improper power application.
- Inappropriate removal of hand from throttle.
- Poor directional control
- Improper use of flight controls.
- Improper pitch attitude during takeoff.
- Failure to establish and maintain proper climb attitude and airspeed.
- Maintaining takeoff slip instead of transitioning to crab after takeoff.

Climb-Out & Engine Shutdown Procedures

The GFM or POH provides useful information about recommended power settings and target airspeeds for best angle of climb, best rate of climb, best cooling performance climb, and cruise performance while in powered flight. Powered gliders may have additional limits that include maximum permitted airspeed with engine extended and maximum airspeed at which to extend or retract the engine. Many self-launching gliders have a time limitation on full throttle operation to prevent overheating and premature engine wear.

The engine may heat up considerably during takeoff and climb, and cooling system mismanagement can lead to dangerously high temperatures in a short time. If the self-launching glider has cowl flaps for cooling, the pilot should set the cowl flaps for high power operations. In some self-launching gliders, operating at full power with cowl flaps closed can result in overheating and damage to the engine in as little as 2 minutes. To minimize the chances of engine damage or fire, the pilot should monitor engine temperatures during high power operations and follow engine operating limitations described in the GFM/POH. If these measures do not reduce high temperatures and since extended overheating could cause an inflight fire, the safest course of action may include shutting down the engine and making a precautionary landing. An inflight fire presents a much greater threat than an emergency landing.

Handling limitations for a given self-launch may include minimum controllable airspeed with power on, minimum controllable airspeed with power off, and other limitations described in the GFM/POH. Self-launching gliders come in many configurations. Those with a top-mounted retractable engine and/or propeller have a thrust line above the longitudinal axis of the glider. Significant power changes may cause substantial pitch attitude changes. For instance, full power setting in these self-launching gliders introduces a nose-down pitching moment.

To counteract this pitching moment, the pilot normally holds the control stick back and uses trim. If the pilot quickly reduces from full power to idle power while maintaining significant control-up stick force, the glider tends to pitch up with the power reduction. This nose-pitching moment could induce a stall. Smooth and coordinated management of power and flight control provides the safest procedure under these conditions.

During climb-out, the pilot should hold a pitch attitude that results in climbing out at the desired airspeed, while adjusting elevator trim as necessary. Pilots should manage climbs in self-launching gliders using smooth control inputs and smooth and gradual throttle adjustments.

When climbing under power, most self-launching gliders exhibit a turning tendency due to P-factor. P-factor results from uneven distribution of thrust caused by the difference between the angle of attack (AOA) of the ascending propeller blade and the descending propeller blade. The pilot uses rudder to counteract P-factor during climbs with power. [*Figure 7-30*]

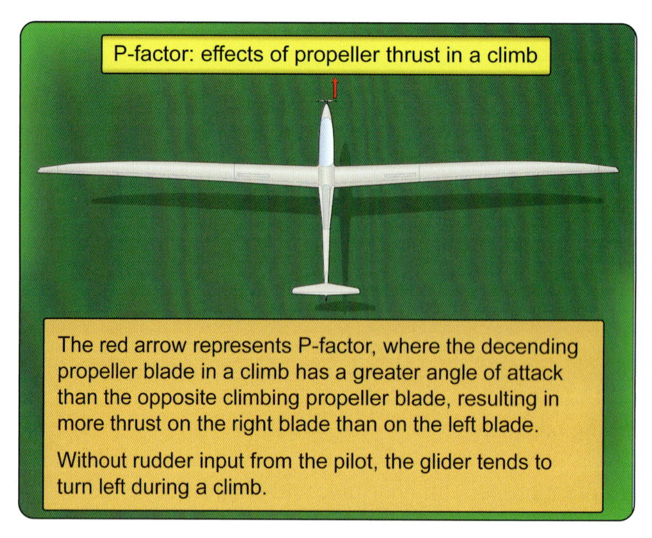

Figure 7-30. *P-factor.*

The pilot should scan for other aircraft traffic before making any turn. Coordinated aileron and rudder control, as well as turns made with a shallow bank angle, result in a more efficient flight and faster climb rate.

Detailed engine shutdown procedures are described in the GFM/POH. The manufacturer provides engine cool-down procedures for reducing engine system temperatures prior to engine shutdown. Lowering the nose to increase airspeed provides faster flow of cooling air to the engine cooling system, and several minutes of reduced throttle and increased cooling airflow may allow the engine to reach an appropriate temperature for shutdown.

When preparing to shut down the engine, the pilot should reduce power slowly to reduce or eliminate shock cooling. If shock cooling occurs, the external parts cool faster and shrink more than the interior components resulting in binding and scuffing of moving parts such as piston rings and valves while the engine continues to operate.

If the engine retracts, additional time after engine shutdown may reduce engine temperature to acceptable limits prior to retracting and stowing the engine in the fuselage. Consult the GFM/POH for details.

Retractable-engine self-launching gliders become aerodynamically more efficient after stowing the engine. The alignment of the propeller blades may need adjustment to prevent interference with the engine bay doors.

When the engine/propeller installation sits aft of the pilot's head, the gliders may have a mirror that enables the pilot to perform a visual propeller alignment check prior to stowing the engine/propeller pod. The GFM/POH contains detailed instructions for stowing the engine and propeller for a particular glider. If a malfunction occurs during engine shutdown and stowage, the pilot may not be able to stow or extend the engine for restart. In this case, glide performance will be reduced by as much as 50%. See GFM/POH for specific information. In anticipation of this eventuality, the pilot should have a landing area within power-off gliding distance.

Some self-launching gliders use a nose-mounted engine/propeller installation that resembles a single-engine airplane. In these self-launching gliders, the shutdown procedure usually consists of operating the engine for a short time at reduced power to cool the engine to shutdown temperature. After shutdown, the pilot should close cowl flaps (if installed) to reduce drag and increase gliding efficiency. The manufacturer may recommend a time interval between engine shutdown and cowl flap closure to prevent excess temperatures buildup in the confined engine compartment. These temperatures may not be harmful to the engine itself, but may degrade structures around the engine, such as composite engine mounts or installed electrical components. Excess engine heat can also result in fuel vapor lock that prevents a subsequent restart.

Some installations have a propeller feathering system that reduces propeller drag during non-powered flight. If the pilot can control the propeller blade pitch while in flight or after shutdown, the pilot should set the propeller as described in the GFM/POH. For any inflight engine restart the pilot should follow the manufacturer's procedures.

Common Errors

Common errors during climb-out and shutdown procedures include:

- Failure to follow manufacturer's recommended procedure for engine shutdown, feathering, and stowing (if applicable).
- Failure to maintain positive aircraft control while performing engine shutdown procedures.
- Failure to follow proper engine extension and restart procedures.

Gliderport/Airport Traffic Patterns & Operations

Gliderports and airport operators within the United States should comply with Federal Aviation Administration (FAA) recommended procedures established in Advisory Circulars (AC), the Aeronautical Information Manual (AIM), and current FAA regulations. These publications serve as good references to help ensure safe glider operations.

Airport managers and glider operators usually establish traffic patterns for their operation to accommodate all the activities that take place. Pilots planning to operate at a gliderport or airport should obtain a thorough briefing or checkout before conducting flights at that facility. The landing surface serves as the primary reference to begin and fly each approach to the landing area. Pilots commonly use an initial point (IP) as shown in *Figure 7-31*, which at the recommended altitude will provide for sufficient gliding distance to reach the landing field with an adequate safety margin. The sequence of a normal approach runs from over the IP to the downwind leg, base leg, final approach, flare, touchdown, rollout, and stop. The IP may be located over the center of the gliderport/airport or at a remote location near the traffic pattern. Once past the IP, pilots normally manage their energy to compensate for wind, traffic, terrain and obstacles. While a rectangular traffic pattern is preferred, pilots may need to modify their traffic pattern since flying a pattern that results in a landing as intended takes precedence over a rectangular pattern structure.

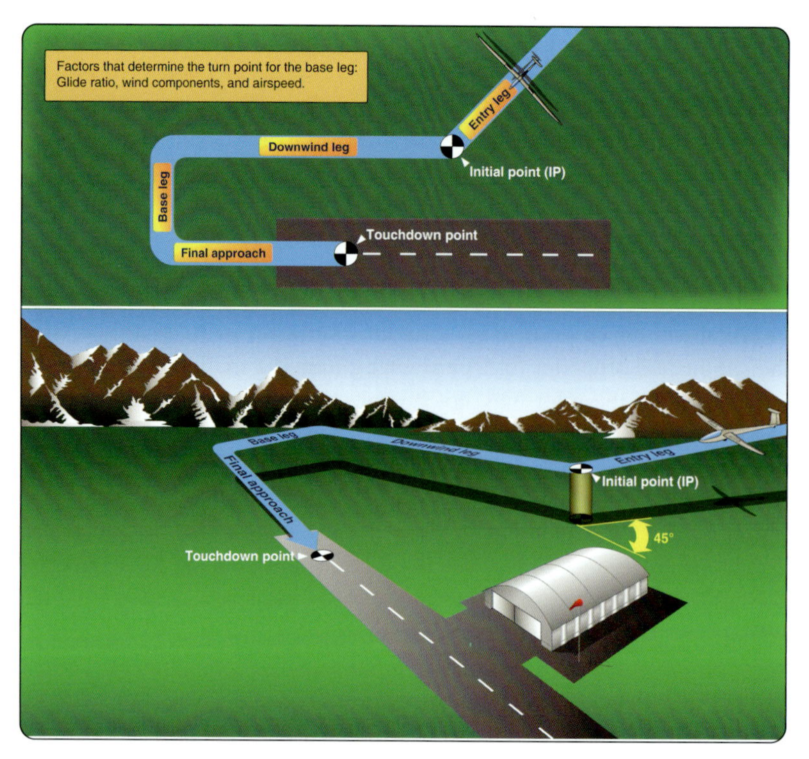

Figure 7-31. *Traffic pattern.*

Pilot should understand the rationale for determining an IP. When approaching an unfamiliar landing area, a pilot may not have a known IP. In this situation, the pilot should use the landing surface as the primary reference and set up a traffic pattern that provides an equivalent margin of safety. Pilots should develop proper placement, altitude, and distances based on wind, traffic, terrain, obstacles, and glider performance. In addition to environmental factors, the pilot should plan for an approach and landing with other participating traffic in the pattern and know the right of way rules.

While glider pilots may compensate for winds and modify the traffic pattern to retain a safe approach angle, the pilot normally plans to fly over the IP, where known, at an altitude of 800 to 1,000 feet AGL or as recommended by the local field operating procedures. Once over the IP, the pilot flies along the downwind leg of the planned landing pattern. During this time, the pilot should look for other aircraft and listen to the radio, if installed, for other aircraft in the vicinity of the gliderport/airport. Glider pilots should plan to make any radio calls early in the pattern and then concentrate on the landing task.

Pilots should complete the landing checklist prior to the downwind leg. A popular landing checklist mnemonic uses the acronym FUSTALL.

- Flaps—set (if applicable)
- Undercarriage—down and locked (if applicable)
- Speed—normal approach speed established (as recommended by the GFM/POH)
- Trim—set
- Air brakes (spoilers/dive brakes)—checked for correct operation
- Landing area—look for wind, other aircraft, and personnel
- Land the glider

Normal Approach & Landing

Planning for an approach begins at some distance from the landing zone. Prior to the IP and downwind leg entry, the pilot should consider the approach angle, distance from the landing area, spacing to accommodate other aircraft, and desired approach airspeed. Pilots normally use the recommended approach speed in the GFM/POH, but they may use 1.5 V_{SO} in the absence of a recommended approach speed. The pilot should use spoilers/dive brakes as necessary to dissipate excess altitude while maintaining the desired approach airspeed. During the entry into the traffic pattern, pilots should manage trim and make coordinated turns limited to no more than 45° of bank, and the pilot should not conduct a 360° turn once established on the downwind leg.

Downwind Leg

When approaching the IP, the pilot should maneuver the glider to enter the downwind leg. The lateral distance from a downwind leg to the landing area should be such that the glider pilot can look down to the centerline of the landing area at a 30- to 45-degree angle. This sight picture equates to a lateral distance to the landing area of 800 to 1200 feet, which allows the glider to fly inside any airplane traffic pattern and gives a better view of the landing area. The exact distance depends on winds, other weather conditions, the type of glider, and the site topography. On a typical downwind leg, the glider should descend to arrive abeam the touchdown point at an altitude between 500 and 600 feet AGL. The pilot may use the spoilers/dive brakes to arrive at this altitude. The pilot should also monitor the glider's position with reference to the touchdown area. If the wind pushes the glider away from or toward the touchdown area, the pilot should establish a wind correction angle, stop the drift, and plan to shorten or lengthen the downwind as needed. On the downwind leg, groundspeed increases with any tailwind and shortens the elapsed time to reach the point at which to turn base.

Base Leg

The base leg normally starts with the touchdown point no more than 45° over the pilot's shoulder looking back at the touchdown area if under a no wind condition. However, pilots should adjust the downwind length based on the glide ratio of the glider flown. Each glider pilot should evaluate the landing conditions, configure the glider for landing under those conditions, and turn to base while keeping the point of intended touchdown within easy gliding range. Performing a slip or extending drag devices can dissipate excess altitude, but nothing on a non- motorized glider will make up for insufficient altitude.

Once established on the base leg, the pilot should scan for and detect any aircraft on long final approach. If using a radio, the glider pilot can broadcast position for turn to final. The pilot should adjust the turn to correct for wind drift encountered on the base leg and roll out on a heading that aligns with the landing area. The pilot should also adjust the spoilers/dive brakes and/or slip as needed to position the glider at the desired approach angle. New pilots should learn to properly scan for another aircraft operating in the traffic pattern. Pilots should also review the current revision of FAA Advisory Circular (AC) 90-48, Pilots' Role in Collision Avoidance.

Final Approach

The turn onto the final approach should not exceed a 45° bank and the pilot should use an appropriate approach angle to start the final descent. The pilot normally makes a coordinated turn from base to final to line up with the centerline of the touchdown area. The pilot should adjust the spoilers/dive brakes, as necessary, to fly the desired approach angle to a specific spot on the ground and establish a stabilized approach at the recommended approach speed.

Stablized Approach

A stabilized approach requires the pilot to judge certain visual clues and then maintain a constant final descent airspeed, approach angle, and configuration. Glider pilots should plan a downwind and base leg that allows them to turn from base to final in position to continue on a stabilized approach. The final approach with spoilers/dive brakes extended approximately half open (not necessarily half travel of the spoiler/dive brake control handle) creates an ideal approach for most gliders. The pilot can devote significant attention toward outside visual cues to fine tune the approach. The pilot should not stare at any one place, but rather scan from one point to another, such as from the aiming point to the horizon, to any objects along the landing surface, to an area well short of the landing surface, and back to the aiming point. This makes it easier to perceive any deviation from the desired glide path and determine if the glider is proceeding directly toward the aiming point. The pilot should also glance at the airspeed indicator periodically and correct for any airspeed deviation.

A glider descending on stabilized final approach travels in a straight line towards a spot on the ground ahead, commonly called the aiming point. If the glider maintains a constant glide path without a flare for landing, it will strike the ground at the aiming point.

To the pilot, the aiming point appears to be stationary. It does not appear to move under the nose of the aircraft and does not appear to move forward away from the aircraft. This feature identifies the aiming point—it does not move. However, objects in front of and beyond the aiming point do appear to move as the distance is closed, and they appear to move in opposite directions! For a constant angle glide path, the distance between the horizon and the aiming point remains constant. If descending at a constant angle and the distance between the perceived aiming point and the horizon appears to increase (aiming point moving down away from the horizon), then the true aiming point is farther down the runway. If the distance between the perceived aiming point and the horizon decreases, meaning that the aiming point is moving up toward the horizon, the true aiming point is closer than perceived.

When the glider is established on final approach, the landing surface normally appears as an elongated rectangle. When viewed from the air during the approach, the phenomenon known as perspective causes the landing surface to assume the shape of a trapezoid with the far end looking narrower than the approach end and the edge lines converging ahead. As a glider continues down the glide path at a constant angle (stabilized), the image the pilot sees keeps the same shape, but with proportionately larger dimensions. In other words, during a stabilized approach, the perceived shape of the landing surface should not change. [*Figure 7-32*]

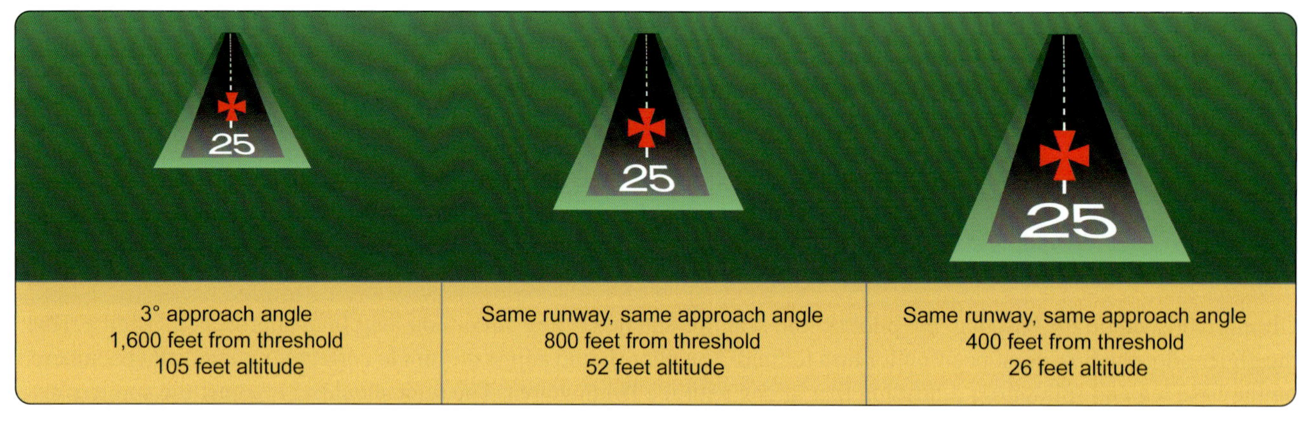

Figure 7-32. *Runway shape during stabilized approach.*

If the approach becomes shallow, the landing surface appears to shorten and becomes wider. Conversely, if the approach is steepened, the landing surface appears to become longer and narrower. [*Figure 7-33*]

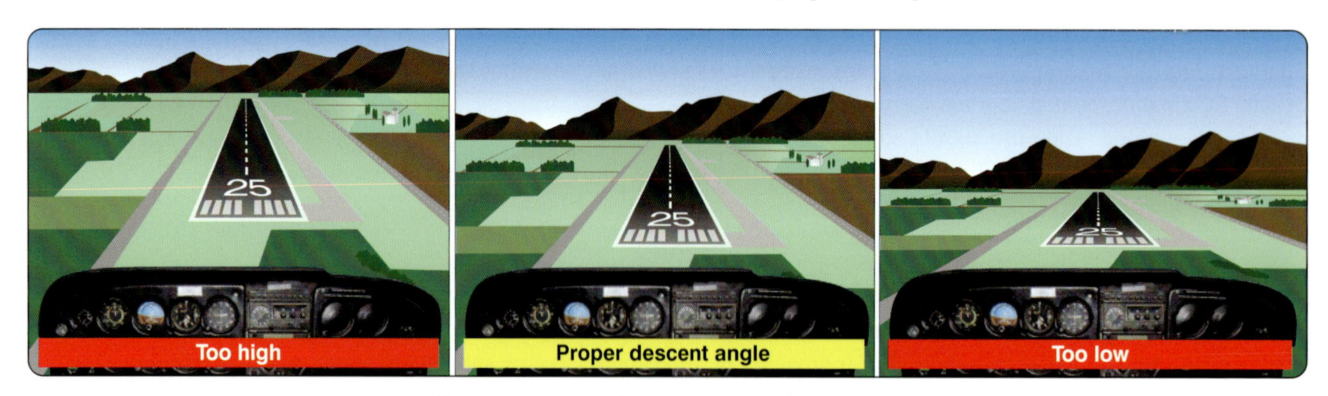

Figure 7-33. *Change in runway shape if approach becomes narrow or steep.*

During instruction and practice, a glider pilot acquires the skill to use visual cues to discern the true aiming point from any distance out on final approach. From this, the pilot determines if the current glide path will result in either an undershoot or overshoot. Note that since the pilot reduces the rate of descent during the flare, the actual touchdown occurs farther down the field. Considering float during round out, a skilled pilot can predict the point of touchdown with some accuracy.

The round out, touchdown, and landing roll are much easier to accomplish when preceded by a stabilized final approach, which reduces the chance of a landing mishap. Therefore, the pilot should detect and correct deviations from the desired glide path early so that the magnitude of corrections is small. The pilot should make appropriate and smooth adjustments in the spoilers/dive brakes to ensure proper glidepath control and should avoid pumping the spoilers/dive brakes from full open to full close. If excess speed develops from a sudden dive at the end of the approach, the glider can float a considerable distance that could preclude touching down in the desired landing area.

Round Out & Flare

When the glider approaches approximately 5 feet above the ground in a normal descent, the pilot begins a slow, smooth transition from a normal approach attitude to a landing attitude, gradually rounding out the flightpath to one that is parallel to and a few inches above the runway. The pilot gradually applies back-elevator pressure to slowly increase the pitch attitude and AOA, which increases at a rate that allows the glider to continue settling slowly as forward speed decreases. This is a continuous process until the glider touches down on the ground.

When the AOA is increased, the lift increases momentarily and gradually decreases the rate of descent and airspeed. During the round out, the airspeed is decreased to touchdown speed while the lift is controlled so the glider settles gently onto the landing surface. The round out is executed at a rate such that the proper landing attitude and the proper touchdown airspeed are attained simultaneously just as the wheels contact the landing surface.

The rate at which the pilot executes the round out depends on the glider's height above the ground, the rate of descent, and the pitch attitude. A round out started excessively high needs to be executed more slowly than one started from a lower height. When the glider appears to be descending very slowly, the increase in pitch attitude should be made at a correspondingly slow rate.

Touchdown

The touchdown is the gentle settling of the glider onto the landing surface. During the round out, the airspeed decays such that the glider touches down on the main gear. Drag devices in a glider such as spoilers or air brakes may also be used to control the touchdown point if needed. As the glider settles, proper landing attitude is attained as necessary. The touchdown should occur with the glider's longitudinal axis parallel to the direction in which the glider is moving along the runway. Failure to accomplish this imposes severe side loads on the landing gear. To avoid these side stresses, the pilot should not allow the glider to touch down while turned into the wind or drifting. After touchdown full deployment of drag devices and use of wheel brakes will increase the deceleration in the landing roll while maintaining directional control.

Pilots should avoid driving the glider into the ground using little or no flare. This type of landing puts excessive loads on the landing gear and wings. Forcing the glider onto the ground at excessive speeds may introduce pilot induced oscillations, such as porpoising. A good glider landing in most gliders with a main wheel and tailwheel (or skid) occurs on the main wheel with the tail wheel just slightly touching or the tail wheel just barely off the surface. The main wheel can withstand the shock of landings, but the tail wheel may not. Pilots should always follow the GFM/POH recommendations of the manufacturer.

After Landing Roll

The landing process should never be considered complete until the glider has been brought to a complete stop. Accidents may occur because of pilots abandoning their vigilance and failing to maintain positive control after getting the glider on the ground. Loss of directional control may lead to an aggravated, uncontrolled, tight turn on the ground, or a ground loop. The combination of centrifugal force acting on the center of gravity (CG) and ground friction of the main wheel resisting it during the ground loop may cause the glider to tip or lean enough for the outside wingtip to contact the ground. Proper directional control needs to be established early on after touchdown. The rudder serves the same purpose on the ground as it does in the air—it controls the yawing of the glider. The effectiveness of the rudder is dependent on the airflow, which depends on the speed of the glider.

The long, low wingtips of the glider are susceptible to damage from runway signage and runway lighting. When landing on a runway or landing strip, the rollout should continue straight along the centerline of the touchdown area, and the pilot should use full back stick on the elevator while keeping the glider wings level with aileron control. Pilots normally ensure that a wing does not contact the ground until the glider reaches a low speed or stops. As control authority decays during the ground roll, the pilot should apply brakes to avoid leaving the runway or landing area. However, if an obstacle becomes a concern (possible in an off-airport landing or landing out), a coordinated turn on the ground may avoid the obstacle. Turning off the runway should be done only when the pilot has the glider under control.

Landings in high, gusty winds or turbulent conditions occur using a higher approach airspeed to improve controllability, provide a safe margin above stall airspeed, and allow better penetration into the headwind on final approach. As a rule of thumb, pilots add one-half the reported gust factor (the difference between the sustained wind and reported gusts) to the normal recommended approach airspeed.

Pilot Induced Pitch Oscillations During Landing

Over controlling the elevator during the landing flare can cause the glider to balloon well above the landing surface even as airspeed decreases toward stalling speed. If the pilot pushes the stick forward after ballooning, the glider will rapidly descend toward the ground. If the pilot pulls the control stick back to arrest the descent, the glider may balloon again or experience a hard or nose-first landing depending on the airspeed. Wind gusts and turbulence increase the likelihood of this type of PIO.

To prevent PIOs during the flare after ballooning, the pilot should stabilize the glider at an altitude of 3 to 4 feet and begin the flare anew. If ballooning occurs at a low airspeed that takes the glider higher than a normal flare altitude, the pilot may reduce the extension of the spoilers/dive brakes to moderate the descent rate.

The spoilers/dive brakes provide significant drag and reduced lift when fully deployed. A sudden extension while landing results in a high sink rate and possible hard contact with the runway. This can lead to a rebound into the air and a series of PIOs. If a wind gust or sudden retraction of the spoiler/dive brakes causes the glider to balloon, the pilot should adjust the spoiler/dive brakes smoothly to reestablish an appropriate flare.

Forward Slip

A forward slip allows the pilot to steepen the approach path without increasing airspeed. While drag devices have reduced the need for forward slips, pilots should know how to use them for short/off-field landings, and approaches over obstacles.

The forward slip retains the glider's direction of motion although the nose of the glider points away from the direction of travel. For a glider in straight flight, the pilot lowers the wing on the windward side using the ailerons. Simultaneously, the pilot yaws the glider's nose by applying opposite rudder to point the glider's longitudinal axis at an angle to its original flightpath. The correct amount of bank and yaw maintains the original ground track. The pilot should also raise the nose if necessary, to prevent the airspeed from increasing.

Pilots should allow an extra margin of altitude for safety as part of an approach. If an arrival with excess altitude occurs when nearing the boundary of the selected field, the pilot can dissipate the excess altitude using a forward slip.

The use of slips has definite limitations. Some pilots may try to lose altitude by erratic slipping rather than by smoothly maneuvering, exercising good judgment, and using only a slight or moderate slip. In off-field landings, erratic or violent slipping may lead to excess speed that could result in landing long or overshooting the entire field.

Sideslip

A sideslip, as distinguished from a forward slip, occurs with the glider's longitudinal axis held parallel to the original flightpath. While keeping the nose aligned using rudder control, the pilot lowers the wing on the upwind side of the glider and adjusts the flightpath direction by varying the bank angle. [*Figure 7-34*]

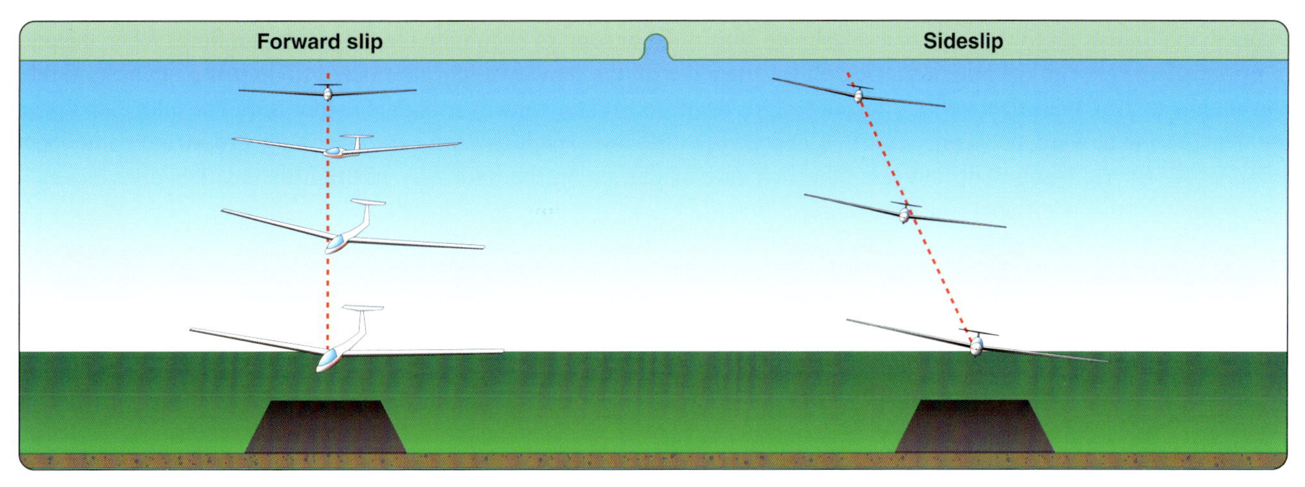

Figure 7-34. *Forward slip and sideslip.*

Common Errors

Common errors when performing a slip include:

- Improper glidepath control.
- Improper use of slips.
- Improper airspeed control.
- Improper correction for crosswind.
- Improper procedure for touchdown/landing.
- Poor directional control during/after landing.
- Improper use of spoilers, air brakes or dive brakes.

Crosswind Landing

Crosswind landings rely on a crab to correct for the effects of the wind on the final approach. A pilot performs a crab by turning the glider to a heading sufficiently into the crosswind to fly a straight track along the desired final approach path. A glider in a crab tracks the extended centerline of the landing area in coordinated flight. [*Figure 7-35A*]

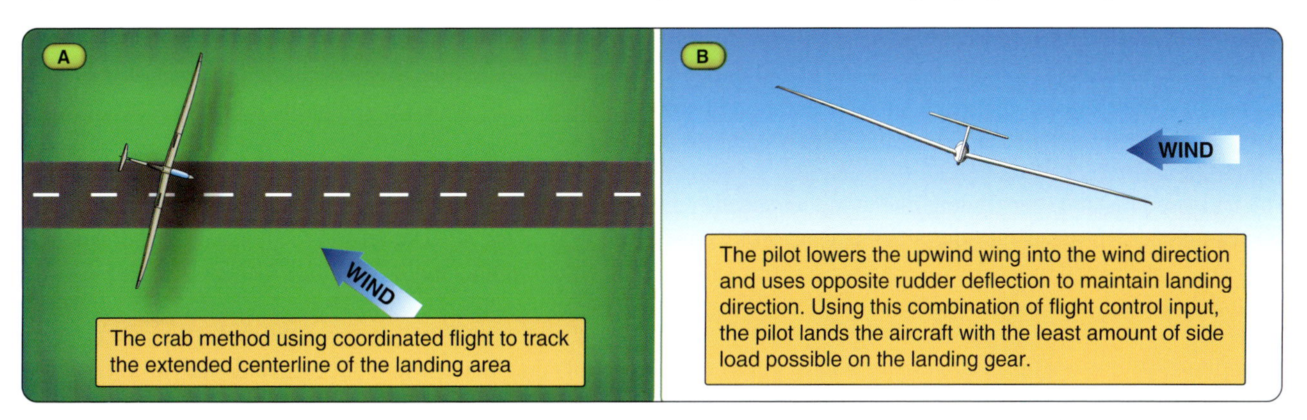

Figure 7-35. *Using the crab method to track the extended centerline of the landing area (A). Controlling side drift by adjusting the glider into the wind before landing (B).*

Pilots should transition from a crab to a sideslip on short final or prior to beginning the round out and flare. When the pilot transitions to a side slip the pilot keeps the glider aligned with the runway in uncoordinated flight using opposing rudder and aileron control. For effectiveness, the pilot aligns the nose with the runway using the rudder first. The pilot then lowers the wing on the upwind side as necessary to oppose the drift that develops from exposure to the crosswind component. [*Figure 7-35B*] Although a slip increases the sink rate of the glider, the pilot may position the spoilers/dive brakes to compensate for this additional sink rate.

Common Errors

Common errors during approach and landing include:

- Failure to complete the landing checklist in a timely manner.
- Inadequate wind drift correction on the base leg.
- An overshooting, undershooting, too steep, or too shallow a turn onto final approach.
- Poor coordination during turn from base to final approach.
- Improper glidepath control.
- Improper use of flaps, spoilers/dive brakes.
- Improper airspeed control.
- Unstabilized approach.
- Improper correction for crosswind.
- Improper procedure for touchdown/landing.
- Poor directional control during/after landing.
- Improper use of wheel brakes.

Downwind Landing

Downwind landings present special hazards and pilots should avoid this type of landing if possible. However, factors like gliderport/airport design, runway slope, or obstacles or high terrain at one end of the runway may dictate downwind takeoff and landing procedures. Emergencies or a launch failure and turn back at low altitude can also result in a downwind landing. On downwind approaches, pilots should plan a shallower approach angle and may use the spoilers/dive brakes or a forward slip, as necessary, to achieve the desired glidepath or to descend more quickly after passing over an obstacle. The pilot should use a normal approach airspeed during a downwind landing and understand that the increase in groundspeed increases the approach area and runway length needed for the landing.

The increased distance for landing due to higher speed can be determined by dividing the actual touchdown speed by the normal touchdown speed and squaring the result. For example, if the tailwind is 10 knots and the normal touchdown speed is 40 knots, the actual touchdown speed is 50 knots. (50/40)*2 = 1.56 and the landing distance increases 56 percent over the normal landing distance.

After touchdown, the pilot should use the wheel brake and all available drag devices to reduce groundspeed and stop within the available distance. Landing with a tailwind means a loss of control effectiveness at a higher groundspeed and requires more braking action.

Common errors during downwind landing include:

- Improper glidepath control.
- Improper use of slips.
- Improper airspeed control.
- Improper correction for wind.
- Improper procedure for touchdown/landing.
- Poor directional control during/after landing.
- Improper use of wheel brakes.

Landing a Self-Launching Glider

If planning a landing under power, the pilot should perform the engine restart checklist, allow sufficient time for the engine to warmup, and ensure that all systems operate properly. The pilot of a self-launching glider should have an alternate landing area available to mitigate the risk associated with decreased performance and higher drag should the engine fail to start and then fail to retract.

The pilot should fly the traffic pattern and land into the wind and with an approach angle that avoids all obstacles. The landing area dimensions should allow for touchdown and roll-out within the performance limitations of the self-launching glider. The pilot should also take into consideration any crosswind conditions and the landing surface. After touchdown, the pilot should maintain directional control, slow the self-launching glider, and clear the landing area. The pilot should complete the after-landing checklist when stopped and clear of the landing area.

Common Errors

Common errors exclusive to self-launching gliders during landing include:

- Attempting engine restart with insufficient time or altitude
- Mismanagement of engine restart.
- Not accounting for decreased glide ratio with the engine pylon extended.
- Failing to have a suitable field available in case the engine does not start.
- Landing with a side load due to parallax in powered gliders with side-by-side seating.
- Failure to use the appropriate checklist.

Nosewheel Glider Oscillations During Launches & Landings

Some gliders feature pneumatic tires on three wheels—a main wheel, tail wheel, and nose-wheel. The large main tire acts like a fulcrum and prevents both the nose tire and the tail tire from contacting the ground at the same time. With the glider in motion and if the pneumatic nose-wheel remains in contact with the ground, any bump compresses the nose-wheel tire. When the nose-wheel tire rebounds after hitting a bump at a fast ground speed, a pitch up can occur that places the tail wheel in contact with the runway, and it may compress and rebound as well.

With sufficient lift, porpoising may result if the nosewheel and tail wheel alternately hit the runway, compress, and rebound. Because this type of porpoising can damage the glider fuselage, the pilot should use elevator control to lift the nosewheel off the runway as soon as practicable during the takeoff roll so that only the glider's main wheel contacts the ground. During the landing, the pilot should hold the glider off during the flare, allow the main wheel and tail wheel touch simultaneously, and then hold the nosewheel off the ground during the rollout.

Tailwheel/Tailskid Equipped Glider Oscillations During Launches & Landings

Many gliders have a tail wheel. When loaded and ready for flight, these gliders have both the main wheel and a tail wheel/tailskid in contact with the ground, and the center of gravity location remains aft of the main wheel(s). While any upward thrust on the main landing gear tends to pitch the nose of the glider upward, the ground contact of the tail wheel or tail skid limits the change in pitch attitude.

A vigorous main wheel bump on the runway surface during the takeoff roll may push the glider into the air momentarily. At low airspeeds and with minimal elevator control, pilot over control of the elevator after an unexpected bounce or launch may result in PIOs. [*Figure 7-36*]

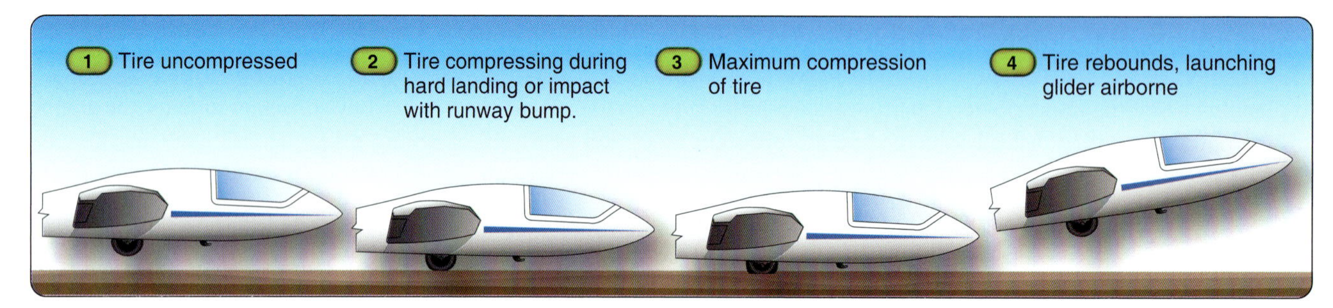

Figure 7-36. *Pneumatic tire rebound.*

During landing, main wheel contact with the ground before the tailwheel or tailskid results in compression and rebound of the pneumatic main tire, which may raise the nose of the glider, increase the angle of attack, and cause resumption of flight. Pilot overcontrol of the elevator may then lead to PIOs.

After Landing & Securing

After landing and stopping, the pilot or crew moves the glider clear of all runways. If parking the glider for a short interval between flights, an appropriate parking spot should not interfere with other gliderport/airport users or create a hazard. Pilots should close glider canopies, which are easily damaged if not secured. Because gliders can suffer wind damage, the glider should be secured if left unattended. Protecting the glider from wind may involve securing a wingtip with a weight or tie down. The manufacturer's handbook should have the recommended methods for securing the glider.

When finished with flying for the day, the pilot or crew should move the glider to the tiedown area and secure it in accordance with the recommendation in the GFM/POH. The tiedown anchors should be strong and secure. Any external control lock should be large, well-marked, and brightly painted. Pitot tube covers should use brightly colored materials for high visibility. Any canopy cover should have a soft inner surface that cannot scuff or scratch the canopy.

If storing the glider in a hangar, persons moving the glider should exercise care and maintain awareness of objects or other aircraft in the hangar. When parked in a hangar, the crew normally chocks the main wheel and tailwheel. Wing stands under each wingtip keep the glider in a wings-level position. If stored with one wing high, a weight should be placed on the lowered wing to hold it down.

If disassembling the glider for storage in a trailer, individuals tow the glider to the trailer area and normally align the fuselage with the long axis of the trailer. The pilot or crew should follow the disassembly checklist in the GFM/POH and stow the glider components securely in the trailer. Storage includes collection and stowage of all tools, closing trailer doors and hatches, and securing the trailer against wind and weather.

Performance Maneuvers

Straight Glides

The glider pilot holds a constant heading and airspeed during a straight glide and uses a prominent point on the ground in front of the glider as a heading reference. During a straight glide, each wingtip should appear at an equal distance relative to the horizon. Straight glides should be coordinated as indicated by a centered yaw string or slip-skid ball. With the wings level, the pilot establishes a pitch attitude relative to a distant point on or below the forward horizon. The pilot can hold a constant pitch attitude and a constant airspeed with little to no control pressure using the elevator trim control.

The glider pilot should listen for airflow noise changes. Any changes in airspeed or coordination cause a change in the wind noise. While gusts cause the sound and airspeed to change momentarily, the pilot can ignore the sound of gusts and hold the glider at a constant pitch attitude to maintain airspeed control.

The glider pilot should learn to fly through a wide range of airspeeds, from minimum controllable airspeed to maximum allowable airspeed. Glider pilots should also note the difference in control pressure with airspeed changes. This provides the pilot with a complete understanding of the feel of the controls of the glider. If the glider equipment includes spoilers/dive brakes and/or flaps, the glider pilot should become familiar with the changes that occur in pitch attitude and airspeed when using these controls.

Common Errors

Common errors during straight glides include:

- Rough or erratic pitch attitude and airspeed control.
- Rough, uncoordinated, or inappropriate control applications.
- Failure to use trim or improper use of trim.
- Improper use of controls when using spoilers, dive brakes, and/or flaps.
- Prolonged uncoordinated flight—yaw or ball not centered.

Turns

Turning involves all three flight controls: ailerons, rudder, and elevator. For purposes of this discussion, turns divide into the following three classes as shown in *Figure 7-37*.

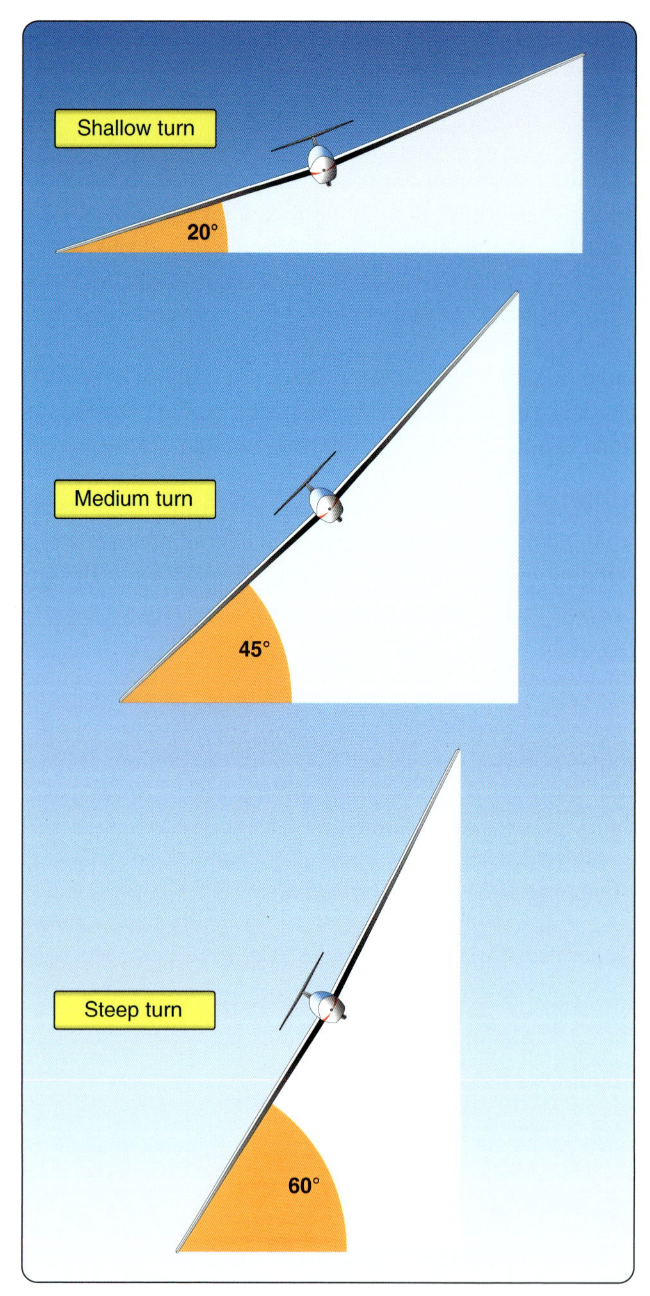

Figure 7-37. *Shallow, medium, and steep turns.*

- Shallow turns (less than approximately 20° of bank) include those in which the inherent lateral stability of the glider levels the wings unless the pilot maintains some aileron pressure to maintain the bank.

- Medium turns result from approximately 20° to 45° of bank. Lateral stability results in little to no aileron control pressure to maintain the bank angle.

- Steep turns result from a degree of bank 45° or more. During a steep turn, the overbanking tendency of a glider overcomes lateral stability, and the bank increases without opposing aileron application.

Most training gliders have a yaw string, typically a piece of yarn taped to the canopy. Pilots refer to the taped end as the head and the free end as the tail. During flight, comparing the head and tail of a yaw string identifies coordinated flight, slips, and skids. During coordinated flight, the yaw string flows straight back on the windscreen (perpendicular to the longitudinal axis) [*Figure 7-38*]. During a slipping turn, the head of the yaw string is to the inside of the turn when compared to the tail. [*Figure 7-39*] During a skidding turn, the head of the yaw string is to the outside of the turn when compared to the tail. [*Figure 7-40*].

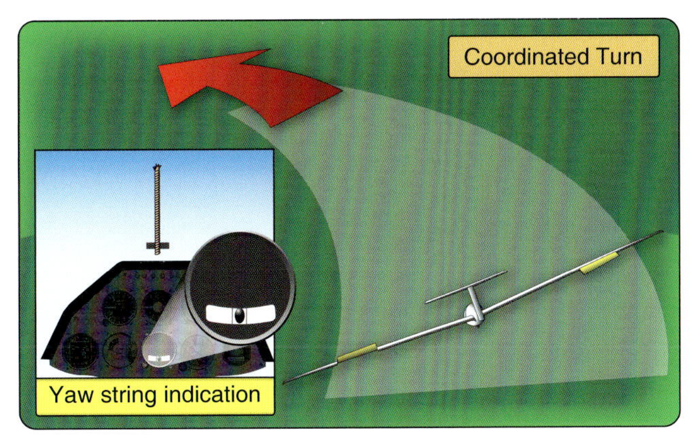

Figure 7-38. *Coordinated turn.*

Turn Coordination

In a slipping turn the glider turns at a lower rate for the bank used due to yaw toward the outside of the turning flightpath. The pilot reestablishes a coordinated turn by decreasing the bank (ailerons), increasing yaw in the direction of the turn (rudder), or a combination of the two. [*Figure 7-39*]

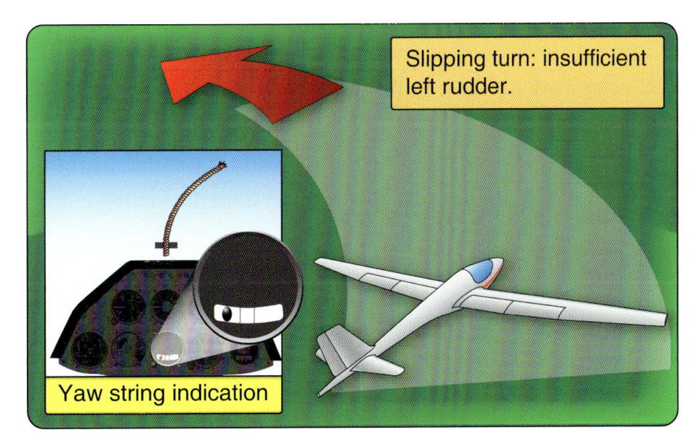

Figure 7-39. *Slipping turn.*

In a skidding turn the glider turns at a higher rate for the bank used due to the yaw toward the inside of the turning flightpath. Correction of a skidding turn involves a decrease in yaw (rudder), an increase in bank (aileron), or a combination of the two. [*Figure 7-40*]

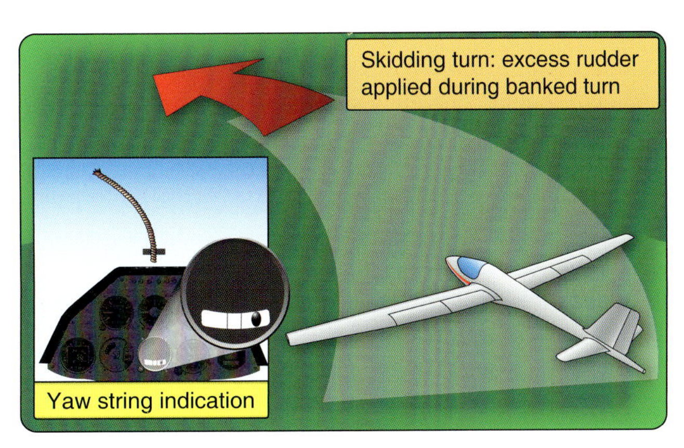

Figure 7-40. *Skidding turn.*

The ball in the slip/skid indicator also indicates coordinated flight, slips, and skids in a similar manner as the head of the yaw string. When using this instrument for coordination, the pilot can apply rudder pressure on the same side as the ball (step on the ball). The pilot's body pressure against the sides, bottom, and back of the seat respond in the same way as the ball, and many pilots can sense the force that pushes to one side or the other rather than straight down into the seat. Pilots can correct for uncoordinated condition by using appropriate rudder and aileron control pressures simultaneously or individually to coordinate the glider.

Roll-In

Before starting any turn, the pilot should clear the airspace in the direction of the turn. When applying aileron to bank the glider, the down aileron deflection on the rising wing generates greater drag while the raised aileron on the lowering wing generates less drag. This difference in drag causes the glider to yaw toward the rising wing or away from the intended direction of turn. When applying pressure to the ailerons to begin a turn and to counteract this adverse yaw, the pilot should also apply rudder pressure in the direction of turn. If using excess rudder pressure, the nose appears to yaw before the pilot establishes the bank. If using insufficient rudder pressure, the nose initially moves in the wrong direction as the pilot begins the turn. If using the correct amount of rudder pressure, the nose starts to move along the horizon, increasing its rate of travel proportionately as the bank increases.

After establishing a medium banked turn, the pilot may relax pressure applied to the aileron control. The glider tends to remain at the selected bank with no further tendency to yaw without aileron deflection. As a result, the pilot may also relax pressure applied to the rudder pedals, and the rudder streamlines itself with the direction of the slipstream. If the pilot maintains constant rudder pressure after establishing the turn, the glider skids to the outside of the turn. If the pilot makes a conscious effort to center the rudder rather than let it streamline itself to the turn, the pilot may inadvertently apply some opposite rudder pressure. This forces the glider to slip to the inside of the turn. The yaw string or ball in the slip/skid moves as described above.

As the angle of bank increases from a shallow bank to a medium bank, the airspeed of the wing on the outside of the turn increases in relation to the inside wing. The additional lift developed on the outside wing balances the lateral stability of the glider. Therefore, the pilot does not use aileron pressure to maintain a medium bank. If the bank increases from a medium to steep, the radius of turn decreases even further. The greater lift of the outside wing will cause the bank to steepen, and the pilot should use opposite aileron to hold the bank constant. Because the outboard wing develops more lift, it also has more induced drag. This causes a slid during steep turns that the pilot corrects with rudder pressure.

To establish the desired angle of bank, the pilot should use visual reference points on the glider, the earth's surface, and the natural horizon. The beginning pilot may lean away from or into the turn but should remain aligned with the seat. Any deviation prevents proper use of visual references. Large application of aileron pressure may produce rapid roll rates and allow little time for corrections before reaching the desired bank. Slower (small control displacement) roll rates provide more time to make necessary pitch and bank corrections.

While establishing the desired angle of bank and during the turn, the pilot should use elevator pressure to maintain the desired airspeed. Throughout the turn, the pilot should cross-check the airspeed indicator to verify the proper pitch. The cross-check and instrument scan should include outside visual references. If the glider gains or loses airspeed, the pilot should adjust the pitch attitude in relation to the horizon. During all turns, the pilot uses aileron, rudder, and elevator control to correct minor variations in pitch and bank.

Roll-Out

The roll-out from a turn involves application of coordinated flight controls in the opposite direction. Aileron and rudder application occur in the direction of the roll-out or toward the high wing. As the angle of bank decreases, the pilot uses elevator pressure, as necessary, to maintain airspeed.

Since the glider continues turning while in a bank, the roll-out should begin before reaching the desired heading. The amount of lead to roll-out on the desired heading depends on the degree of bank used in the turn. Normally, pilots use one-half the bank angle. For example, if the bank is 30°, the pilot leads the roll-out by 15°. As the wings become level, the pilot can smoothly relax control pressures, so the controls neutralize as the glider returns to straight flight. As the roll-out occurs, outside visual references provide indications that the wings have leveled, and the turn stopped.

Common errors during a turn include:

- Failure to clear turn.
- Nose movement before the bank starts—rudder is being applied too soon.
- Significant bank before the nose moves, or nose movement in the opposite direction—the rudder is being applied too late.
- Up or down nose movement when entering a bank—excessive or insufficient elevator is being applied.
- Rough or abrupt use of controls during the roll-in and roll-out.
- Failure to establish and maintain the desired angle of bank.
- Overshooting/undershooting the desired heading.

Steep Turns

In thermaling flight, small-radius turns can keep the glider in or near the core of a thermal updraft. To keep the radius of turn small, the pilot can establish a steep bank while maintaining an appropriate airspeed, such as minimum sink or best glide speed. The pilot should also understand that as the bank angle increases, the stall speed increases.

A steep turn results in a rapid heading change, and the pilot should clear the area for other traffic. While banked steeply, the rudder may act like an elevator. A little top rudder helps keep the nose-up attitude. If the pilot does not add sufficient back pressure or top rudder to maintain the desired airspeed as the bank angle steepens, the glider may enter a spiral dive. In summary, during a coordinated steep turn, the pilot uses back pressure on the elevator for airspeed control, aileron pressure against the raised wing for bank control, and top rudder pressure to maintain pitch attitude.

Common Errors

Common errors during steep turns include:

- Failure to clear turn.
- Uncoordinated use of controls.
- Loss of orientation.
- Failure to maintain airspeed within tolerance.
- Inappropriate division of attention inside and outside the glider.
- Unintentional stall, spin, or spiral dive.
- Excessive deviation from desired heading during roll-out.

Slow Flight

Maneuvering during slow flight demonstrates the flight characteristics and degree of controllability of a glider at reduced speeds. Pilots should develop awareness of the flight characteristics of any glider they fly to recognize and avoid stalls that may inadvertently occur during low airspeed flight used in takeoffs, climbs, thermaling, and approaches to landing.

Maneuvering during slow flight develops the pilot's sense of feel and ability to use the controls, and to improve proficiency. Pilots should use outside visual references to maintain the desired pitch attitude as well as periodically scan the airspeed indicator.

The maneuver starts from either best glide speed or minimum sink speed at a safe altitude. The pilot smoothly and gradually increases pitch attitude. While the glider airspeed decreases, the position of the nose in relation to the horizon will rise as the angle of attack increases. Since lift diminishes with the square of airspeed, the increase in angle of attack to keep lift constant becomes more pronounced as airspeed decays. As speed continues to decrease and approach the airspeed at which any further increase in angle of attack or load factor would result in a stall, the glider reaches the edge of its flight envelope at a minimum controllable airspeed. In smooth air and with no turning, minimum controllable airspeed is lower than in rough air. If in turbulence, the pilot should fly with a sufficient speed margin above the minimum controllable airspeed to avoid a stall. During slow flight or during flight at minimum controllable airspeed, the pilot should continually use the horizon to maintain desired pitch attitude and glance at the airspeed indicator to maintain the target airspeed. Trimming the glider, as necessary, compensates for changes in control pressure. The diminished effectiveness of the flight controls during slow unaccelerated flight should familiarize the pilot with the characteristics and feel of flight near the 1G stalling speed.

After establishing a slow airspeed in straight flight, turns further demonstrate the glider's characteristics at that selected airspeed. During turns, the pilot should decrease pitch attitude as needed to maintain airspeed. Otherwise, as bank steepens, the increase in load factor may result in a stall. A stall may also occur in a turn because of abrupt or rough control movements or turbulence, which increase load factor. Abruptly raising the flaps during minimum controllable airspeed flight also results in sudden loss of lift and may cause a stall. The actual speed at which a stall occurs also depends upon conditions such as the gross weight and CG location.

Pilots should also practice slow flight with the glider in different configurations such as with spoilers/dive brakes, flaps, and landing gear extended and retracted. This provides additional understanding of the changes in pitch attitude caused by the increase in drag in different configurations.

Common Errors

Common errors during slow flight maneuvers include:

- Failure to clear the area.
- Failure to establish or to maintain desired airspeed.
- Improper use of trim.
- Rough or uncoordinated use of controls.
- Excessive deviation from desired heading during roll-out.
- Failure to recognize indications of a stall.

Stall Recognition & Recovery

A stall can occur at any airspeed or attitude. In a powered glider, a stall can also occur at any power setting. Intentional stalls familiarize the pilot with the conditions that produce stalls, assist in recognizing an approaching stall, and develop the skills necessary to prevent or recover from a stall. The pilot should learn the stall characteristics and recovery procedures of the glider being flown.

Stall accidents usually result from an inadvertent stall at a low altitude in which a recovery was not accomplished prior to contact with the surface. The longer it takes to recognize the approaching stall, the more complete the stall becomes, and the greater the expected altitude loss. To mitigate the risk involving loss of altitude during recovery, pilots should practice stalls at an altitude that allows recovery within gliding distance of a landing area and no lower than 1,500 feet AGL.

Many gliders do not have an electrical or mechanical stall warning device. Pilots should recognize an approaching stall by sight, sound, and feel. The following cues should warn the pilot of an approaching stall.

1. Vision—visualizing the relative wind and angle of attack. Yaw string (if equipped) movement from normal flight position.

2. Hearing—a change in sound due to loss of airspeed.

3. Feeling—As a stall begins, the pilot starts to feel airframe buffeting or aerodynamic vibration.

   A. Kinesthesia, or the sensing of changes in direction or speed of motion, if properly developed, warns of a decrease in speed or the beginning of a settling, or mushing, of the glider.

   B. As speed decreases, the resistance to pressure on the controls becomes progressively less. The ailerons, elevator, and rudder have significantly less authority and require more movement to control the glider. Under low-speed stalling conditions, the lag between control movements and the response of the glider becomes more pronounced.

Pilots should always make clearing turns before performing stalls. During the practice of intentional stalls, the major objective is not to learn how to stall a glider, but rather to learn how to recognize an approaching stall and take prompt corrective action. The recovery actions involve a coordinated recovery.

First, at the indication of a stall, the pilot should immediately lower the pitch attitude and AOA by releasing the back-elevator pressure or by moving the elevator control forward. This lowers the nose and returns the wing to an effective AOA. The amount of elevator control pressure or movement to use depends on the design of the glider, the severity of the stall, and proximity to the ground. In some gliders, a moderate movement of the elevator control—perhaps slightly forward of neutral—suffices, while others may require a forcible push to the full forward position. However, an excessive negative load on the wings caused by excessive forward movement of the elevator may impede, rather than hasten, the stall recovery. The object is to reduce the AOA sufficiently to allow the wing to regain lift. [*Figure 7-41*]

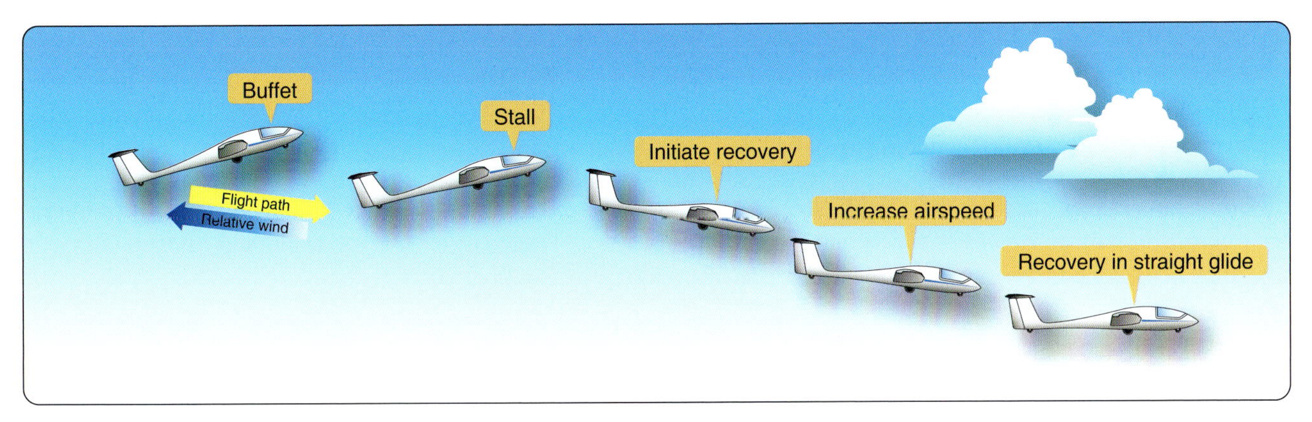

Figure 7-41. *Stall recovery.*

When practicing stalls in a powered glider, the pilot should experience the stall and recovery with and without the engine running. In a stall with power available, the pilot should smoothly and promptly apply maximum allowable power during the stall recovery while lowering the pitch attitude to increase the self-launching glider's speed and assist in reducing the AOA. The applied power reduces the loss of altitude. Maximum allowable power applied at the instant of a stall does not usually cause overspeeding of an engine equipped with a fixed-pitch propeller, due to the heavy air load imposed on the propeller at low airspeeds. However, the pilot may reduce the power as airspeed increases so the airspeed or rpm does not become excessive.

Introduction to stalls should consist of approaches to stalls with recovery initiated at the first airframe buffet or when the pilot recognizes partial loss of control. Using this method, pilots become familiar with the initial indications of an approaching stall without fully stalling the glider. Whether in an unpowered or powered glider, stall recovery occurs by reducing the angle of attack, leveling the wings with coordinated control inputs, and returning to straight flight. Whenever practicing stalls while turning, the pilot should maintain a constant bank angle until the stall occurs.

Stalls in most gliders move progressively outward from the wing roots (where the wing attaches to the fuselage) to the wingtips. This occurs because the wings have a smaller angle of incidence at the tips than at the wing roots. When an exceedance of the critical angle of attack results in a stall, the outer part of the wing can retain some aerodynamic effectiveness. During recovery from a stall, the return of lift begins at the tips and progresses toward the roots, thus giving the ailerons authority to level the wings.

Using the ailerons requires finesse to avoid an aggravated stall condition. For example, if the right wing drops during the stall and the pilot uses excessive aileron control to the left to raise the wing, the aileron that deflects downward on the right wing would change the camber of that portion of the wing. The increased AOA could cause the wing to stall at the tip. The increase in drag created by the high AOA on that wing might cause the airplane to yaw in that direction. This adverse yaw could result in a spin unless the pilot maintains directional control with rudder or reduces aileron deflection.

Even with application of excessive aileron pressure, a spin does not occur if the pilot maintains directional (yaw) control using timely application of coordinated rudder pressure. Therefore, the pilot should use rudder properly during both entry and recovery from a stall. The rudder in stall recovery counteracts any tendency of the glider to yaw. A pilot using correct stall recovery technique decreases the pitch attitude by applying forward elevator pressure to reduce the AOA and simultaneously maintains directional control with coordinated use of the aileron and rudder.

Advanced stalls include secondary, accelerated, and crossed-control stalls. These stalls expand a pilot's stall/spin awareness.

Secondary Stalls

A secondary stall occurs during a recovery from a preceding stall. It may occur when the pilot attempts to complete a stall recovery with abrupt control input or before the glider has regained sufficient flying speed, which results in a repeated stall. When this stall occurs, the pilot should release back-elevator pressure as in a normal stall recovery.

Accelerated Stalls

Actual accelerated stalls occur most frequently during turns in the traffic pattern close to the ground while maneuvering the glider for the approach. A glider pilot should recognize signs of an imminent accelerated stall and take prompt action to prevent a completely stalled condition to avoid loss of altitude or a spin.

Performing intentional accelerated stalls can show the pilot how these stalls occur, enhance pilot recognition of conditions that can cause a stall, and reinforce timely and proper recovery action. During training at a safe altitude, pilots should learn how to recover at the first indication of a stall or immediately after the stall occurs.

A glider at a given weight consistently stalls at the same indicated airspeed in unaccelerated 1G flight. However, the glider stalls at a higher indicated airspeed when the pilot imposes a maneuvering load as in a steep turn or pull-up. These maneuvers rely on an increased angle of attack to generate the lift used to change the path of the glider. If the demand for additional lift exceeds the critical angle of attack, an accelerated stall occurs and could surprise a pilot. Depending on the wing configuration and quality of coordination, one wing may stall prior to the other wing. If the wings have a slight or pronounced sweep, one wing can rapidly develop more lift than the other, and a spin could occur before the pilot can react. For this reason, pilots should avoid turning too tightly in the traffic pattern to prevent exceeding the critical angle of attack and any resulting accelerated stall at a low altitude.

Pilots should not perform accelerated maneuver stalls in any glider if the GFM/POH prohibits this maneuver. If permitted, training for this maneuver occurs with a bank of approximately 45° and never at a speed greater than the manufacturer's recommended airspeed for the maneuver or the design maneuvering speed specified for the glider. At the design maneuvering speed, the glider will stall before application of full aerodynamic control can exceed the glider's limit load factor. The stall unloads the wings, cuts off the acceleration, and prevents structural damage.

A glider slipping toward the inside of the turn at the time the stall occurs tends to roll rapidly toward the outside of the turn as the nose pitches down and the outside wing stalls first. A glider skidding toward the outside of the turn tends to roll to the inside of the turn because the inside wing stalls first. During a stall in a coordinated turn, the glider's nose pitches away from the pilot just as it does in wings-level stall since both wings stall simultaneously. The configuration of the wings has a strong influence on exactly how a glider reacts to different airflows. A pilot should fly the specific glider into these situations at a safe altitude to determine how the glider will react. This training should condition the pilot to avoid an accelerated stall that could result in an accident.

As with any deliberate stall, the area should be clear of other aircraft. From straight flight at maneuvering speed or less, the pilot should roll the glider into a steep banked (45° maximum) turn and gradually apply back-elevator pressure. After establishing the bank, the pilot smoothly and steadily increases back-elevator pressure. The resulting apparent centrifugal force pushes the pilot's body down in the seat, increases the wing loading, and decreases the airspeed. The pilot should firmly increase back-elevator pressure until a definite stall occurs.

When the glider stalls, recovery involves the prompt release of back-elevator pressure. In an uncoordinated turn, one wing may tend to drop suddenly, causing the glider to roll in that direction. If this occurs, the pilot should lower the nose and use coordinated control pressure to return glider to wings-level, straight flight.

An accelerated stall could occur any time the pilot applies excessive back-elevator pressure or increases the AOA too rapidly. Although demonstrated from a steep turn, the maneuver allows the pilot to experience accelerated stall characteristics and develop the ability to recover instinctively at the onset of a stall at any other-than-normal stall speed or flight attitude.

Crossed-Control Stalls

A crossed-control stall demonstration maneuver shows the effect of improper control technique and emphasizes the importance of coordinated control pressures whenever making turns. This type of stall occurs with the controls crossed—aileron pressure applied in one direction and rudder pressure in the opposite direction while exceeding the critical AOA. [*Figure 7-42*]

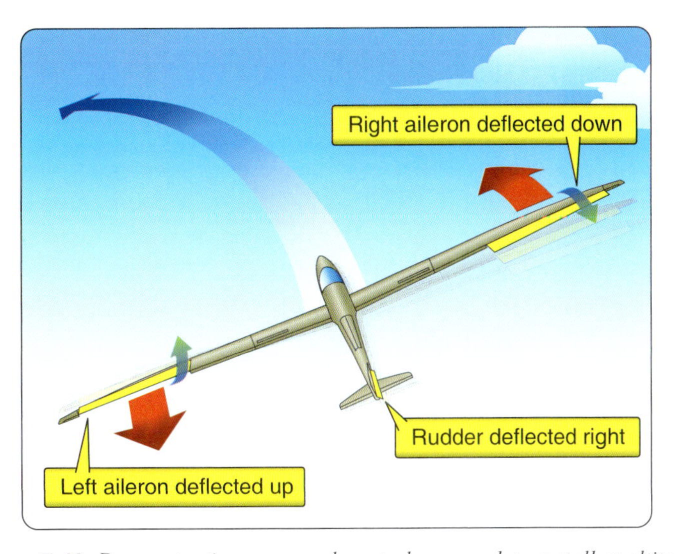

Figure 7-42. *Demonstrating a crossed-control approach to a stall at altitude.*

This stall most commonly occurs during a poorly planned and executed base-to-final turn when overshooting the centerline of the runway during the turn. Normally, the pilot should correct by increasing the rate of turn using coordinated aileron and rudder. At the relatively low altitude of a base-to-final approach turn, improperly trained pilots sometimes fear steepening the bank to increase the rate of turn and incorrectly use excessive rudder pressure to yaw the airplane.

The addition of rudder pressure on the inside of the turn causes the speed of the outer wing to increase, creating greater lift on that wing. To keep that wing from rising and to maintain a constant angle of bank, the pilot applies opposite aileron pressure. The added inside rudder pressure also causes the nose to lower in relation to the horizon. Consequently, the pilot adds additional back-elevator pressure to maintain a constant pitch attitude. The resulting turn uses rudder applied in one direction, aileron in the opposite direction, and excessive back-elevator pressure—a pronounced crossed-control condition. The down aileron on the inside of the turn helps drag that wing back, slowing it and decreasing its lift. This further causes the glider to roll. The roll may be so fast that it is possible the bank will be vertical or past vertical before the pilot can stop and reverse it.

The demonstration of the maneuver should occur at a safe altitude because of the possible extreme nose-down attitude and loss of altitude that may result. Before demonstrating this stall, the pilot should clear the area for other air traffic. As the pilot establishes the gliding attitude and airspeed, the glider should be retrimmed. With the glide established, the pilot rolls the glider into a medium banked turn to simulate a final approach turn that would overshoot the centerline of the runway. During the turn, the pilot applies excessive rudder pressure in the direction of the turn while holding bank constant with opposite aileron pressure. At the same time, increased back-elevator pressure keeps the nose from lowering.

All these control pressures increase until the glider stalls. When the stall occurs, releasing the control pressures and simultaneously decreasing the AOA initiates the recovery. In a crossed-control stall, the glider often stalls with little warning. The nose may pitch down, the inside wing may suddenly drop, and the glider may continue to roll to an inverted position. This is usually the beginning of a spin.

The pilot should recover before the glider enters an abnormal attitude (vertical spiral or spin) by returning to wings-level, straight flight using coordinated control inputs. The pilot should recognize imminent stall and take immediate action to prevent a completely stalled condition. This type of stall during an actual approach to a landing would likely result in ground contact before recovery.

Common Errors

Common errors during advanced stalls include:

- Improper pitch and bank control during straight-ahead and turning stalls.
- Rough or uncoordinated control procedures.
- Failure to recognize the first indications of a stall.
- Premature recovery when demonstrating a full stall.
- Poor recognition and recovery procedures.
- Excessive altitude loss, excessive airspeed, or encountering a secondary stall during recovery.

Chapter Summary

Regardless of the launch method, pilots and ground personnel should have the ability to communicate effectively. This not only includes knowing appropriate signals, but also knowing when to use them. Briefing of all personnel before takeoff enhances the safety of a glider operation. Takeoffs normally occur with the assistance of a wing runner; however, the chapter also discusses unassisted takeoff techniques that experienced pilots may use. The chapter discusses various maneuvers including level glides, turns, steep turns, release procedures, slack line avoidance and recovery, and boxing the wake. A pilot can discern turn coordination using a yaw string or inclinometer, and the chapter gives practical advice regarding the use of these instruments when adjusting from a slip or skid. Pilots should understand the traffic pattern procedures for every field they use. A minimum altitude for the initial point (IP) of the traffic pattern should allow the glider pilot to maneuver to the landing field and make a successful landing in a variety of conditions including normal, crosswind, and downwind. A pilot should know how to make a stable final approach to an aiming point, use drag devices and slips to control the descent angle, and land in a crosswind with no side load. Pilots should recognize various stalls and know how to avoid conditions that could result in any unintentional stall. Flight at minimum controllable airspeed and stall training at a safe altitude build pilot awareness of and resistance to unsafe operating conditions.

Chapter 8: Abnormal & Emergency Procedures

Introduction

Glider pilots train for abnormal and emergency procedures in case of control problems, instrument failures, or equipment malfunctions. In addition, since forced landings may occur due to inadequate lift on a cross-country flight, glider pilots should understand how to use emergency equipment and survival gear as a practical necessity.

Aerotow Abnormal & Emergency Procedures

Environmental factors, pilot error, and mechanical equipment failure can cause an abnormality during aerotow. The tow and glider pilots should avoid situations that would place the other pilot at risk. Should a situation arise where the glider position threatens the safety of the tow plane, the glider pilot may have no other choice but to release the tow rope.

Environmental Factors

Environmental factors for terminating the tow include encountering clouds, mountain rotors (area of turbulence created by wind and mountainous terrain), or restricted visibility. Any of these factors may require the glider pilot to release.

Pilot Error

Examples of tow-pilot errors include starting the takeoff before the glider pilot gives the ready signal, using steep banks during the aerotow without prior consent of the glider pilot, or frivolous use of aerotow signals, such as "release immediately!" For the glider pilot, examples of pilot error include allowing the glider to get out of position during the tow, losing sight of the towplane, or leaving air brakes open during takeoff and climb.

Kiting

A hazardous situation may occur during takeoff if the glider pilot flies high above and loses sight of the towplane. If the tension on the tow line pulls the towplane tail up, the tow pilot may run out of up-elevator authority. In addition to a the towplane experiencing a nose down pitching moment, it is also loosing airspeed. The kiting glider creates a huge amount of drag which slows the towplane. It can take 1,500 feet or more for the towplane to recover once the rope breaks or the glider pilot releases.

Unintended Spoiler Deployment

Spoiler-related emergencies often result from a pilot's failure to lock the spoilers in the closed position prior to takeoff. As the glider accelerates down the runway, or soon after it becomes airborne, the airflow deploys the spoilers, greatly increasing the glider's drag. The pilot may not notice the spoilers open, but the increased drag can make it impossible for the tow pilot to climb and clear obstacles. A release by the tow pilot leaves the glider low to the ground and in a high drag configuration. If the glider pilot does not recognize the situation, and close the spoilers, an unsafe landing becomes likely. If conditions permit a continued climb the towpilot can waggle the rudder to indicate something is wrong with the glider. The glider pilot should differentiate between the towpilot rocking the wings indicating release immediately, and the towpilot fanning the rudder indicating something is wrong with the glider, avoiding an unnecessary release.

If time allows, the tow pilot can use the radio and warn the glider pilot to close the spoilers. Without using radio, the tow pilot can use the "Rudder Waggle" signal for the glider pilot to check the glider. The glider pilot should not confuse this signal with the "Wing Rock" signal for the glider pilot to release the tow rope. This mistake would leave the glider pilot at low altitude with the spoilers open and make a safe landing very difficult.

Mechanical Failures

Mechanical equipment failures include tow hook failures, towplane mechanical or airframe failures, and glider airframe failures. The tow pilot rocks the wings of the towplane to request an immediate release by the glider pilot. [*Figure 8-1* right panel] The glider pilot may deliberately terminate the tow or launch anytime it seems safer than continuing. For example, if the glider pilot discovers control binding once air pressure builds on the surfaces, releasing the tow line seems less risky than continuing. If a mechanical failure occurs, the glider pilot should assess the situation and determine the best course of action. Options include remaining on the aerotow, planning to release at a point in the future, or releasing immediately.

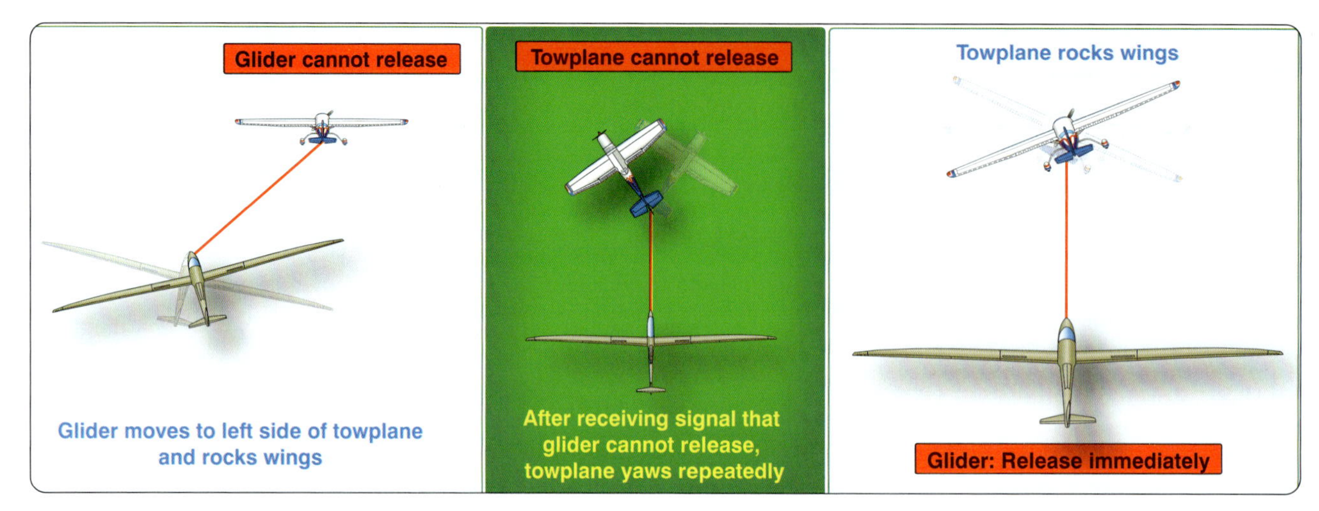

Figure 8-1. *Towplane and glider signals due to release mechanism failure (left and center panels) and towplane signal commanding glider release (right panel).*

blank

Tow release failure can occur in either the towplane or the glider tow hook system. If the glider release mechanism fails, the glider pilot should notify the tow pilot either by radio or tow signal as depicted in the left panel of *Figure 8-1* above, and the glider should maintain the high tow position. The tow pilot should tow the glider over the gliderport or airport and release the glider from the towplane. The towline should fall back and below the glider. The design of some tow hook mechanisms allows the line to pull free from the glider by its own weight. After release by the towplane and since some glider hooks do not back release, the glider pilot should pull the release to increase the likelihood of a towline release from the glider. In the rare event the tow pilot also cannot release, the tow pilot will yaw the tow plane repeatedly as shown in the enter panel of *Figure 8-1* above.

tow planes often use a variant of the Schweizer or Tost glider tow hitch. [*Figure 8-2*] and [*Figure 8-3*] The hitch connects to the tow plane fuselage below the rudder. The specific rings that attach tow lines to these hitch types can be seen in *Figure 8-4*. The wing runner should know the correct method of attachment. If the tow plane has the Schweizer tow hitch, the tow ring may rotate forward and trap the sleeve that locks the tow hitch in place if the glider flies too far above the tow plane during the tow. This may prevent the tow pilot from releasing the tow line. [*Figure 8-5*]

Figure 8-2. *A Schweizer tow hitch.*

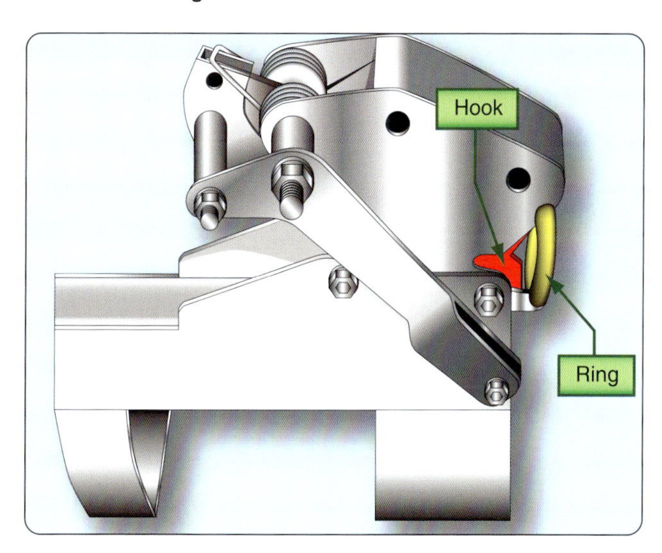

Figure 8-3. *A Tost tow hitch.*

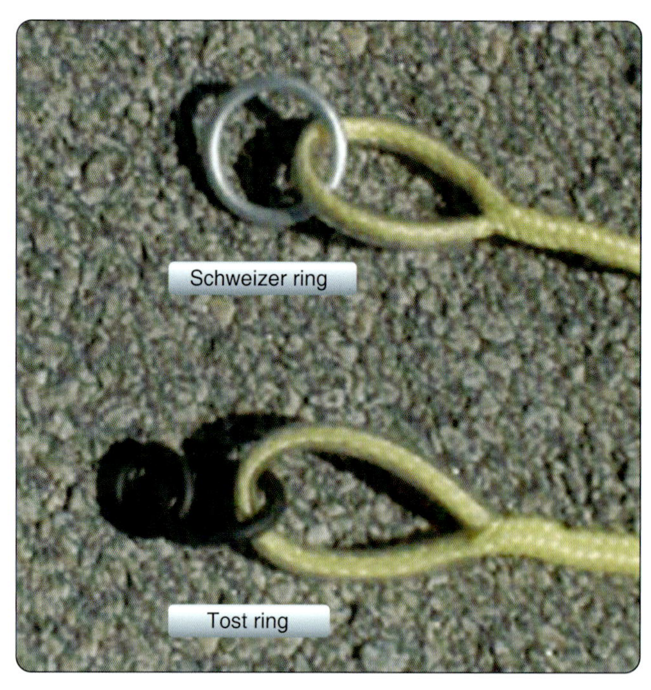

Figure 8-4. *Examples of Schweizer and Tost tow rings.*

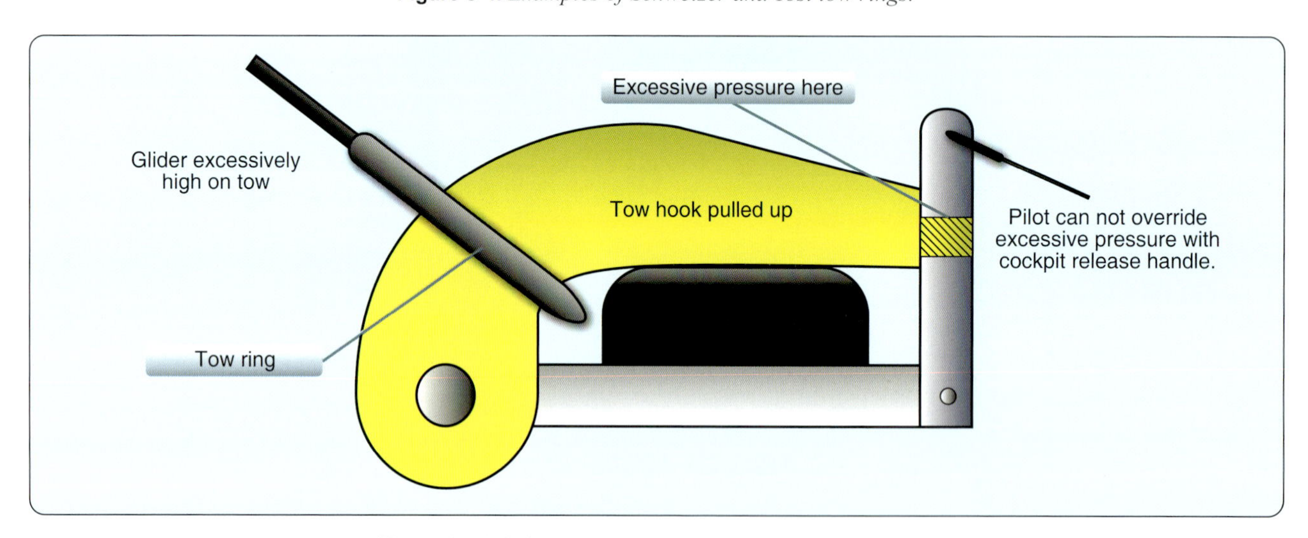

Figure 8-5 A Schweizer tow hitch under upward tension.

Failure of the release mechanism in both towplane and glider occurs very rarely. If it does occur, radio or tow signals between glider pilot and tow pilot should confirm the situation. One option has the glider and towplane landing without releasing from one another. Once a descent to the gliderport or airport begins in this scenario, the glider pilot should move to and maintain the low-tow position and use spoilers/dive brakes to avoid overtaking the towplane. The tow pilot should plan a shallow approach that avoids obstacles and allows the glider to touch down first. The towplane could experience a hard landing if the glider pilot uses the wheel brake excessively and slows the towplane below its flying speed. The glider pilot should use the spoilers or dive brakes to stay on the runway and use the wheel brake only as necessary to avoid overtaking the towplane.

A second option in a dual release failure situation involves the glider pilot breaking the towline while within gliding distance of the landing field; however, using this slack-line procedure increases risk of entanglement and could damage the glider or towplane if the line does not break as designed. The pilot flies the glider above the towplane, dives down to develop slackline, and then fully extends the dive brakes and/or spoilers to induce a tension overload. As the towplane accelerates without the load of the glider and the glider decelerates due to the increase in drag, the towline should come under sudden tension and break free. If the glider pilot cannot verify that the line break occurred at the glider's weak link, the glider pilot should land using the landing procedure for an attached line in the GFM/POH.

The following sections discuss five tow failure situations for which a pilot should have a plan. While the best course of action depends on many variables, such as runway length, airport environment, density altitude, and wind, all tow failures or emergency releases have one thing in common: the need to maintain control of the glider. For example, the pilot should avoid stalling the glider or dragging a wingtip on the ground during a low altitude turn.

Tow Failure with Sufficient Runway To Land and Stop

If a tow failure or deliberate release occurs prior to towplane lift-off, the towplane either continues the takeoff or aborts the takeoff. If the towplane aborts the takeoff, the tow pilot should maneuver the towplane to the left side of the runway. For a glider still on the runway, the glider pilot should pull the release, decelerate using the wheel brake, and maneuver to the right side of the runway. If the glider becomes airborne, the glider pilot should pull the towline release, land ahead, and be prepared to maneuver to the right side of the runway. [*Figure 8-6*, panel 1] Pulling the towline release in either case ensures that the rope remains clear of the glider. Since local procedures vary, both the glider pilot and tow pilot should know any specific gliderport or airport procedures.

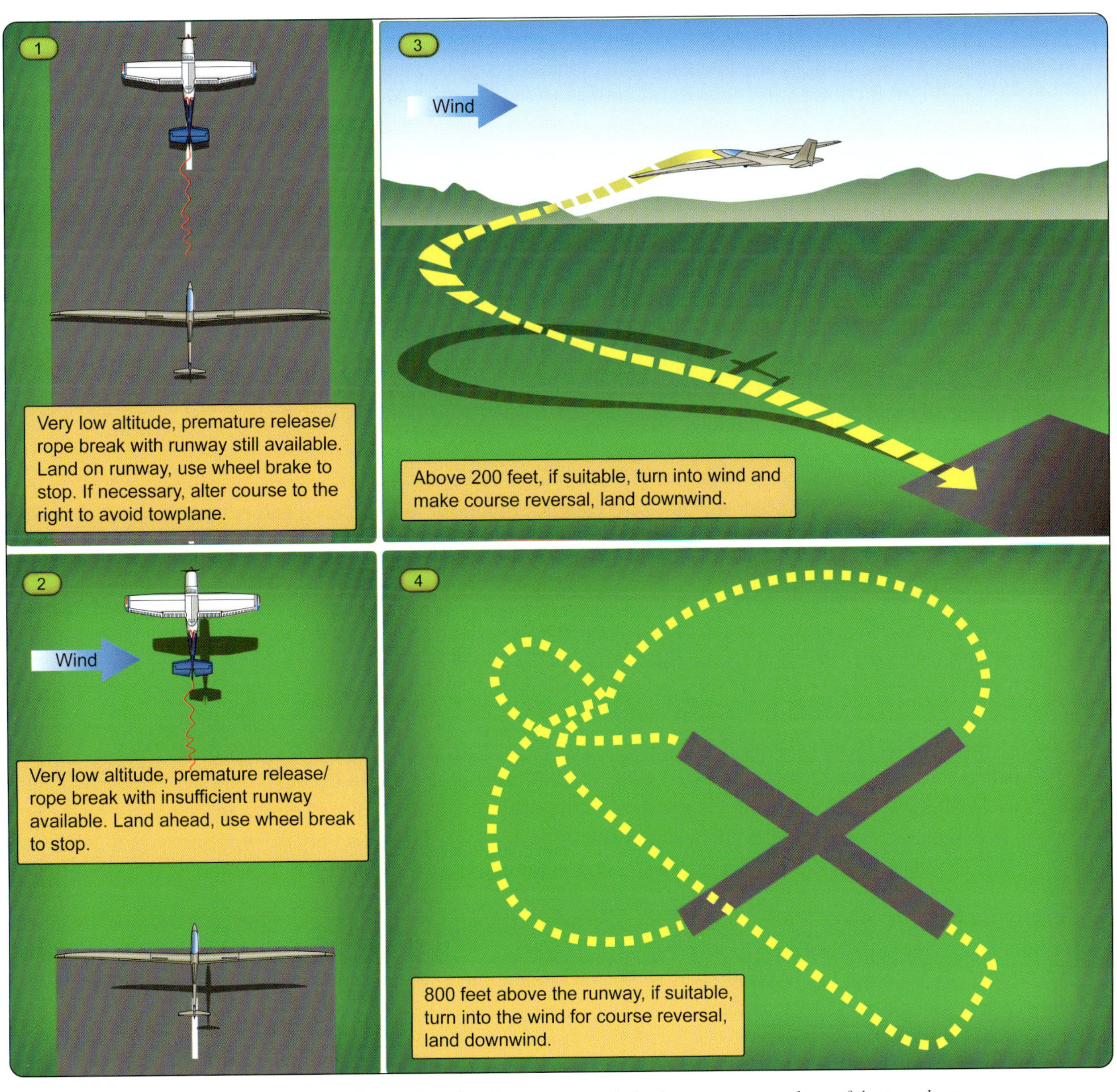

Figure 8-6. *Situations for tow line break, uncommanded release, or power loss of the towplane.*

Tow Failure after Takeoff—Glider Unable to Return to Airport

When an inadvertent release, tow line break, or a signal to release from the towplane occurs at a point at which the glider has neither sufficient runway directly ahead to land nor sufficient altitude (typically 200 feet above ground level) to make a safe turn back to the field, the pilot should land the glider ahead. [*Figure 8-6*, panel 2] Glider pilots should consider any attempt to turn at low altitude prior to landing as high risk because of the likelihood of dragging a wingtip on the ground or stalling the glider. Landing ahead and slowing the glider as much as possible prior to touchdown and rolling onto unfamiliar terrain often provides the best outcome. Low speed means low impact forces, which reduce the likelihood of pilot injury or damage to the glider. Glider pilots should look on both sides and ahead when choosing the best area for an off-field landing. While more altitude provides more range during an emergency, landing under control supersedes getting to a "perfect" landing area almost within glide distance.

Tow Failure above Return to Runway Altitude

A glider pilot may consider and attempt a downwind landing on the departure runway if the glider possesses sufficient altitude to make a course reversal. [*Figure 8-6*, panel 3]

The pilot should only attempt a course reversal and downwind landing option if within gliding distance of the airport or landing area. In ideal conditions, this maneuver typically requires a minimum altitude of 200 feet above ground level. If there are strong headwinds, the pilot should consider that the glider will reach 200 feet AGL with less distance traveled across the ground. This may allow for landing-ahead options on the existing runway compared to reversing course, landing with less return runway available, at a considerably higher groundspeed, and possibly risking overshooting the entire runway on return.

The glider pilot has responsibility to avoid the towplane if the tow terminates due to a towplane emergency. Since, the tow pilot deals with that emergency and may maneuver the aircraft abruptly, the glider pilot should never follow the towplane down.

After releasing from the towplane at low altitude and if the glider pilot chooses to make a turn of approximately 180° for a downwind landing, the pilot should maintain flying speed. The pilot should immediately lower the nose to a pitch attitude that maintains the appropriate airspeed. If a rope break occurred, the glider pilot should release the rope portion still attached to the glider to avoid any entanglement.

During a course reversal, the pilot should make the initial turn into any crosswind using a 45° bank angle. This bank angle provides a safe margin above stall speed, incurs an acceptable amount of altitude loss, and completes a course reversal in a timely manner. Using a shallow bank angle may not allow enough time for the glider to align with the landing area. An excessively steep bank angle may result in an accelerated stall or wingtip ground contact. Provided the departure drifted downwind and the pilot makes the turn back to the field into the crosswind, the glider may only need minor additional course corrections to align with the intended landing area. Throughout the maneuver, the pilot should maintain the appropriate approach speed and proper coordination to maximize range.

While downwind landings result in higher groundspeeds, the glider pilot should maintain the appropriate approach airspeed. Landing downwind results in a shallower than normal approach, and the pilot may use spoilers or dive brakes to control the descent path during the straight-in portion of the approach. Higher groundspeed becomes especially noticeable during the flare. After touchdown, the pilot should use the spoilers or dive brakes and use wheel brakes to slow and stop the glider before any loss of directional control. During the latter part of the roll-out, the glider may feel unresponsive to the controls even though rolling along the runway at a higher than normal groundspeed.

Tow Failure above 800' AGL

When the emergency occurs at or above 800 feet above the ground, the glider pilot has more time to assess the situation. Depending on the gliderport or airport environment, the pilot may choose to land on a cross runway, into the wind on the departure runway, or on a taxiway. [*Figure 8-6*, panel 4] In some situations, an off-gliderport or off-airport landing may be safer than attempting to land on the gliderport or airport.

Tow Failure above Traffic Pattern Altitude

If an emergency occurs above the traffic pattern altitude, the glider pilot should maneuver away from the towplane, and release the tow line if still attached. The glider pilot should turn toward the gliderport or airport while evaluating the situation and then decide whether to search for lift or make an immediate return to the gliderport or airport for landing. Pilots should remember their obligation when dropping objects (such as a tow line) from an aircraft according to Title 14 of the Code of Federal Regulations part 91, section 91.15, and not create a hazard to persons and property on the ground. [*Figure 8-12*, panel 4]

Slack Line

Slack line is a reduction of tension in the tow line. If the slack is severe enough, it might entangle the glider or cause damage to the glider or towplane. The following situations may result in a slack line:

- Abrupt power reduction by the towplane
- Aerotow descents
- Glider turns inside the towplane turn radius [*Figure 8-7*]
- Updrafts and downdrafts
- Abrupt recovery from a wake box corner position
- Glider dives toward towplane [*Figure 8-8*]

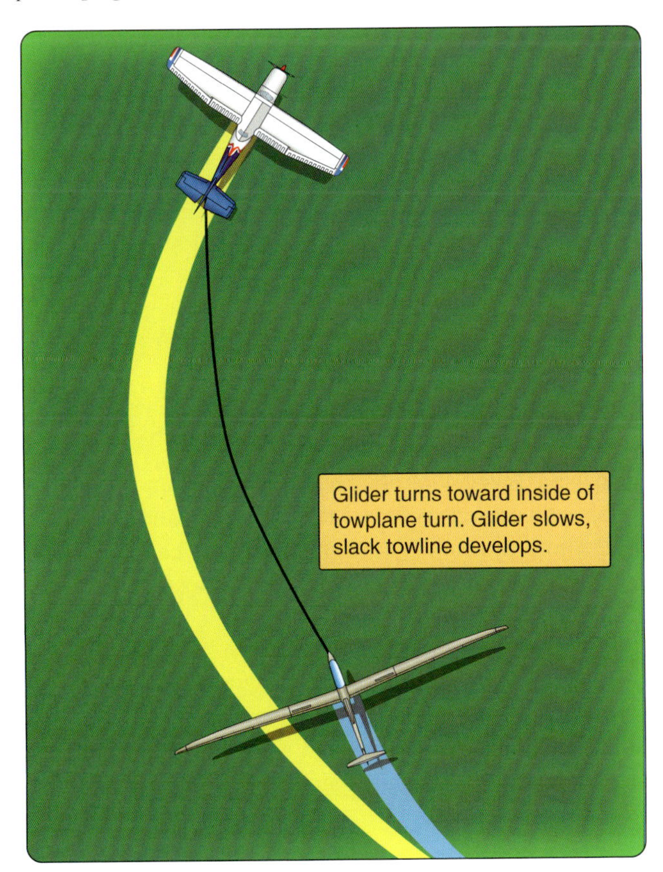

Figure 8-7. *Glider bank steeper than that of towplane, causing slack in tow line.*

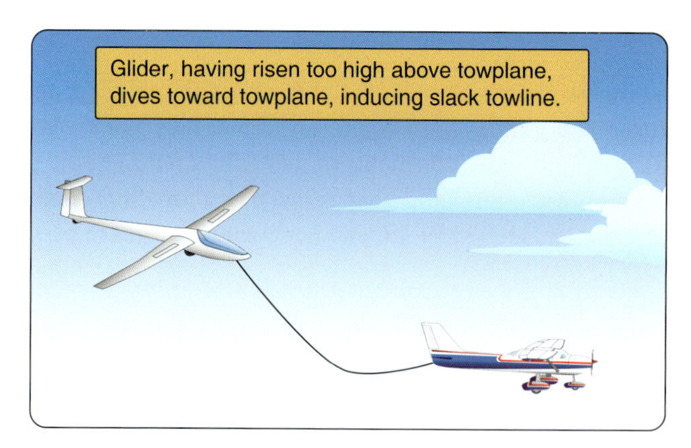

Figure 8-8. *One method of generating tow line slack.*

When the towplane precedes the glider into an updraft, the glider pilot perceives the towplane as climbing faster and higher than the glider. Then, as the glider enters the updraft, it climbs higher and faster than the towplane did. As a result, the glider pilot pitches the glider over to regain the proper tow altitude, gains airspeed more quickly than the towplane, and creates a slack tow line.

A glider pilot should initiate slack tow line recovery procedures as soon possible by increasing drag. For example, the glider pilot may slip back into alignment with the towplane. If slipping motion fails to reduce the slack sufficiently, careful use of spoilers/dive brakes can decelerate the glider and take up the slack. As the tow line tightens and the tow stabilizes, the glider pilot removes the extra drag and resumes the desired aerotow position. The glider pilot should immediately release from the aerotow if the slack in the tow line becomes excessive or gets beyond the pilot's capability to recover safely.

Common errors regarding a slack line include:

- Failure to take corrective action at the first indication of a slack line.
- Improper procedure to correct slack line causing excessive stress on the tow line, towplane, and the glider.
- Failure to decrease drag as tow line slack decreases.

Ground Launch Abnormal & Emergency Procedures

Abnormal Procedures

The launch equipment operator manages a speed-controlled winch some distance from the glider, and the initial tow speed and tension could vary. A tow line speed too great could exceed glider limitations while a speed too low may make liftoff difficult, prevent further climb after liftoff, or result in stall after takeoff. The pilot should use appropriate radio calls and augment with visual signals if necessary, to direct the launch operator to increase or decrease speed. The pilot should release the tow line and land ahead if an abnormal situation develops.

A launch mechanism malfunction may interrupt a ground launch. A gradual deceleration in rate of climb or airspeed may be an indication of this type of malfunction. If suspecting a launch mechanism malfunction, the glider pilot should release and land ahead. However, a pilot might confuse an unintended opening of the spoilers with a winch malfunction if the pilot does not notice the spoilers opening. The pilot should quickly check the status of the spoilers if performance appears degraded during the tow.

Wind gradient (a sudden increase in windspeed with height) can have a noticeable effect on ground launches. A significant or sudden wind gradient may increase indicated airspeed and exceed the maximum ground launch tow speed. [*Figure 8-9*] When encountering a wind gradient, the pilot should push forward on the stick to reduce the tension on the tow line, which reduces indicated airspeed. The only way for the glider to resume climb without exceeding the maximum ground launch airspeed involves signaling the launch operator to reduce tow speed. After the reduction of towing speed, the pilot can resume normal climb. If launching using a winch with automatic tension control, the glider pilot uses conventional pitch changes to control airspeed and climbs at a higher rate if wind speed increases in a gradient.

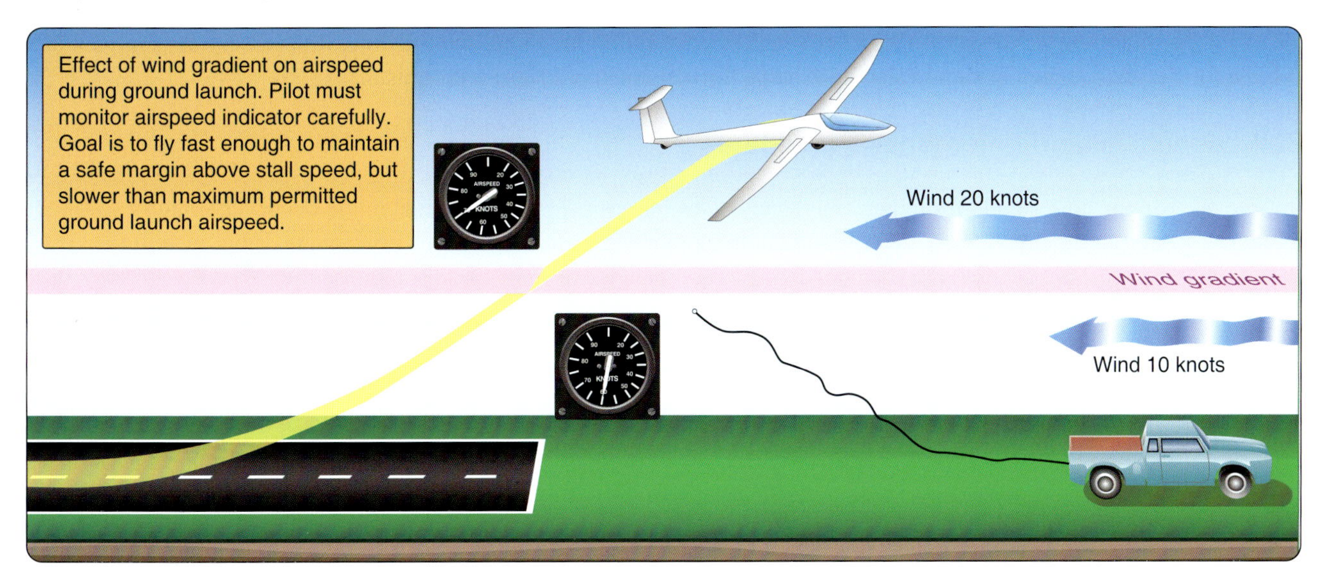

Figure 8-9. *Ground launch wind gradient.*

Emergency Procedures

A broken tow line may cause an emergency during a ground launch. [*Figure 8-10*] When a tow line failure occurs, the glider pilot should pull the release handle, immediately lower the nose of the glider, and maintain a safe airspeed. Distinguishing features of the ground launch include nose-high pitch attitude and a relatively low altitude for a significant portion of the launch and climb. If a tow line break occurs and the glider pilot fails to respond promptly, the nose-high attitude of the glider may result in a stall.

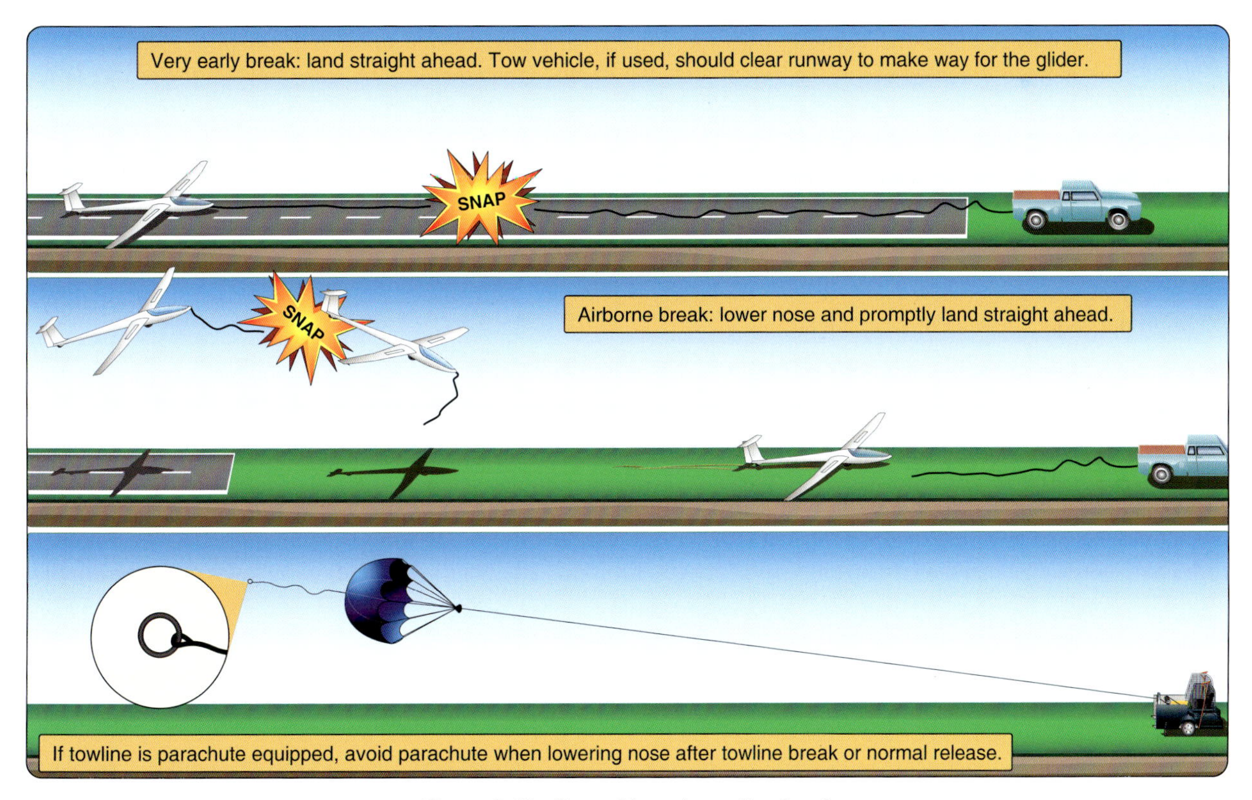

Figure 8-10. *Ground launch tow line break.*

If the glider tow release mechanism fails, the pilot should fly at airspeeds no lower than best lift over drag (L/D) airspeed. The pilot should fly over and then past the ground launch equipment. This method allows the glider tow hook back release to activate or the tow line weak link to fail. The ground launch equipment also uses an emergency release mechanism in the event the glider tow release fails. Winches should have a guillotine to cut the tow line, if necessary. If using a motor vehicle for a ground launch, it should also have some form of backup release mechanism.

Self-Launch Takeoff Emergency Procedures

Prior to takeoff, the pilot should formulate emergency plans for any type of failure that might occur. Thorough knowledge of aircraft performance data, normal takeoff and landing procedures, and emergency procedures as outlined in the GFM/POH help bring about the successful management of any emergency.

Mismanagement of the aircraft systems through lack of knowledge may cause serious difficulty. For instance, if the spoilers or dive brakes remain open during takeoff and climb or open inadvertently, the self-launching glider may not generate sufficient excess power to continue climbing. Other emergency situations may include inflight fire, structural failure, encounters with severe turbulence, wind shear, canopy failure, and inadvertent encounter with instrument meteorological conditions (IMC).

Possible options for handling emergencies depend on altitude above the terrain, wind, and weather conditions. As a part of preflight planning, pilots should review the effects of density altitude on glider performance, the takeoff runway length, landing areas near the gliderport, and potential air traffic that could affect the pilot's approach and landing decisions. Emergency options may include landing ahead on the remaining runway, landing off field, or returning to the gliderport to land on an available runway. The appropriate emergency procedures may be found in the GFM/POH for the specific self-launching glider.

Spiral Dives

An excessive low-nose attitude during a steep turn may result in a significant increase in airspeed and loss in altitude, which indicates a spiral dive. If the pilot attempts to recover from this situation by applying back elevator pressure only, the limiting load factor may be exceeded, causing structural failure. To recover from a spiral dive, the pilot should first reduce the angle of bank with coordinated use of the rudder and aileron, and then smoothly increase pitch to the proper attitude.

Common errors during spiral dives include:

- Failure to recognize when a spiral dive develops.
- Rough, abrupt, or uncoordinated control application during recovery.
- Improper sequence of control applications.

Spins

During a spin, the glider follows a downward corkscrew path. A spin entry may occur after an aggravated stall of one or both wings that results in rotation around the vertical axis. As the glider rotates around the vertical axis, the outer wing develops more lift and less drag than the inner stalled wing, creating a rolling, yawing, and pitching motion. [*Figure 8-11*] The pilot may not realize that the critical angle of attack has been exceeded until the glider yaws toward the lowering wing.

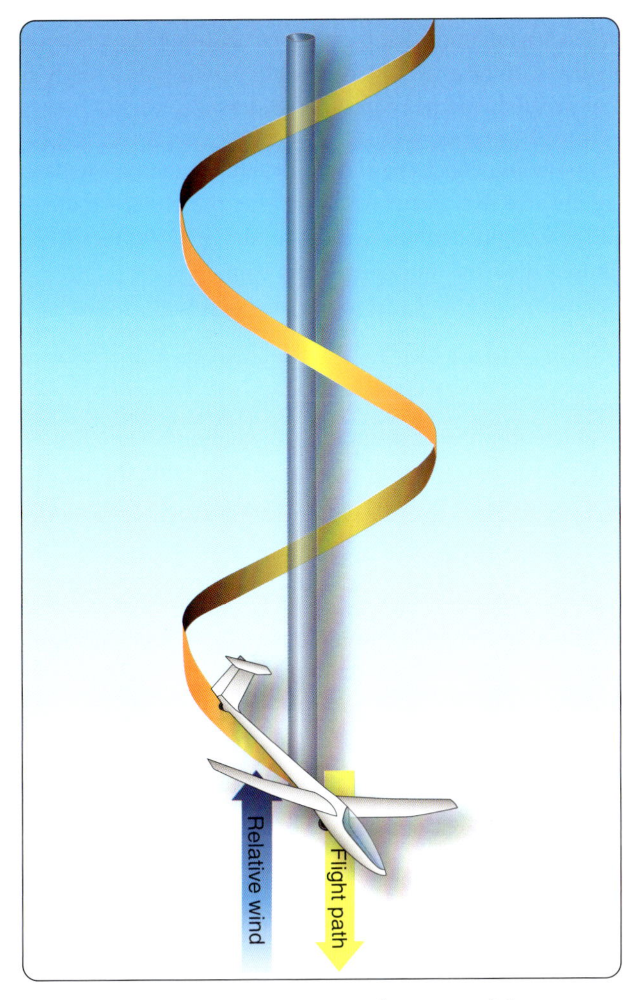

Figure 8-11. *Autorotation of spinning glider.*

Many gliders need considerable effort to enter a spin and require good judgment and technique to intentionally enter a spin. However, these same gliders may enter a spin accidentally if the pilot mishandles the controls in turns, stalls, and flight at minimum controllable airspeeds. This fact explains why pilots should practice stalls and develop the ability to recognize and recover from them.

Continued practice of stall recognition and recovery helps the pilot develop an instinctive and prompt reaction to prevent a spin. The pilot should apply immediate corrective action any time the glider approaches spin conditions, and the pilot should immediately execute spin recovery procedures if a spin occurs.

A flight instructor may demonstrate spins and spin recovery techniques with emphasis on any special spin procedures or techniques required for a particular glider. Before beginning any spin operations, the following items should be reviewed:

- GFM/POH limitations section, placards, or type certification data sheet, to determine if the glider is approved for spins.
- Weight and balance limitations.
- Proper recommended entry and recovery procedures.
- Any requirements for parachutes as given in Title 14 of the Code of Federal Regulations (14 CFR) part 91.

Before any flight, and especially one with intentional spins planned, pilots should check for excess or loose items that may affect the weight, CG, and controllability of the glider. Slack control cables (particularly rudder and elevator) could prevent full control deflections and delay or preclude recovery in some gliders.

Prior to initiation of a spin, the pilots should check the flight area above and below the glider for other air traffic and not spin if any conflict exists. Clearing the area may occur while slowing the glider for the spin entry. All spin training should initiate at an altitude high enough for a completed recovery at or above 1,500 feet AGL and within gliding distance of a landing area. The following paragraphs describe four phases of a spin.

Entry Phase

In the entry phase, the pilot provides the necessary elements for the spin. The spin demonstration entry procedure begins much like a stall. However, as the glider approaches a stall, the pilot smoothly applies full rudder in the direction selected for spin rotation and full back (up) elevator to the limit of travel. The pilot should maintain the ailerons in the neutral position during the spin procedure unless the GFM/POH specifies otherwise.

Incipient Phase

The incipient phase begins as the glider stalls and rotation begins. As an incipient spin develops, the indicated airspeed should read near or below stall airspeed. This phase can run until the spin develops fully, which may take up to two turns for most gliders. Instructors commonly use an incipient spin for introduction to spin training and begin recovery prior to the first of 360° of rotation. In this phase, the aerodynamic and inertial forces have not achieved a balance.

Developed Phase

The developed phase occurs when the glider's angular rotation rate, airspeed, and vertical speed stabilize with the glider in a nearly vertical flightpath. Equilibrium occurs in this phase as aerodynamic forces and inertial forces balance and the attitude, angles, and self-sustaining motions about the vertical axis become constant or repetitive.

Recovery Phase

This phase may last for a quarter turn to several turns. To accomplish spin recovery, pilots should follow the manufacturer's recommended procedures. In the absence of the manufacturer's recommended spin recovery procedures, pilots should follow the following steps for general spin recovery:

1. Ailerons to neutral. Ailerons may have an adverse effect on spin recovery. Aileron control in the direction of the spin may increase the rate of rotation and delay the recovery. Aileron control opposite the direction of the spin may cause the down aileron to move the wing deeper into the stall and aggravate the situation. Retract flaps, if extended, as soon as possible after spin entry.

2. Apply full opposite rudder against the rotation. Ensure full (against the stop) opposite rudder until rotation stops. As spin rotation stops, neutralize the rudder. If not neutralized at this time, the deflected rudder may cause a yawing effect in the opposite direction.

3. When rotation stops, apply a positive and brisk, straightforward movement of the elevator control past neutral to recover from the stall. Slow and overly cautious control movements during spin recovery may result in the glider continuing to spin indefinitely, even with anti-spin inputs. A brisk and positive technique, on the other hand, results in a more positive spin recovery. Hold the controls firmly in this position until airspeed begins increasing.

4. With rotation stopped, angle of attack below the critical angle, and nose-down attitude, airspeed increases rapidly. Make smooth pitch control input of sufficient magnitude. Pulling up aggressively may cause a second stall and spin during the recovery. Waiting too long to pull up could lead to high airspeed and excessive G-loading.

The FAA recommends these recovery procedures for use only in the absence of the manufacturer's procedures. Before any pilot begins spin training, the pilot should understand any spin recovery procedures provided by the manufacturer.

The most common problems in spin recovery include pilot confusion when determining the direction of spin rotation and whether the maneuver constitutes a spin or a spiral dive. An inclinometer does not indicate the direction of a spin. A high or increasing airspeed indicates a spiral dive. In a spin, the airspeed reads at or below stalling speed.

Common errors when encountering/practicing spins include:

- Failure to clear area before a spin.
- Failure to establish proper configuration prior to spin entry.
- Failure to correct airspeed for spin entry.
- Failure to recognize conditions leading to a spin.
- Failure to achieve and maintain stall during spin entry.
- Improper use of controls during spin entry, rotation, or recovery.
- Disorientation during spin.
- Failure to distinguish a spiral dive from a spin.
- Excessive speed or secondary stall during spin recovery.
- Failure to recover before descent below safe altitude.
- Failure to recover with a landing area within gliding distance.

Off-Field Landing Procedures

Off-field landings may occur in the vicinity of the launching airport due to unexpected rapid weather deterioration, a significant change in wind direction, unanticipated amounts of sinking air, disorientation, lack of situational awareness, tow failures, and other emergencies. In these situations, a precautionary off-field landing may present less risk than an approach back to the airport. If the pilot loses sight of the airport or if the glide back to the airport comes up short for any reason, the attempt to return to the airport may result in damage to the glider or injury to the pilot.

A glider pilot should always be prepared for off-field landings as the absence of sufficient lift may require an off-field landing. Even when flying in a self-launching glider, the engine or power system could fail.

The basic ingredients for a successful off-field landing include an awareness of wind direction, wind strength at the surface, and obstacles along the approach path. The glider pilot should select suitable landing areas while airborne with sufficient height and time to plan and perform a safe approach and landing, and then make an accurate landing to the actual selected field.

These basic ingredients for a successful off-field landing can be summarized as follows:

- Recognizing the possibility of imminent off-field landing.
- Selecting a suitable area, then a suitable landing field within that area.
- Planning the approach with wind, obstacles, and local terrain in mind.

- Executing the approach, landing, and stopping the glider as soon as possible.
- Contacting ground crew and notifying them of the off-field landing location.

Denial represents the most common off-field landing planning failure. The pilot experiencing denial finds it easier to focus on continuing the flight and attempting to find a way to climb back up and fly away. This false optimism leaves little or no time to plan an off-field landing if the attempt to climb fails.

When planning an off-field landing, flying downwind offers more range and a greater area to search than flying upwind. After selection of a landing zone, wind awareness allows the pilot to plan the orientation and direction of a landing approach into the wind to shorten the landing roll. Pilots should visualize the wind flowing over and around the intended landing area including how low altitude turbulence may exist in the area downwind of hills, buildings, and other obstructions.

Pilots should use a methodical approach to an off-field landing based on a set of decision heights. The pilot should select a general landing area no lower than 2,000 feet above ground level (AGL) and select the intended landing field no lower than 1,500 feet AGL. At 1,000 feet AGL, the pilot should commit to flying the approach and landing.

Pilots should consider safety rather than an easy retrieval as the highest priority when selecting a landing site. During an off-field landing approach, the pilot will likely not know the precise elevation of the landing site. Since this makes the altimeter less useful, the pilot should fly the approach and assess the progress by recognizing and maintaining the angle that brings the glider to the intended aiming point for the landing site. When landing with a tailwind (due to slope or one-way entry into the selected field due to terrain or obstacles), the pilot should use a shallower approach angle.

A good approach clears each visible obstacle, including clearing any poles and wires by a safe margin. From the air, wires may not appear until right in front of the pilot, whereas towers supporting wires appear from a greater distance. The pilot should assume that wires run between telephone poles, supporting structures, and buildings. The pilot should plan to overfly any wires that may be present, even if not actually seeing them.

The pilot should select a field of adequate length and one with no visible slope, when possible. Any slope visible from the air becomes steep when close to the ground. Color may assist in assessment of slope. High spots often appear lighter in color than low spots because soil moisture tends to collect in low spots and darkens the color of the soil. If the landing will occur on a slope, landing uphill works better than downhill. With a slight downhill grade, the glider will stay airborne longer and experience a longer landing roll, which may result in a collision with objects at the far end of the selected field.

A pilot familiar with the colors of local seasonal vegetation can identify crops and other vegetation from the air. Tall crops generally present more danger than low crops. Pilots should also avoid discontinuities such as lines or crop changes since these discontinuities often indicate the presence of a fence, ditch, irrigation pipe, or some other obstacle or machinery that could damage a glider.

The recommended approach procedure should include the following legs:

- Crosswind leg on the downwind side of the field
- Upwind leg
- Crosswind leg on the upwind side of the field
- Downwind leg
- Base leg
- Final approach

This approach procedure provides the opportunity to see the intended landing area from all sides. Pilots should use every opportunity while flying an off-field approach to inspect the landing area and look for obstacles or other hazards. [*Figure 8-12*]

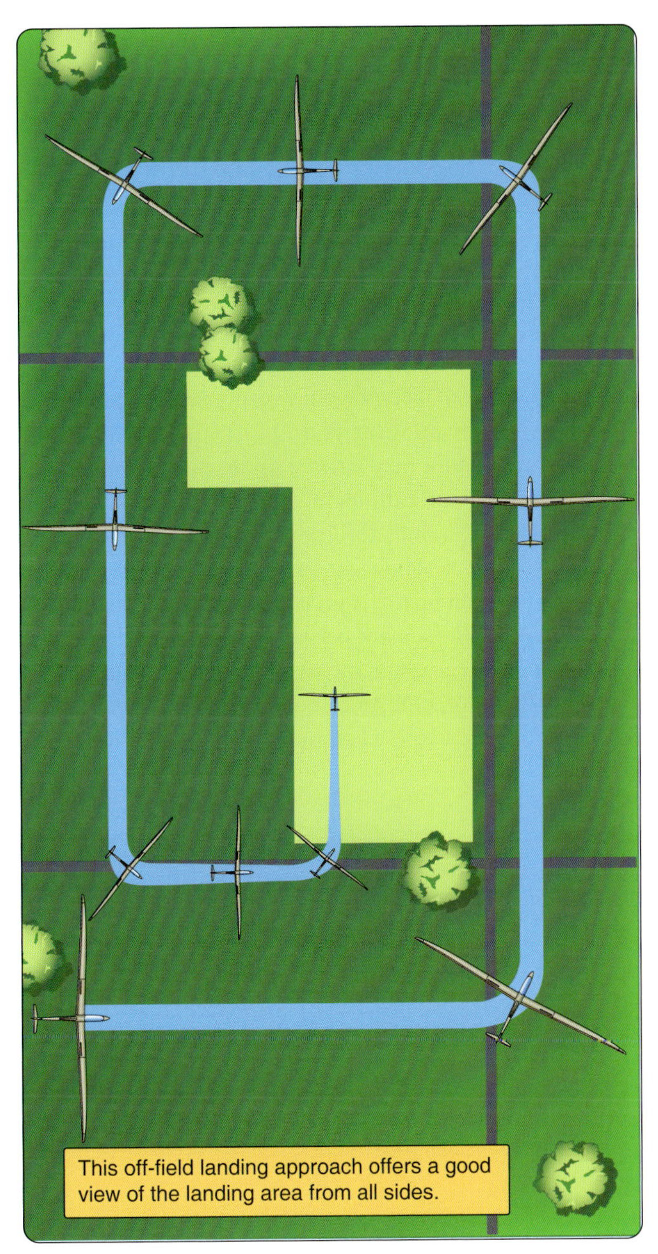

Figure 8-12. *Off-field landing approach.*

The pilot should also consider how obstacles affect the length of landing area available for touchdown, roll out, and stopping the glider. Flying over an obstacle 50 feet high means the glider will overfly the first 500 feet or so of the landing area during the descent to flare and landing. If the selected field has obstacles on the final approach path, the field should also allow for the descent, flare, and landing roll after clearing the obstacle.

Aerodynamic drag works better than wheel brakes at flying speeds. Pilots who hold the glider off during the flare and touch down at the lowest safe speed will land within a shorter distance. After touchdown, using the wheel brake immediately will stop the glider and help prevent collision with any unseen obstacles.

Afterlanding Off Field

Off-Field Landing without Injury

A pilot should tend to personal needs first, secure the glider, and then contact the retrieval crew. If cell service exists, the pilot can use a cell phone to call the retrieval crew. To help identify position, the pilot should relay the GPS coordinates. Pilots should write down the coordinates in case the GPS loses power. If able, calling other glider pilots in the area on the glider-to-glider radio frequency or calling the tow plane can assist with retrieval.

Once contact has been made to arrange for retrieval, the pilot may collect any special tools needed for glider de-rigging or installing gust locks on the glider's flight controls. Since a normal retrieval depends on location, weather and time of day, the pilot may need to set up equipment to handle the current or expected environmental conditions.

Off-Field Landing with Injury or Emergency

A pilot should address any critical injuries and contact emergency response personnel, other aircraft, or any other source of identifiable assistance. If cell coverage exists, the pilot can dial 911, speak to an operator, and provide a clear description of the location. If cell service does not work, a pilot may use the glider radio, if operable, to broadcast a Mayday distress call on emergency frequency 121.5 MHz. Many aircraft, including civil airliners, routinely monitor this frequency. Their altitude gives the line-of-sight aviation transceiver enhanced range when transmitting or receiving. The pilot may try any other frequency likely to elicit a response such as glider air-to-air frequency or the tow plane frequency. Some gliders have an Emergency Locator Transmitter (ELT) on board. A pilot needing emergency assistance should activate the ELT, if available. While 14 CFR, part 91, section 91.207 does not pertain to gliders, an ELT or EPIRB adds safety when flying a glider on a cross-country. Personal locator devices offered by several companies use the 406 MHz satellite signal, and GPS technology to accurately track and relay the pilot's location in the event of an off-field landing that requires emergency assistance. Additionally, pilots may elect to carry satellite communications and tracking devices that allow communication even outside of cell coverage.

If the glider comes to rest in a precarious position, the pilot should secure the aircraft if able. An injured pilot should stay with the glider. Rescue personnel can locate the glider more easily from the air than locate an individual. The pilot might obtain protection from the elements by crawling into the fuselage, crawling under a wing, or using any parachute canopy to rig a makeshift tent around the glider structure. After attending to medical needs and contacting rescue personnel, the pilot should attend to clothing, food, and water issues. The pilot should make every attempt to conserve energy. If unsafe to stay with the glider, the pilot should move to a nearby location for shelter but leave clear written instructions in a prominent location in the glider detailing the shelter location.

System & Equipment Malfunctions

Flight Instrument Malfunctions

Instrument failures can result from improper maintenance practices, internal instrument failure, or an external cause. Improper airspeed indicator maintenance might involve failure to connect the instrument correctly to pitot and static lines. External failures include clogging from insects or ice.

Pilots should practice setting normal attitudes for all flight regimes using outside cues including the skills needed to make a safe approach without a functioning airspeed indicator or altimeter. In fact, many older and vintage gliders do not require an operational altimeter. A flight review or periodic instruction provides an excellent opportunity to review these procedures.

Static line contamination affects both the altimeter and the airspeed indicator. If either instrument malfunctions because of static line contamination, the indications of the other instrument may also be incorrect. The pilot should use external cues and evaluate the indications of any instrument connected to the static port. If in doubt about the accuracy of the instruments, the pilot should not rely on the instrument indications. After landing and prior to the next flight, an aviation maintenance professional should evaluate the instrument system.

Altimeter Malfunctions

During the approach to land without a functioning altimeter, the pilot should assess the angle to the target area frequently. Entering the approach from an apparent normal height, or even from a higher-than-normal height provides a margin of safety. If the current descent angle would overfly the target, the pilot applies spoilers or dive brakes to steepen the descent angle. If necessary, a forward slip or turning slip dissipates excess altitude. If the approach angle will result in a landing short of the target, the pilot should close the spoilers or dive brakes and can modify the approach path to shorten the distance to the targeted landing area as needed.

Airspeed Indicator Malfunctions

If the airspeed indicator appears to be erratic or inaccurate, the pilot should fly the glider by pitch attitude for best glide or minimum sink airspeed. Additional airspeed cues include control response and wind noise. At very low airspeeds, controls feel mushy and wind noise is generally low. At higher airspeeds, control becomes crisper and wind noise takes on a more insistent hissing quality. The pilot may amplify the sound of the relative wind by opening the sliding window installed in the canopy and by opening the air vent control. Turbulent or gusty wind conditions generally require additional airspeed to ensure adequate control authority. If in doubt, flying slightly faster than optimum airspeed provides a safety margin above stall speed. However, a speed higher than best glide airspeed increases the rate of descent and decreases range.

Variometer Malfunctions

Variometer failure makes it difficult for the pilot to locate and exploit sources of lift. If near an airport, the pilot may elect to make a precautionary landing so troubleshooting and repair can take place. Without a nearby airport, the pilot can look for clues to sources of lift. The pilot can use the altimeter to gauge rate of climb or descent in the absence of a functioning variometer. Tapping the altimeter with a finger often overcomes internal friction in the altimeter, allowing the hand to move upward or downward. The direction of the movement gives an idea of the rate of climb or descent over the last few seconds.

Compass Malfunctions

If the compass performs poorly or not at all, the pilot should cross-check current position with aeronautical charts and with electronic methods of navigation, such as GPS, if available. The position of the sun, combined with knowledge of the time of day, can also help with orientation. Section lines, major roads, and prominent landmarks often provide helpful cues for orientation and the direction of flight.

Glider Canopy Malfunctions

Glider Canopy Opens Unexpectedly

Canopy-related emergencies often result from pilot failure to lock the canopy in the closed position prior to takeoff. If the canopy opens in flight, the pilot should focus on flying the glider. The pilot should maintain adequate airspeed while selecting a suitable landing area. An open canopy causes higher than normal drag, and the pilot should plan a steeper-than-normal descent path.

If the canopy opens while on aerotow, the pilot should maintain a normal flying attitude and tow position. The glider pilot should not attempt to close the canopy or release prematurely. After continuing to climb with the tow plane to several thousand feet above the ground and releasing the tow rope, the pilot may try to close the canopy only if able to maintain glider control. If flying with a passenger on board and conditions allow, the pilot can direct the passenger to close and lock the canopy.

Broken Glider Canopy

If the canopy breaks during flight, the best response involves landing as soon as practicable. If the canopy shatters, drag increases and the pilot should plan a steeper-than-normal descent path during the approach.

Frosted Glider Canopy

During flight at high altitude or in low ambient temperatures, frost formation or frozen condensation on the inside surface of the canopy can obstruct vision. The pilot should open the air vents and the side window to ventilate the cabin and to evacuate moist air before this occurs. Descending to lower altitudes for warmer air or flying in direct sunlight may help defrost the canopy.

Water Ballast Malfunctions

One example of ballast failure involves asymmetrical wing tank draining. With one wing heavier than the other, the glider may become difficult to control at low airspeeds and during the landing rollout. Another failure involves wing tanks that drain using a central pipe that passes through the fuselage. A leak in this system may trap water in the fuselage. The pilot should determine if a means exists to evacuate the water from the fuselage. If the water collects far enough forward or aft, it may cause an out of CG condition and degrade pitch control. Pilots can regain elevator effectiveness by flying at mid-to-high airspeeds. If pitch control degrades significantly, the pilot with sufficient altitude and a parachute might consider a bailout as the safest choice.

Retractable Landing Gear Malfunctions

During flight, the pilot cannot generally resolve landing gear failures related to mechanical malfunctions. The pilot should fly the approach at normal airspeed and may need to use more spoiler or dive brake than normal during the approach due to the reduced drag. The pilot should land on the smoothest surface available, preferably an area that has good turf to reduce damage to the glider. A full stall, hard, or tailwheel first landing increases the chances of injury and damage, and the pilot should make a soft touchdown slightly above stall speed.

In a gear-up landing, the glider makes considerable noise as it slides along the runway, and the wingtips travel closer to the ground. The pilot should keep the wings level while possible and keep the glider path as straight as possible while using the rudder to avoid collisions with objects on the ground or along the runway border. The pilot should focus on personal safety since no method exists to prevent glider skin damage during a gear-up landing.

Primary Flight Control Systems

Failure of any primary flight control system presents a serious threat to safety. Incomplete preflight assembly represents the most frequent cause of control system failure. The crew should use a written checklist to verify each assembly operation and avoid interruptions during assembly. If interruptions occur, the crew should rerun the checklist from the beginning. A critical assembly check aids in confirming the glider has been assembled correctly and a positive control check prior to flight verifies control system continuity.

Elevator Malfunctions

The most serious control system malfunction involves a failure of the elevator. Causes of elevator failure include the following:

- An improper connection of the elevator actuators during assembly.
- An elevator control lock that was not removed before flight.
- Separation of the elevator gap seal tape.
- Interference with free and full travel of the control stick or system caused by a foreign object such as a water bottle, camera, or unsecured rear-seat cushion.
- A control stick secured by a lap belt or shoulder harness in the back seat.
- A structural failure of the glider due to overstressing or flutter.

If the pilot detects elevator irregularity or failure early in the takeoff roll, release of the tow line (or power reduction to idle in a self-launching glider), may allow obstacle avoidance. The pilot should use the brakes firmly to stop the glider as soon as possible.

In an aerotow launch, the glider pilot should consider the effect the glider has on the safety of the tow pilot and if the glider pilot has a parachute. If the elevator control irregularity becomes apparent after takeoff and if close to the ground with a flat or slightly nose-low pitch attitude, the pilot should release the tow line. If sufficient elevator control exists during climb, staying attached and achieving a high altitude gives the pilot time to abandon the glider and deploy a parachute, if available.

If continuing the climb, the pilot can experiment with the effect of other flight controls on the pitch attitude of the glider. These include the effects of various wing flap settings, spoilers or dive brakes, elevator trim system, and raising or lowering the landing gear. If flying a self-launching glider, the pilot can also experiment with the effect of power settings on pitch attitude.

If aileron control functions, the pilot can bank the glider and use the rudder to moderate the attitude of the nose relative to the horizon. When approaching the desired pitch attitude, adjusting the bank angle can maintain the desired pitch attitude. Forward slips may have a predictable effect on pitch attitude and can be used to moderate it. Usually, a combination of these techniques allows some control of pitch attitude. Although difficult to use, these techniques allow some control. Achieving an altitude sufficient to permit bailing out usually ends in survival as parachutes rarely fail.

Elevator gap seal tape, if in poor condition, can degrade elevator responsiveness. If the adhesive that bonds the gap seal leading edge to the horizontal stabilizer begins to fail, the airflow may lift the leading edge of the gap seal. This provides, in effect, a small spoiler that disturbs the airflow over the elevator just aft of the lifted seal. In extreme cases, this effect can remove all elevator authority.

Aileron Malfunctions

Aileron failures can cause serious control problems. Causes of aileron failure include:

- Improper connection of the aileron control actuators during assembly.
- Aileron control lock not removed before flight.
- Separation of the aileron gap seal tape.
- Interference of a foreign object with free and full travel of the control stick or aileron circuit.
- A control stick secured by a lap belt or shoulder harness in the back seat.
- Structural failure and/or aileron flutter.

The pilot might counteract this failure successfully because each wing has an aileron. If one aileron becomes disconnected or locked by an external control lock, the degree of motion still available in the other aileron may exert some influence on bank control. A glider with limited aileron movement may control more easily at high airspeeds than at low airspeeds.

If both ailerons malfunction and compromise roll control, the pilot may use the secondary effect of the rudder to make gentle bank adjustments while maintaining a safe margin above stall speed. If the pilot applies left rudder in wings-level flight, the nose yaws to the left. If the pressure is held, the wings begin a gentle bank to the left. If the pilot applies and holds right rudder pressure, the glider yaws to the right, then begins to bank to the right. The pilot can use this secondary banking effect of the rudder for limited roll control. However, if the bank angle becomes excessive, recovery to wings-level flight using the rudder alone may become impossible. If the bank becomes too steep, the pilot should use any aileron influence available, as well as all available rudder to level the wings. If a parachute is available and the glider becomes uncontrollable at low airspeed, the best chance to escape serious injury may be to bail out from a safe altitude.

Rudder Malfunctions

An actual rudder failure rarely occurs because removing and installing the vertical fin/rudder combination does not occur as part of the sequence of rigging and de-rigging the glider. The pilot should recognize any obvious directional control issue caused by a rudder malfunction at the very beginning of a launch and abort immediately.

Rudder malfunctions may occur if the pilot forgets to remove the rudder control lock or when an unsecured object interferes with the free and full travel of the rudder pedals. Preflight preparation should include safe stowage of all items on board and removal of all flight control locks. The pretakeoff checklist encompasses all primary flight controls for correct and full travel prior to launch.

During flight, if an object interferes with or jams the rudder pedals, the pilot should attempt to remove it. If removal fails, the pilot can attempt to deform, crush, or dislodge the object by applying force on the rudder pedals. Varying the load factor may dislodge the object but could also result in moving the object and jamming the elevator or aileron controls. If the object cannot be retrieved and stowed, the pilot should consider a precautionary landing.

In the air, the pilot may obtain some degree of directional control by using adverse yaw. During rollout from an aborted launch or during landing rollout without rudder control, the pilot can deliberately ground the wingtip toward the direction of desired yaw. Putting the wingtip on the ground for a fraction of a second causes a slight yaw in that direction; however, holding the wingtip firmly on the ground may cause a ground loop in the direction of the grounded wingtip.

Commonly misplaced objects that can cause rudder control interference include:

- Water bottles.
- Cameras.
- Electronic computers.
- Containers of food and similar items.
- Clothing.
- Sunglasses.

Secondary Flight Controls Systems

Secondary flight control systems include the elevator trim system, wing flaps, and spoilers or dive brakes. Malfunction of these systems may present a serious challenge.

Elevator Trim Malfunctions

When compensating for a malfunctioning elevator trim system, the pilot should apply pressure on the control stick to maintain the desired pitch attitude, and then bring the flight to safe conclusion.

Spoiler/Dive Brake Malfunctions

Spoiler or dive brake system failures arise from rigging errors or omissions, environmental factors, and mechanical failures. Without proper connection, one or both spoilers or one or both dive brakes could deploy without the possibility of retraction. Spoiler or dive brake deployment during the launch or the climb may cause a launch emergency and possible tow failure. Spoilers or dive brakes that deploy asymmetrically, result in yaw and roll tendencies.

If an asymmetric spoiler or dive brake extension occurs and the pilot cannot retract the extended spoiler or dive brake, the pilot may attempt to deploy the other spoiler or dive brake to restore the symmetry, which protects against stalling or a spin. If this condition arises during launch or climb, the pilot should abort the launch, extend the other spoiler or dive brake to restore symmetry, and land.

Environmental factors include low temperature or icing during long, high altitude flights. Lower temperatures cause contraction of all glider components. Uneven contraction may bind the spoilers or dive brakes and make them difficult or impossible to deploy. Any moisture trapped in the system may freeze and interfere with operation of the spoilers or dive brakes. On the other hand, a rise in temperature causes all glider components to expand, which could bind the spoilers or dive brakes in the closed position. This could occur with the glider parked on the ground in direct summer sunlight.

Mechanical failures can cause asymmetrical spoiler or dive brake extension during flight. Causes may include a broken weld in the spoiler or dive brake actuator mechanism, a defective control connector, or other mechanical failure. The glider yaws and banks toward the wing with the extended spoiler or dive brake. Aileron and rudder counteract these tendencies. To eliminate any possibility of a stall or spin entry, the pilot should maintain a safe margin above stall airspeed. While deploying the other spoiler or dive brake relieves the asymmetry, it reduces gliding range. This may be a significant concern if over rough terrain. Nevertheless, a controlled landing in rough terrain has much less associated risk than an asymmetric stall or spin.

Miscellaneous Flight System Malfunctions

Towhook Malfunctions

Failure modes include failure to release and uncommanded release. The pilot should form an emergency plan prior to launch for either condition. A pilot who cannot release the tow line should alert the tow pilot and follow appropriate emergency procedures.

Oxygen System Malfunctions

If a suspected or known failure of the oxygen system occurs, the pilot should descend immediately to an altitude that does not require supplemental oxygen. A pilot deprived of sufficient oxygen, even for a short interval, loses critical thinking capability and may develop a false sense of wellbeing. After descending, the pilot should avoid hyperventilation and breathe normally to restore oxygen to the bloodstream.

For high altitude flights, such as a wave flight, the oxygen bailout bottle should be in good condition and within easy reach if a high-altitude escape becomes necessary. Pilots who make high altitude flights should train for an event requiring glider abandonment, use of oxygen, and proper use of a parachute.

Drogue Chute Malfunctions

Some gliders have a drogue chute to add drag during the approach to landing and enhance a steep approach. The drogue chute stows in the aft tip (tail cone) of the fuselage or in a special compartment in the base of the rudder. If the chute deploys accidentally or inadvertently during the launch, the pilot normally jettisons it.

During an approach, an improperly packed or damp drogue chute may fail to deploy. If this happens, the pilot may use the rudder to sideslip briefly or use the rudder to yaw the glider to attempt deployment. Either technique can increase the drag force on the component that pulls the parachute out of the compartment. If neither technique deploys the drogue chute, the drogue canopy may deploy spontaneously. If this occurs, the pilot has the option to jettison the chute.

Another malfunction involves failure of the drogue chute to inflate. If this happens, the canopy "streams" like a twisting ribbon of nylon, providing only a fraction of the expected drag. Full inflation, although unlikely after streaming occurs, would increase drag substantially. Rather than face any sudden increase in drag and if in doubt as to the status of deployment, the pilot can jettison the chute. Regardless of the malfunction type, the pilot should review approach and landing options for the drogue chute conditions.

Self-Launching Gliders

Self-launching gliders have multiple systems to support powered flight. The pilot should review the GFM/POH for the self-launching glider flown, develop an understanding of the glider systems, and work with a glider instructor as necessary to develop proficiency. Additional systems in a self-launching glider may include the following:

- Fuel tanks, lines, and pumps.
- Engine or propeller extension and retraction systems.
- Electrical system including engine starter system.
- Lubricating oil system.
- Engine cooling system.
- Engine throttle controls.
- Propeller blade pitch controls.
- Engine monitoring instruments and systems.

An engine in a self-launching glider may fail. Failures range from a very slight power loss at full throttle to catastrophic and sudden failure during a full-power takeoff. Fuel contamination or fuel exhaustion cause a significant number of these failures.

Full power operation of an internal combustion engine cannot continue without a supply of fuel, source of ignition, internal cooling, and lubrication. The pilot should monitor the engine temperature, oil pressure, fuel pressure, and revolutions per minute (rpm) carefully to ensure the desired engine performance. Warning signs of impending difficulty include excessively high engine temperatures, abnormal engine oil temperatures, low oil pressure, low rpm despite high throttle settings, low fuel pressure, or abnormal engine operation that includes surging, backfiring, or missing. Abnormal engine performance may indicate complete engine failure will occur within a short time. Even if total engine failure does not occur, an engine that cannot produce full power may create an inability to hold altitude or climb. The best course of action, if airborne, involves landing followed by appropriate maintenance and repair.

Regardless of the type of engine failure, the pilot should maintain flying airspeed and control the glider. Pilots flying self-launching gliders with a pod-mounted external engine above the fuselage need to lower the nose more aggressively than pilots flying a glider with an engine mounted in the nose. In the former, the thrust of the engine during full power operations tends to provide a nose-down pitching moment. If power fails, a nose-up pitching moment occurs due to the substantial parasite drag of the engine pod high above the longitudinal axis of the glider. At low altitudes, the pilot may not have time to stow the engine and will land with the engine extended. The GFM/POH contains authoritative information regarding the correct sequence of pilot actions in the event of power failure.

A power failure during launch or climb may provide limited time to maneuver. The pilot should concentrate on flying the glider, selecting a suitable landing area, and making a safe landing. Troubleshooting should only occur if safe to do so. Even if the pilot restores power, full power may not come back. Flight with partial power may result in an inability to clear obstacles, such as wires, poles, hangars, or nearby terrain.

Inability to Restart a Self-Launching/Sustainer Glider Engine While Airborne

Nearly all self-launching gliders have a procedure designed to start the engine while airborne. This procedure allows the pilot to fly home safely during a flight where soaring flight conditions deteriorate. However, engines may not start in some situations. The reasons include lack of fuel, ignition malfunction, low engine temperature due to cold soak, insufficient battery output, fuel vapor lock, lack of propeller response to blade pitch controls, and other factors. This becomes a serious problem with unsuitable terrain below for a safe off-field landing.

The pilot should attempt the engine restart at an altitude high enough to complete a safe power-off approach and landing in case the restart fails. Self-launching glider pilots should not allow themselves to get into a situation at altitude where the only acceptable resolution involves relying on the engine.

Self-Launching Glider Propeller Malfunctions

Propeller failures include propeller damage or disintegration, propeller drive belt or drive gear failure, or failure of the variable blade pitch control system. To perform an air-driven engine restart many self-launching gliders require placing propeller blades in a particular blade pitch position. In the absence of this adjustment, the propeller blades cannot deliver enough torque to start the engine.

Self-Launching Glider Electrical System Malfunctions

An electrical system failure in a self-launching glider may render electrically controlled propeller pitch inoperative or prevent deployment of a pod engine for an air restart. Certain electrical failures prevent activation of an electric starter if used for an air restart. However, if able to maintain a suitable landing spot and with sufficient rising air, the pilot can continue to fly the glider without electrical power. If the pilot cannot reach an airport or suitable landing field, an off-airport landing will occur.

Pilots fly some self-launching gliders at night using engine power for cruising. The glider must have the appropriate aeronautical lighting required for night operations (14 CFR part 91, section 91.209). If carrying passengers, the pilot must meet the recent flight experience to for night takeoffs and landings in accordance with 14 CFR part 61, section 61.57(b).

Flight at night increases accident risk. If an electrical system failure during night operations extinguishes the position lights, pilots of nearby aircraft cannot see the self-launching glider and the glider pilot must assume responsibility for collision avoidance. The glider pilot may also have difficulty seeing the flight instruments or electrical circuit breakers. It makes good sense to have a working flashlight for such an emergency.

The pilot should not reset any circuit breakers if smoke or the smell of smoke is present as recommended in CE- 10-11R1, Special Airworthiness Information Bulletin, dated January 14, 2010, available for download on the FAA's DRS website.

Resetting a circuit breaker in flight may increase the risk of an electrical overload and fire. [*Figure 8-13*] If electrical smoke fills the cabin, the pilot should consider ventilating the cabin and head directly for the nearest suitable airport and land. The aviation transceiver installed in the instrument panel may not function, but the pilot may use a portable battery-operated aviation two-way radio if available. The pilot can also receive landing instruction through air traffic control (ATC) light-signals. The pilot should review 14 CFR part 91, section 125, and the Aeronautical Information Manual (AIM) section 4-2-13, Traffic Control Light Signals.

Figure 8-13. *Self-launching glider circuit breakers.*

Inflight Fire

If a fire ignites, the pilot should do everything possible to reduce the spread of the fire and land as soon as possible. The self-launching glider GFM/POH is the authoritative source for emergency response to suspected in-flight fire. In general, the response includes the following steps:

- Reduce throttle to idle.
- Shut off fuel valves.
- Shut off engine ignition.
- Turn off the electrical system or any device contributing to the fire.
- Consider a slip to keep flames away from the fuselage, if applicable.
- Land immediately and stop as quickly as possible.
- Evacuate the self-launching glider.

After landing, the pilot and any passenger should exit the glider, move upwind and away from the glider, and keep onlookers away. The principal danger after evacuating the glider comes from fuel ignition and explosion, with the potential for serious injury.

Modern gliders contain composite materials and resins that can produce poisonous fumes when overheated or burned. The glider pilot should avoid breathing the fumes and may consider jettisoning the canopy while in flight. This same modern construction also means a fire can spread very quickly. A quick landing or bailing out may save the pilot's life. If the fire spreads to critical structures, the airframe may fail before the landing.

Emergency Equipment & Survival Gear

Emergency equipment and survival gear enhance safety for all cross-country flights.

Survival Gear Checklists

Checklists help the pilot assemble survival equipment in an orderly manner. The essentials for survival include reliable and usable supplies of food and water. Equipment including blankets and clothing helps maintain a safe body temperature, which could become difficult to manage in extreme cold or extreme heat.

Food & Water

During cross-country flight, pilots should carry an adequate supply of water and food (especially high-energy foods such as energy bars, granolas, and dried fruits). If necessary for survival, pilots may drink water from ballast tanks if accessible and free of contaminants, such as antifreeze.

Clothing

Pilots should wear clothing appropriate to the local environment, including hat or cap, shirts, sweaters, pants, socks, walking shoes, space blanket, and gloves or mittens. Layered clothing provides flexibility to meet the demands of the environment since removal or addition of layers helps to regulate body temperature. A parachute canopy can be used as an effective layered garment to conserve body heat or to provide relief from excessive sunlight, and sunglasses can protect eyes from that light. Desert areas may be very hot in the day and very cold at night. Prolonged exposure to either condition can debilitate a stranded pilot.

Communication

Pilots can use radios, telephones, or cell phones to summon assistance. An aviation transceiver can be tuned to broadcast and receive on the emergency frequency 121.5 MHz, the frequency used by a tow plane, the glider-to-glider frequency, or any other usable frequency that elicits a response. Newer ELTs provide a continuous signal on the 406 MHz Search and Rescue (SAR) system and activate either automatically or manually to transmit a unique digital identification code to the first satellite that comes into range. The satellite receives the signal and relays it to a ground station. If there is no ground station in view, the satellite sends it to the first available ground station. The ground station processor measures the Doppler shift of the ELT signal and calculates its position. This calculation is usually accurate to within 1.5 nautical miles on the first satellite pass and becomes more accurate with each pass. If the beacon has an integrated GPS or access to one on board the aircraft, the ELT transmits GPS position with the digital data.

After the ground station has completed processing, it transmits the identification and position to the United States Mission Control Center (USMCC). The USMCC attaches the information contained in the 406 MHz beacon registration database for that ELT and generates an alert message. If the location lies within the continental U.S., the alert is sent to the Air Force Rescue Coordination Center (AFRCC) at Langley Air Force Base, Virginia. The AFRCC then takes the registration data and attempts to ascertain the aircraft's disposition. By calling the emergency contact numbers, or by calling flight service stations with the N-number, they can quickly determine whether the aircraft is safe on the ground and not in need of SAR.

Since most activations result from false alarms, resolution over the phone saves SAR assets for actual emergencies. If the AFRCC is unable to verify the aircraft is safe on the ground, it launches an SAR mission. This normally involves assigning the search to the USAF Auxiliary Civil Air Patrol and may include requesting assistance from the local SAR responders or law enforcement personnel.

The unique digital code of each 406 MHz beacon associates each beacon with a particular aircraft. The beacon registration contains information such as tail number, home airport, type and color of aircraft, and several emergency points of contact. This provides rapid access to flight plans and other vital information and speeds the search effort.

Pilots can supplement the use of electronic devices with signal mirrors, smoke, or prominent parachute canopy displays, which provide a good visual signal during daylight. The pilot can use a parachute canopy and case to lay out a prominent marker for searching aircraft. Flashlights and light beacons work well at night. Fires and flames from combustible material appear visible by night and provide smoke that may be seen during daylight. Signal flares can work during day or night. A whistle provides a good method for making a loud sound, but all audible signals including shouting and other noisemaking activities have limited range.

Navigation Equipment

Pilots use aviation charts during flight planning and to navigate during flight. Chart data can pinpoint the location when a pilot makes an off-airport landing. Sectional charts have a useful scale for most cross-country flights. GPS coordinates also help the ground crew equipped with a GPS receiver and appropriate charts and maps. Commercially available detailed GPS maps make navigation easier for the ground crew.

Medical Equipment

Medical kits routinely include bandages, medical tape, disinfectants, a tourniquet, matches, a knife or scissors, bug and snake repellent, and other useful items. Pilots should check that the kit contains medical items suitable for the current operating environment and replace any used or expired component. Glider occupants should have access to the kit after any emergency landing.

Stowage

Loose items may shift when encountering inflight turbulence, low-G maneuvers, or during a hard landing. Stowing equipment properly protects occupants and maintains integrity of all flight controls and glider system controls.

Parachute

Any parachute should be clean, dry, and stored in a cool place when not in use. Contaminants could reduce or destroy the integrity of the parachute material. The pilot has responsibility to ensure that the parachute meets any required FAA inspection criteria.

Chapter Summary

This chapter presents a variety of abnormal and emergency scenarios. If an emergency should occur during an aerotow, each pilot should act in a manner that allows the pilot of the other aircraft to achieve a safe outcome. Emergencies can also occur during a launch using a winch or vehicle. These emergencies can occur in a matter of seconds, which require split-second decision making. Various emergencies during a self-launch can happen, and the pilot should know how to handle the various emergency scenarios discussed in this chapter. Gliders often fly in steep banks and at slow speeds. Pilot error during these conditions can result in a spiral dive or a spin. Pilots should understand the difference between these two scenarios and know how to recover from each one. Various system and equipment malfunctions can lead to an off-airport landing. Pilots prepare for this eventuality by carrying emergency equipment and survival gear for the weather and terrain they could encounter.

Chapter 9: Glider Flight & Weather

Introduction

Glider pilots face a multitude of decisions, starting with the decision to take to the air. Pilots must determine if weather conditions are safe and if current conditions support a soaring flight. Gliders, being powered by gravity, are always sinking through the air. Therefore, glider pilots must seek air that rises faster than the sink rate of the glider to enable prolonged flight. Glider pilots refer to rising air as lift, not to be confused with the lift created by the wing. This chapter focuses on weather that commonly affects glider flights. However, all pilots should understand weather theory, weather hazards to aviation, and the technical details of aviation weather products. The Aviation Weather Handbook (FAA-H-8083-28) provides comprehensive information.

The Atmosphere

Without the atmosphere, wind, clouds, precipitation, and protection from solar radiation would not exist. The height of the atmosphere represents a small distance when compared to the 3,438 nautical mile radius of the earth. There is no specific upper limit to the atmosphere—it simply thins to a point where it fades away into space. The atmosphere below about 27 NM (164,000 feet) contains 99.9 percent of atmospheric mass. At that altitude, atmospheric density drops to approximately one-thousandth of its value at sea level. [*Figure 9-1*]

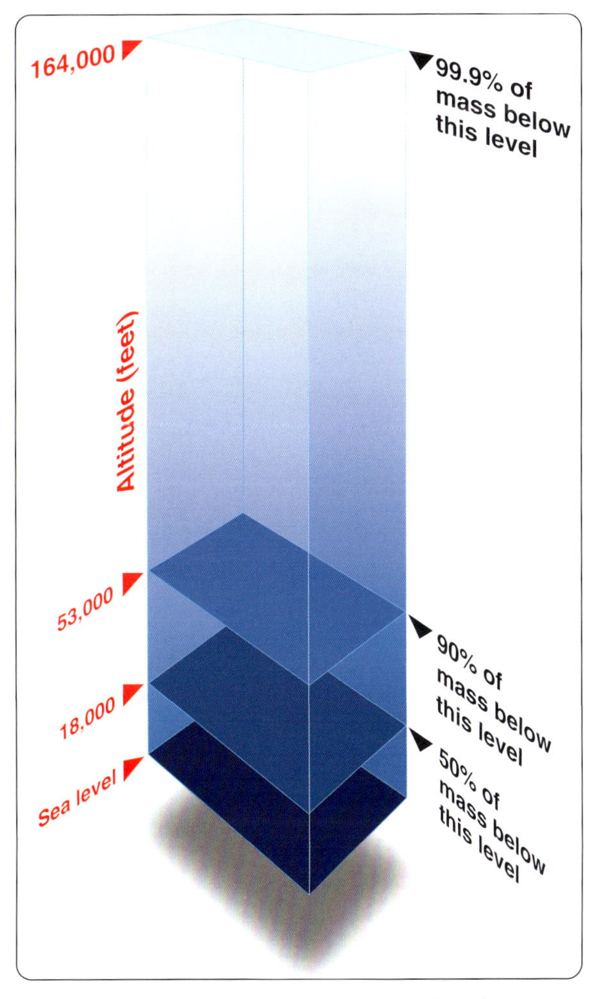

Figure 9-1. *Atmospheric mass by altitude.*

Composition

Two gases, nitrogen (N~2) and oxygen (O~2), comprise 99 percent of the volume of the total atmosphere on average. The remaining volume contains various trace gases and small amounts of water, ice, dust, and other particles. While the proportion of nitrogen to oxygen remains the same to approximately 260,000 feet, the amount of water vapor (H~2O) in the air can vary. For example, water vapor content above tropical areas and oceans accounts for up to 4 percent of the atmosphere by volume and displaces some nitrogen and oxygen gas. Conversely, the water vapor in the atmosphere over deserts and at high altitudes consists of much less than 1 percent of the total volume. [*Figure 9-2*]

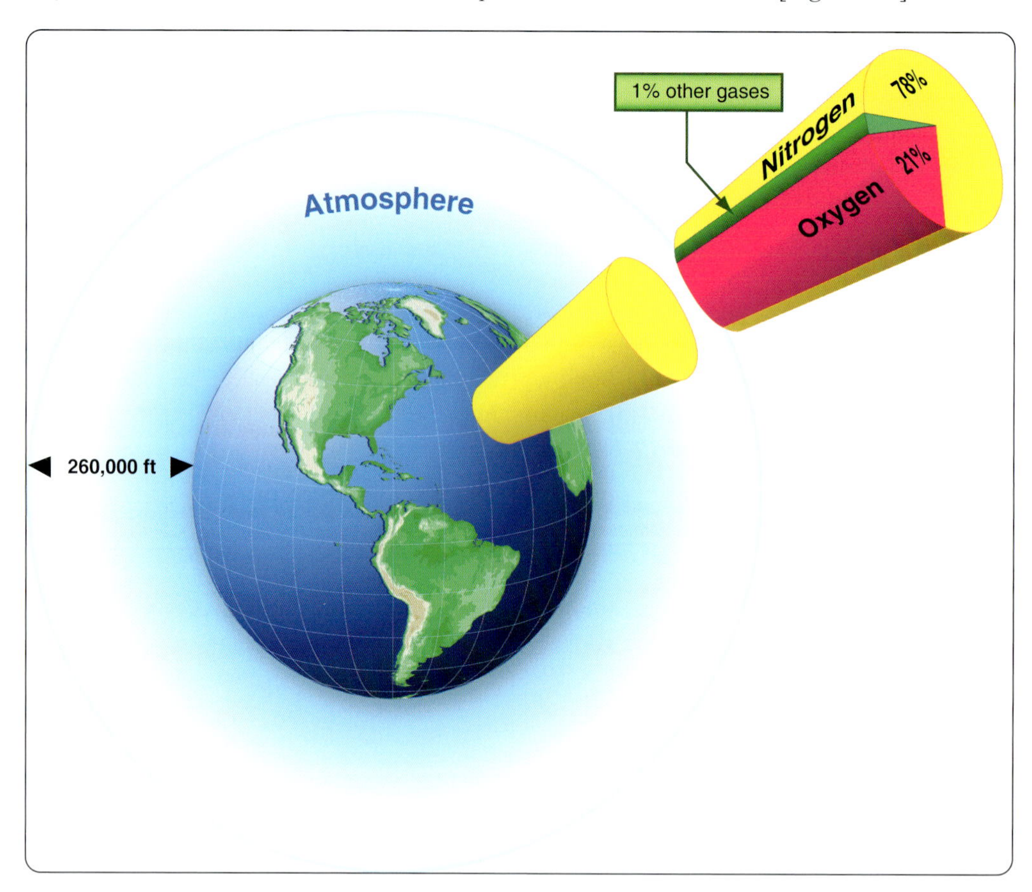

Figure 9-2. *The composition of the atmosphere.*

Although water vapor exists in the atmosphere in small amounts as compared to nitrogen and oxygen, it has a significant impact on weather. The additional physical states of water vapor as a liquid and solid contribute to the formation of clouds, precipitation, fog, and ice.

Atmospheric Measurements

Temperature, density, and pressure measurements provide information about the atmosphere. These variables change over time, and combined with vertical and horizontal differences, the measurements and trends comprise data for weather reports and forecasts.

Temperature

People often describe air temperature in terms of whether the air feels hot or cold. In aviation, quantitative measurements use the Celsius (°C) scale or the Fahrenheit (°F) scale. Temperature of the atmosphere depends on the average kinetic energy of molecules. Fast-moving molecules have high kinetic energy and higher temperatures. Conversely, slow-moving molecules have lower kinetic energy and lower temperatures.

Density

The density of any substance gives its mass per unit of volume. Low air density means a smaller number of air molecules (or less massive molecules on average) in a specified volume, while high air density means a greater number of air molecules (or more massive molecules on average) in a specified volume.

Pressure

Molecules in a volume of air not only possess a certain temperature and density, but they also collide with other gas molecules and push on nearby objects. The collisions result in measurable pressure or a force per unit of area. Since the force created by moving gas molecules acts equally in all directions, localized measurements of gas pressure using calibrated equipment should equal each other. Common units of measure for pressure include pounds per square inch (lb/in*2), inches of mercury ("Hg), and the equivalent units of millibars (mb) or hectopascals (hPA).

Ideal Gas Law

How do temperature, pressure, and density relate to each other? Dry air behaves almost like an ideal gas, meaning it obeys the ideal gas law P/DT = R, where P is pressure, D is density, T is temperature, and R is a constant. In general, the density, pressure, and temperature of an air parcel change predictably in accordance with the variables in the formula.

Standard Atmosphere

To provide a common reference used for temperature and pressure, scientists established the International Standard Atmosphere (ISA). This standard atmosphere uses a representative vertical distribution of temperature and pressure variables for pressure altimeter calibrations. The standard conditions are also a starting point for most aircraft performance data. At sea level, the standard atmosphere consists of a barometric pressure of 29.92 "Hg, 1,013.2 mb, or 14.7 lb/in*2, and a temperature of 15 °C or 59 °F.

Since temperature normally decreases with altitude at a predictable rate, a standard lapse rate calculation gives the standard temperature at various altitudes. Below 36,000 feet, the standard temperature lapse rate is 2 °C (3.5 °F) per 1,000 feet of altitude change. Pressure does not decrease linearly with altitude, but a 1 "Hg decrease for each 1,000 feet of increased altitude approximates the rate of pressure change below 10,000 feet. Pilots can use the standard lapse rates for flight planning purposes with the understanding that variations from standard conditions exist in the atmosphere. [*Figure 9-3*]

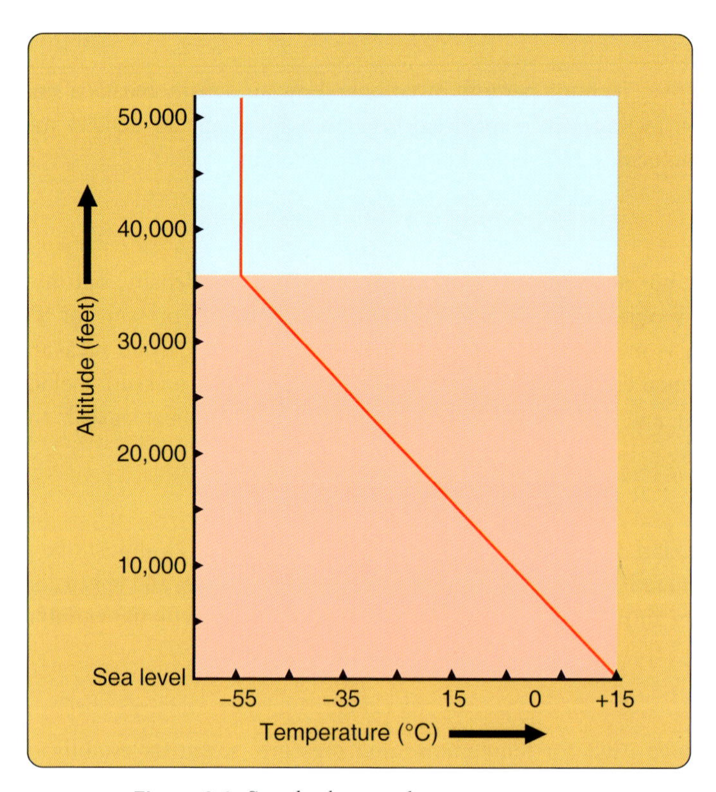

Figure 9-3. *Standard atmosphere temperatures.*

Layers of the Atmosphere

Scientists divide earth's atmosphere into five layers: troposphere, stratosphere, mesosphere, thermosphere, and exosphere. [*Figure 9-4*] The rate of change in temperature as altitude increases defines these layers. The lowest layer, called the troposphere, exhibits an average decrease in temperature from the earth's surface to about 36,000 feet above mean sea level (MSL). The troposphere extends to a higher altitude over the tropics and a lower altitude over the polar regions. It also varies seasonally, being higher during the summer and lower during the winter.

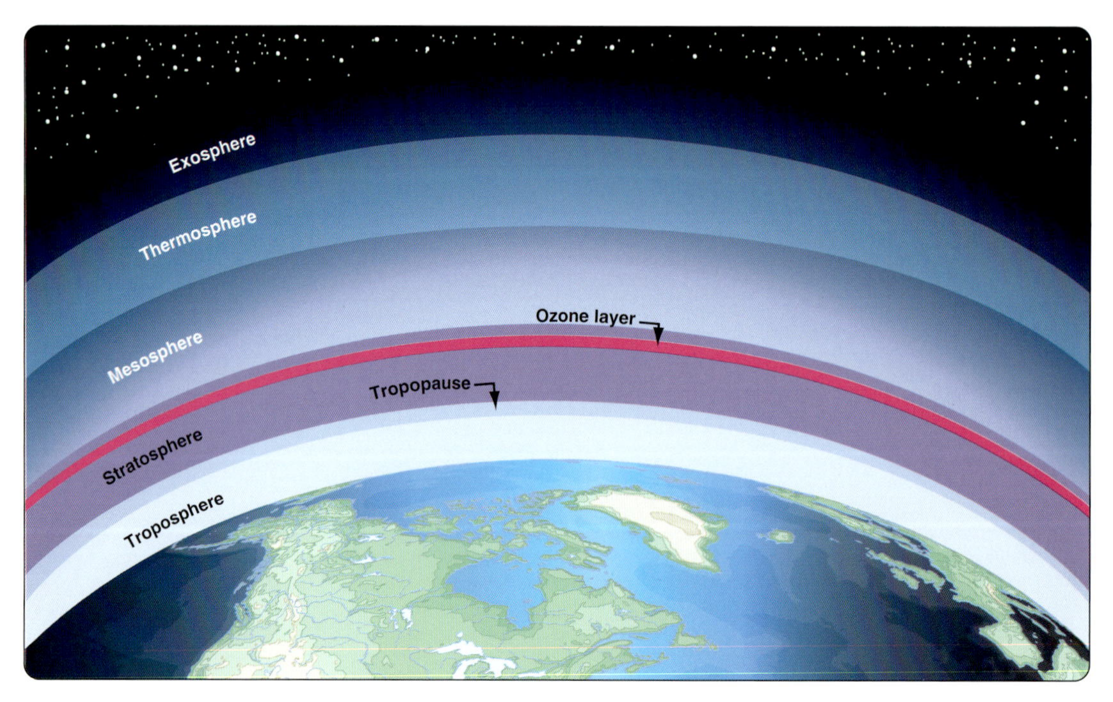

Figure 9-4. *Layers of the atmosphere.*

Almost all of earth's weather occurs in the troposphere as most of the water vapor and clouds are found in this layer. The lower part of the troposphere interacts with the land and sea surface, providing thermals, mountain waves, and sea-breeze fronts. Although temperatures decrease as altitudes increase in the troposphere, local areas where temperature increases with altitude (inversions) commonly occur.

The top of the troposphere or tropopause has a pressure of about ten percent of MSL pressure (0.1 atmosphere) and density drops to about 25 percent of its sea-level value. Temperature reaches its minimum value at the tropopause, approximately –55 °C (–67 °F). For pilots, this is an important part of the atmosphere because it is associated with a variety of weather phenomena, such as thunderstorm tops, clear air turbulence, and jet streams. The vertical limit of the tropopause varies with the height of the troposphere.

The tropopause separates the troposphere from the stratosphere. With increasing height in the stratosphere, the temperature tends to change very slowly at first. However, as altitude increases the temperature increases to approximately 0 °C (32 °F) reaching its maximum value at about 160,000 feet MSL. Unlike the troposphere in which the air moves freely both vertically and horizontally, the air within the stratosphere generally moves horizontally.

Gliders have reached into the lower stratosphere using mountain waves. At high altitudes, supplemental oxygen requirements become mandatory. Layers above the stratosphere have some interesting features that are normally not of importance to glider pilots. However, interested pilots may refer to any general text on weather or meteorology.

Scale of Weather Events

When preparing forecasts, meteorologists consider atmospheric circulation on different scales. To aid the forecasting of short- and long-term weather, various weather events have been organized into three broad categories or scales of circulation. The size and lifespan of the phenomena in each scale are roughly proportional, so that larger scales coincide with longer lifetimes. The term "microscale" refers to features with spatial dimensions of 0.1 to 1 NM, which last for seconds to minutes. An example is an individual thermal. The term "mesoscale" refers to the horizontal dimensions of 1 to 1,000 NM, which last minutes to weeks. Examples include mountain waves, sea breeze fronts, thunderstorms, and fronts. Research scientists break down the mesoscale into further subdivisions to better classify various phenomena. The term "macroscale" refers to the horizontal dimensions greater than 1,000 NM, which last weeks to months. These include the long waves in the general global circulation and the jet streams embedded within those waves. [*Figure 9-5*]

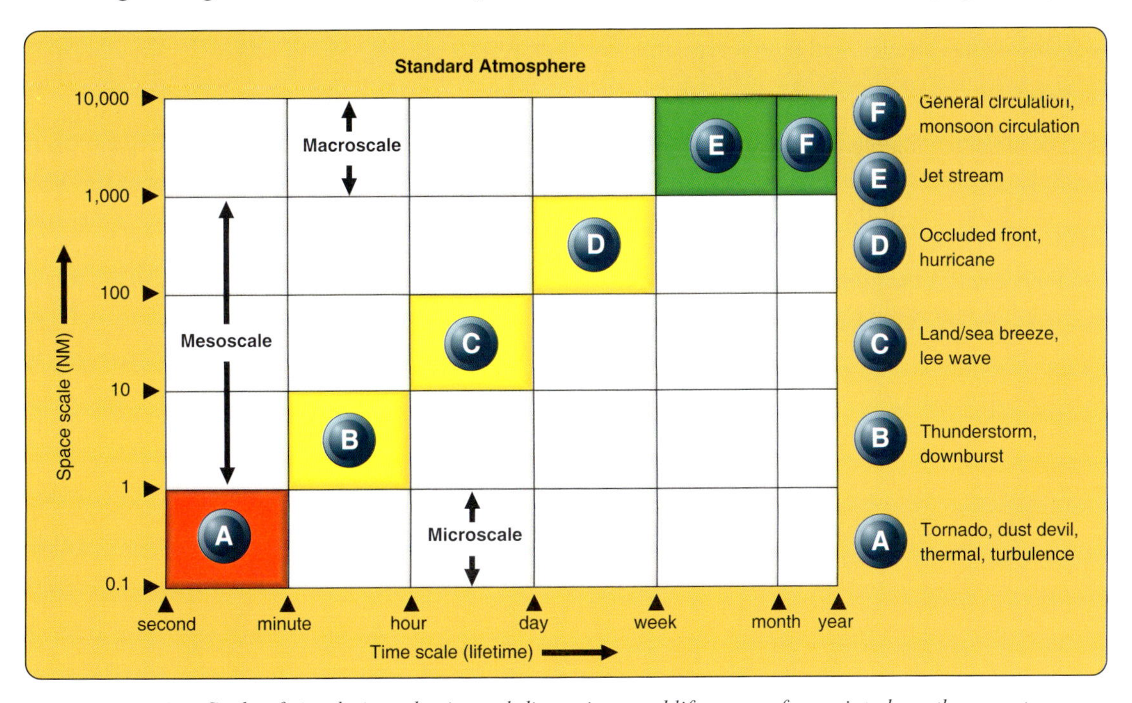

Figure 9-5. *Scale of circulation—horizontal dimensions and life spans of associated weather events.*

Smaller scale features are embedded in larger scale features. For instance, a microscale thermal may be just one of many thermals in a mesoscale convergence line, like a sea breeze front.

The sea breeze front may occur only under certain synoptic (i.e., simultaneous) conditions controlled by the macroscale circulations. The scales interact, with feedback from smaller to larger scales and vice versa, in ways not fully understood by atmospheric scientists. Generally, the behavior and evolution of macroscale features are more predictable, with forecast accuracy decreasing as scale diminishes. For instance, forecasts of up to a few days for major events, such as a trough with an associated cold front, have become increasingly accurate. However, no one can forecast the exact time and location of an individual thermal an hour ahead of time. Since most of the features of interest to soaring pilots lie in the smaller mesoscale and microscale ranges, prediction of gliding weather presents a significant challenge.

Pilots interpreting forecasts should begin with the macroscale, which identifies large-scale patterns that may produce good gliding conditions. This varies from site to site and depends on whether the goal is thermal, ridge, or wave soaring. Then, mesoscale features should be considered. This may include items such as the cloudiness and temperature structure of the air mass behind a cold front, as well as the amount of rain produced by the front. Developing an understanding of lift types and environments in which they form, can help a pilot predict local weather conditions that affect glider flights.

Thermals

Thermals are the most common form of rising air used to sustain glider flight. The paragraphs in this section explore topics related to thermals, including thermal structure, atmospheric stability, and air masses conducive to flight in thermals.

Thermal Shape & Structure

Convection describes a form of heat transfer involving the movement or flow of mass in a fluid (gas or liquid). Rising convection currents of air or "thermals" are one means by which the atmosphere transfers heat energy vertically. Thermals do not necessarily develop on a warm sunny day. Subtle differences in the atmosphere make the difference between a warm, sunny day with plenty of thermals and a warm, sunny day that produces no thermals. Glider pilots who understand the conditions that create thermals can use their own forecasting skills to predict thermal activity.

Two conceptual models exist for the structure of thermals: the bubble model and the column or plume model. These models simplify a complex and often turbulent phenomenon, so pilots should expect many exceptions and variations while flying in thermals. Many books, articles, and Internet resources provide further reading on this subject.

The bubble model describes an individual thermal resembling a vortex ring, with rising air in the middle and descending air on the sides. The air in the middle of the vortex ring rises faster than the entire thermal bubble. The model fits occasional reports from glider pilots. At times, two glider pilots in the same thermal will find different amounts of lift. For example, one glider may be at the top of the bubble climbing only slowly, while a lower glider climbs rapidly in the stronger part of the bubble below. [*Figure 9-6*] Often, a glider flying below another glider circling in a thermal can contact the same thermal and climb, even if the gliders are displaced vertically by 1,000 feet or more. This suggests the column or plume model of thermals is more common. [*Figure 9-7*]

Figure 9-6. *The bubble or vortex ring model of a thermal.*

Figure 9-7. *The column or plume model of a thermal.*

The applicability of the models may depend on the supply of warm air near the surface. Within a small, heated area, one single bubble may rise and take all the warmed surface air with it. On the other hand, if a large area becomes warm and one spot acts as the initial trigger, surrounding warm air can flow into the relative void left by the initial thermal. The in-rushing warm air follows the same path, creating a thermal column or plume. Since all the warmed air near the surface does not usually have the exact same temperature, a column could exist with a few or several imbedded bubbles. Individual bubbles within a thermal plume may merge, while at other times, two adjacent and distinct bubbles may exist side by side.

Whether behaving as a bubble or column, the air in the middle of the thermal rises faster than the air near the edges of the thermal. A horizontal slice through an idealized thermal provides a bull's-eye pattern; however, cross sections of real-world thermals exhibit dissymmetry. [*Figure 9-8*]

Figure 9-8. *Cross-section through a thermal. Darker green is stronger lift; red is sink.*

A typical thermal cross-section has a diameter of 500–1,000 feet, though the size can vary considerably. Typically, due to mixing with the surrounding air, thermals expand as they rise. Thus, the thermal column may resemble a cone, with the narrowest part near the ground. Thermal plumes also tilt in a steady wind and can distort in the presence of vertical shear. In strong vertical shear, thermals can become very turbulent or become completely broken apart. *Figure 9-9* shows a schematic of a thermal lifecycle in windshear.

Figure 9-9. *Lifecycle of a typical thermal with cumulus cloud.*

A stable atmosphere hinders vertical motion, while an unstable atmosphere promotes vertical motion. A certain amount of atmospheric instability supports development of thermals. However, moist air and strong atmospheric instability may lead to thunderstorm formation. Thus, an understanding of atmospheric stability promotes recognition of favorable flight conditions as well as recognition of weather hazards and associated risk.

When discussing atmospheric stability, a layer of air in the atmosphere represents the dynamic system and a parcel of air represents the displaced element. In a stable dynamic system, a displaced element returns to its original position. In an unstable dynamic system, a displaced element continues to move farther away from its original position. In a neutral dynamic system, a displaced element neither returns to nor moves farther away from its original position.

A parcel of dry or unsaturated air moving upward in the atmosphere expands and cools as it rises due to decreasing pressure. By contrast, a descending parcel of dry or unsaturated air compresses and warms due to increasing pressure. When no transfer of heat between the displaced parcel and the surrounding ambient air occurs, the process is called adiabatic. During this adiabatic process, a rising unsaturated parcel cools at a lapse rate of 3 °C (5.4 °F) per 1,000 feet. This dry adiabatic lapse rate (DALR) approximates what happens in nature although some mixing of air occurs as thermals rise.

Figure 9-10 below demonstrates one means to predict whether a layer of the atmosphere will function like a stable, unstable, or neutral dynamic system. Panels A and B represent two scenarios with the same temperature of 20 °C at the surface, but with different air layer temperatures at 3,000 feet above ground level (AGL). Using the DALR for both scenarios, a parcel of 20 °C air that lifts from the surface cools to 11 °C by the time it reaches 3,000 feet AGL. In scenario A, the lifted parcel is still warmer than the surrounding air and will continue to rise by convection—a condition of instability that could produce a good thermal. In scenario B, the lifted parcel at 3,000 feet AGL has cooled to a lower temperature than the surrounding air and will descend. In this case, the layer exhibits system stability. See the Aviation Weather Handbook (FAA-H-8083-28) for more information on atmospheric stability.

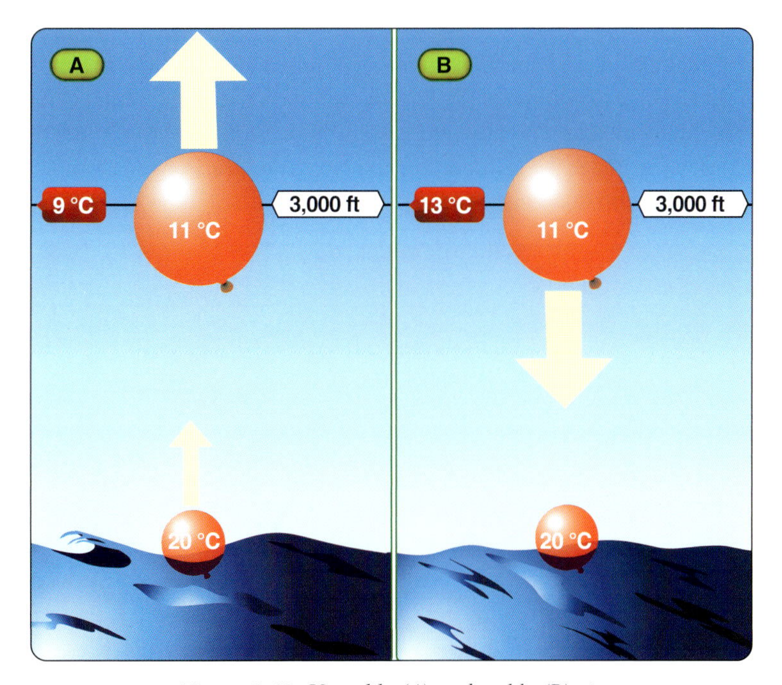

Figure 9-10. *Unstable (A) and stable (B) air.*

Changing the values in *Figure 9-10* illustrates factors that would affect atmospheric stability. A stable layer can turn unstable in one of two ways. In scenario B, if the surface temperature warms by more than 2 °C (to greater than 22 °C), or if the air at 3,000 feet cools by more than 2 °C (to less than 11 °C), the atmospheric layer to 3,000 feet becomes unstable. Warming of lower layers or cooling of higher layers of the atmosphere with no other changes reduces stability and leads to a better environment for thermals. If the layer aloft and at the surface warm or cool by the same amount, then the stability of the layer remains unchanged. The layer exhibits greater stability if the air temperature aloft remains constant, but the surface air cools.

An inversion occurs when the troposphere warms as altitude increases. Inversions can occur at different altitudes and vary in strength. In strong inversions, the temperature can rise as much as 10 °C in a few hundred feet of altitude gain. Along with trapping haze or pollution below, an inversion also effectively caps any thermal activity.

Although moisture in the form of water vapor makes up a small percentage of the atmosphere, it can affect the temperature lapse rate of a rising parcel of air. A rising parcel of air cools at the DALR until it reaches its dewpoint, at which time the water vapor in the parcel begins to condense. The condensation process releases heat, referred to as latent heat, within the rising parcel of air. Therefore, a rising parcel of saturated air cools at a rate lower than the DALR. This saturated adiabatic lapse rate (SALR), varies substantially with altitude. At lower altitudes, it approximates of 1.2 °C per 1,000 feet, whereas at middle altitudes it increases to 2.2 °C per 1,000 feet. Above approximately 30,000 feet, little water vapor exists to condense, and the SALR approaches the DALR.

Air Masses Conducive to Thermal Soaring

Generally, the best air masses for thermals are those with cool air aloft, with conditions dry enough to allow the sun's heating radiation to warm the surface and to limit extensive formation of cumulus clouds. This cool air aloft can originate after passage of a Pacific cold front in the Western United States or from polar continental regions such as interior Canada in the Eastern United States. In both cases, high pressure building into the region often includes an inversion aloft, which keeps cumulus from growing into rain showers or thunderstorms. However, as the high pressure builds after the second or third day, the inversion often lowers to the point that thermals suitable for gliding no longer form. This can lead to warm and sunny, but very stable conditions. Fronts that arrive too close together can also cause poor postfrontal soaring, as high clouds from the next front keep the surface from warming enough. Very shallow cold fronts from the northeast direction (with cold air only one- or two- thousand feet deep) often have a stabilizing effect along the plains directly east of the Rocky Mountains. This is due to cool low-level air undercutting warmer air aloft flowing from the west.

In the desert southwest, the Great Basin, and intermountain west, good summertime thermals often result from intense heating from below, even in the absence of cooling aloft. This dry air mass with continental origins produces cumulus bases 10,000 feet AGL or higher. At times, this air spreads into eastern New Mexico and western Texas. Later in the summer, however, some of these regions come under the influence of the North American Monsoon, which can lead to widespread and daily late morning or early afternoon thundershowers. [*Figure 9-11*]

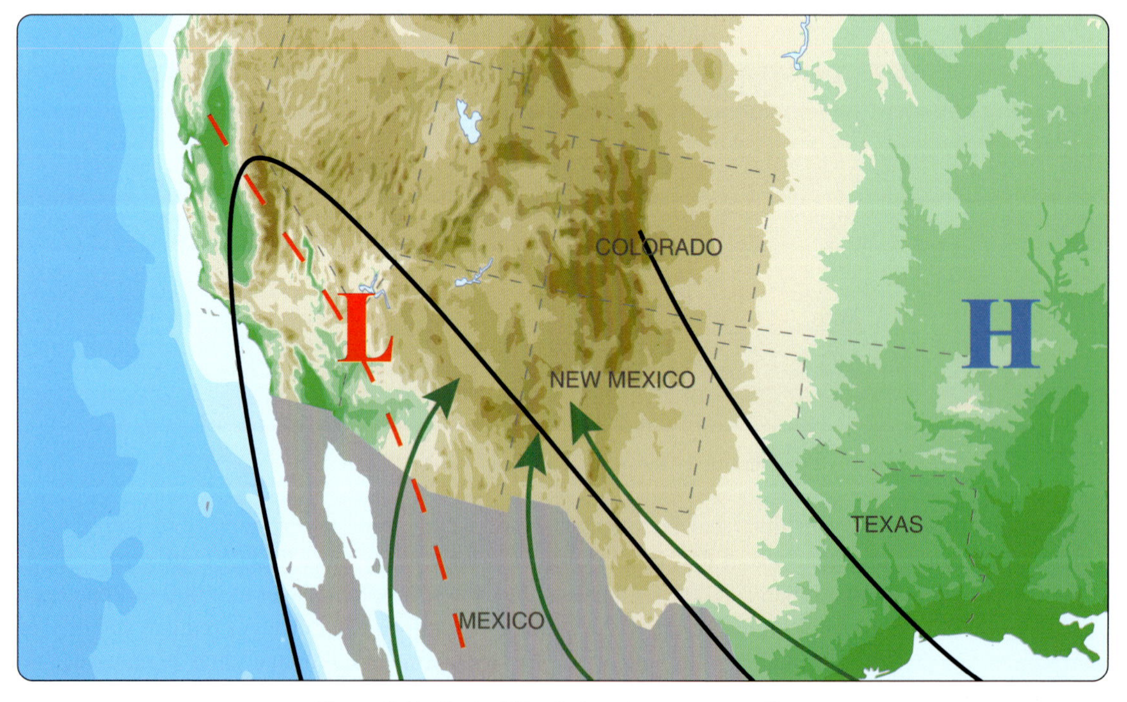

Figure 9-11. *Typical North American monsoon flow.*

Cloud Streets

Cumulus clouds often appear randomly distributed across the sky, especially over relatively flat terrain. Under the right conditions, however, cumulus clouds can align in a long band, called a cloud street. An individual cloud street can extend 50 miles or more while an entire field of cloud streets can extend hundreds of miles. The spacing between streets is typically three times the height of the clouds. Cloud streets align parallel to the wind direction and indicate the pattern of rising and descending air. Glider pilots can often fly many miles with little or no circling, sometimes achieving glide ratios far exceeding the still-air value by flying near and parallel to the clouds while avoiding the space between the clouds. Thus, cloud streets mark an ideal location for flying a downwind cross-country flight.

A cross-section of an idealized cloud street formation illustrates a distinct circulation, with updrafts under the clouds and downdrafts in between. [*Figure 9-12*] Due to the circulation, sink between streets may be stronger than typically found the same distance away from random cumulus clouds.

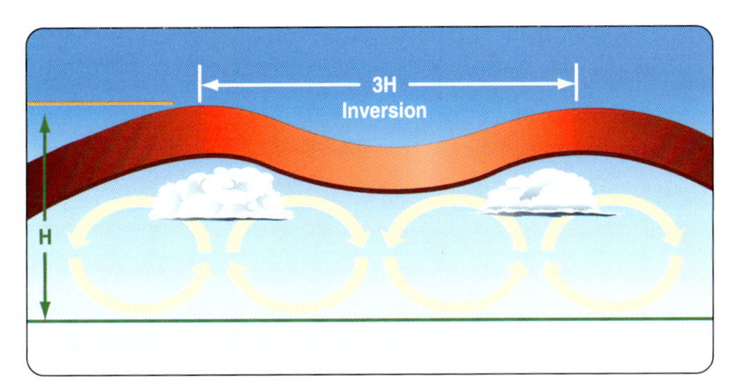

Figure 9-12. *Circulation across a cloud street.*

Cloud streets usually occur over land with cold air outbreaks, for instance, following a cold front. Brisk surface winds and a wind direction remaining nearly constant up to the cloud base are favorable cloud street conditions. Windspeed should increase by 10 to 20 knots between the surface and cloud base, with a maximum somewhere in the middle of or near the top of the convective layer. Thermals should be capped by a notable inversion or stable layer.

Thermal streets, with a circulation like *Figure 9-12*, may exist without cumulus clouds. Without clouds as markers, use of such streets becomes difficult. A glider pilot flying upwind or downwind in consistent sink should alter course crosswind to avoid inadvertently flying along a line of sink between thermal streets that may exist.

Cloud Streets

Figure 9-13 shows a wavelike form for the inversion capping the cumulus clouds. If winds above the inversion run perpendicular to the cloud streets and increase at 10 knots per 5,000 feet or more, cloud street waves can form in the stable air above. Though usually relatively weak, thermal waves can produce lift of 100 to 500 fpm and allow smooth flight along streets above the cloud base.

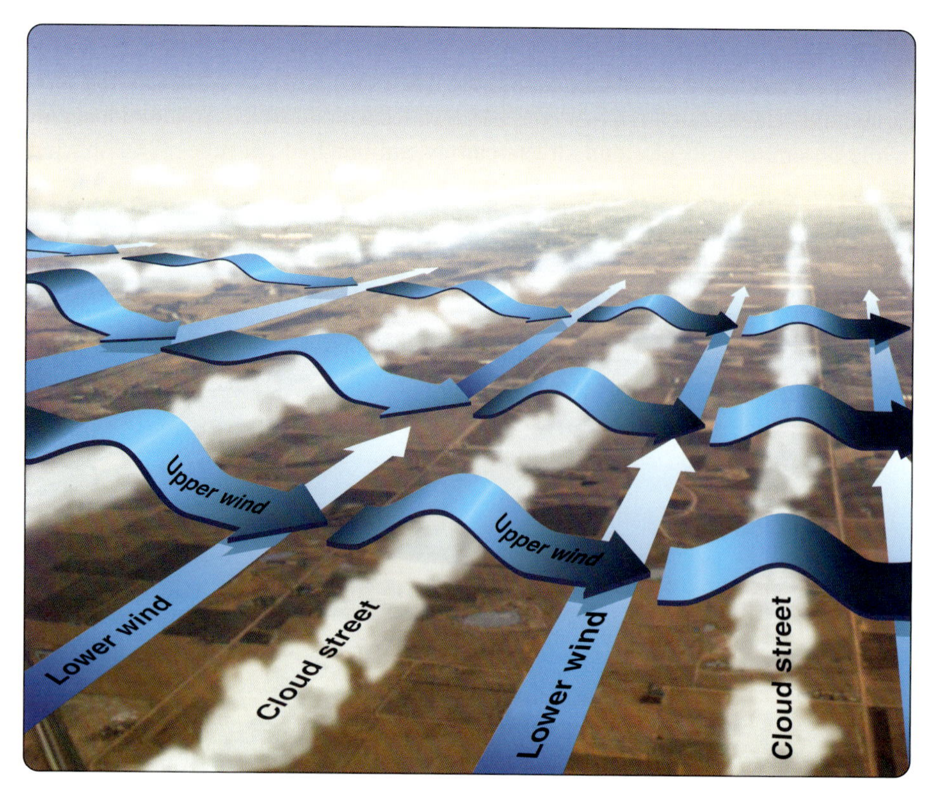

Figure 9-13. *Cloud street thermal wave.*

So-called cumulus waves may exist where the cumulus clouds do not organize in streets. Cumulus waves require a capping inversion or stable layer and increasing wind above cumulus clouds. However, directional shear is not necessary. Cumulus waves may also be short lived, and difficult to work for any length of time. An exception occurs when the cumulus cloud anchors to some feature, such as a ridge line or short mountain range. As a final note, thermal waves can also form without clouds. Without clouds, the possible influence of a ridge or mountain in creating the wave lift becomes difficult to determine.

Thunderstorms

Forecasters sometimes use the term "deep convection" to refer to convection that rises to high levels, which usually means thunderstorms, and they use the term "convective activity" to refer to thunderstorms. The tremendous amount of energy associated with cumulonimbus clouds stems from the release of latent heat as condensation occurs within the growing cloud. While an unstable atmosphere can provide great conditions for thermal formation, an atmosphere that is moist and unstable can create cumulonimbus (Cb) or thunderclouds. Cb clouds are the recognized standard marker of thunderstorms. When Cb builds sufficiently, it changes from rainstorm to thunderstorm status. Not all precipitating, large cumulus formations are accompanied by lightning and thunder, but the presence of these clouds indicates that hazardous conditions exist or may intensify.

Thunderstorms can occur any time of year, though they are more common during the spring and summer seasons. They can occur anywhere in the continental United States but are not common along the immediate West Coast, where an average of only about one per year occurs over a given location. During the summer months, the desert southwest locations, extending northeastward into the Rocky Mountains and adjacent Great Plains, experience an average of 30 to 40 thunderstorms annually. Additionally, in the southeastern United States, especially Florida, between 30 and 50 thunderstorms occur in an average year per location. [*Figure 9-14*] Thunderstorms in the cool seasons usually occur in conjunction with some forcing mechanism, such as a fast- moving cold front or a strong upper-level trough.

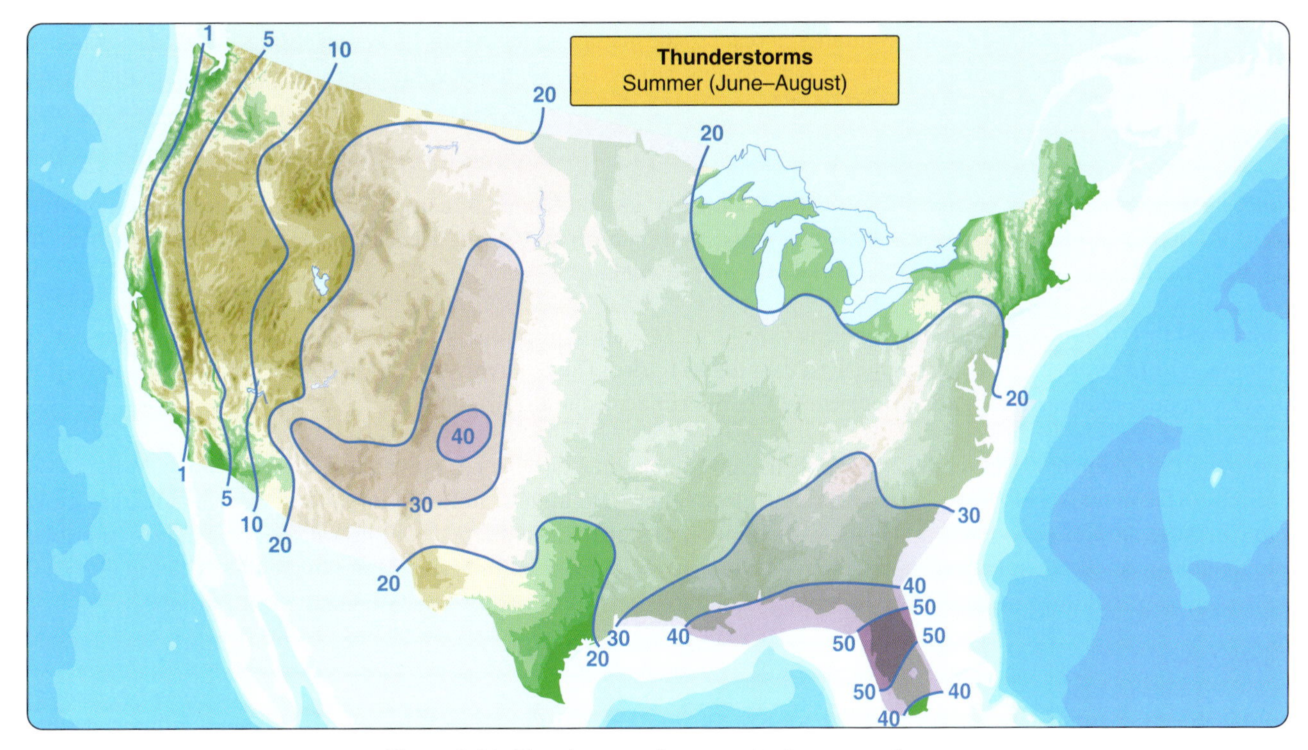

Figure 9-14. *Thunderstorm frequency in the summertime.*

The lifecycle of an airmass or ordinary thunderstorm consists of three main stages: cumulus, mature, and dissipating. The term "ordinary" describes the type of thunderstorm consisting of a single Cb, since individual thunderstorms can develop in a uniform large-scale air mass. The entire lifecycle of an ordinary thunderstorm takes on the order of an hour, though a remnant cloud from the dissipated Cb can last substantially longer.

The cumulus stage of a thunderstorm is characterized by a cumulus cloud growing to a towering cumulus (Tcu). As air rises within the cloud during this stage, the intensity of the updraft increases, and the cloud base broadens to a few miles in diameter. [*Figure 9-15*] As the cloud increases in size, the strong updraft in the middle of the cloud does not entrain or carry along dryer air surrounding the cloud, and general downward motion of air around the Tcu may suppress other smaller cumulus in the vicinity. Toward the end of the cumulus stage, downdrafts and precipitation begin to form within the cloud. On some days, small cumulus can be around for hours, before Tcu form, while on other days, unstable air allows any cumulus cloud that forms to rapidly transform into a Tcu.

Figure 9-15. *A cumulus cloud becoming a towering cumulus.*

As the development of a thunderstorm continues, it reaches the mature stage. By this time, downdrafts known as downbursts or microbursts reach the ground and spread out creating strong and sometimes damaging surface winds. Glider pilots should avoid flight toward these downdrafts and their associated windshear, as the glider might lose the capability to reach a selected landing area.

Note: hazardous windshear creates a significant risk for any ground launch or aerotow, and these operations should not occur if the pilot suspects a windshear encounter associated with Cb may occur. Depending on the size of windshear, the towplane could be in a tailwind while the glider is in a headwind.

Pilots need to watch dissipating thunderstorms closely for new dark, firm bases that indicate formation of a new cell. In addition, outflow from a Cb may cause the air it encounters to rise. The relatively cool air in the outflow can provide nearby air with a boost, leading to formation of a nearby Cb, not connected to the original Cb.

The risk from thunderstorms involves several hazards, including turbulence, strong updrafts and downdrafts, strong shifting surface winds, hail, icing, poor visibility or low ceilings, lightning, and even tornadoes. Once a cloud has grown into Cb, these hazards may develop, with or without obvious signs. Since thermal soaring weather can rapidly deteriorate into thunderstorm weather, understanding the risk associated with these hazards should prompt a glider pilot to remain on the ground or avoid them in the air. The following paragraphs provide more information about these hazards.

Turbulence

A pilot should never intentionally fly into a Cb since severe or extreme turbulence leading to structural failure can occur anywhere within the thunderstorm. Violent updrafts can be followed a second or two later by violent downdrafts, with occasional side gusts. Severe turbulence commonly occurs close to the storm, and moderate turbulence may exist within several miles of a thunderstorm. Below the base of the Cb, moderate to severe turbulence can occur along the boundary between the cool outflow and warm air feeding the Cb. Pilots should expect turbulence near the surface from a gust front as cool outflow spreads from the storm. Unpredictable smaller scale turbulent gusts can occur anywhere near a thunderstorm and avoiding the gust front does not guarantee avoidance of severe turbulence.

Updrafts and Downdrafts

Large and strong updrafts and downdrafts accompany thunderstorms in the mature stage. Updrafts feeding the Cb from under the base can exceed 1,000 fpm. Near the cloud base, a pilot may have difficulty determining the distance to the edge of the cloud, and strong updrafts could suck a glider into the storm cloud. During the late cumulus and early mature stage, updrafts feeding the cloud can cover many square miles. As the storm enters its mature stage, downbursts or microbursts can occur even without very heavy precipitation present. Downbursts can also cover many square miles with descending air of 2,000 fpm or more. A pilot flying under a forming downburst, which may not be visible, could encounter sink of 3,000 fpm or greater in extreme cases. Such a downburst encountered at pattern altitude can cut the normal time available to the pilot for executing an approach. While a normal pattern from 800 feet AGL to the ground could span 3 minutes, contact with the ground occurs in 19 seconds in 2,500 fpm sink.

Outflow Winds

When a downburst or microburst hits the ground, the downdraft spreads out, leading to strong surface winds, known as thunderstorm outflow. Typically, the winds strike quickly and give little warning of their approach. While flying, pilots should keep a sharp lookout between any storm and the intended landing spot for signs of a wind shift. Blowing dust, smoke, or wind streaks on a lake caused by wind from the storm may indicate a rapidly approaching gust front. Thunderstorm outflow winds usually travel at speeds of 20 to 40 knots for a period of 5 to 10 minutes before diminishing. However, winds can easily exceed 60 knots, and in some cases, with a slow- moving thunderstorm, strong winds can last substantially longer. Although damaging outflow winds usually do not extend more than 5 or 10 miles from the Cb, winds of 20 or 30 knots can extend 50 miles or more from large thunderstorms.

Hail

Hail associated with any thunderstorm can exist within part of the main rain shaft. Hail can also occur many miles from the main rain shaft, especially under the thunderstorm anvil. Pea-sized hail does not usually damage a glider, but the large hail associated with a severe storm can dent metal gliders or damage the gelcoat on composite gliders, both in the air or on the ground.

Icing

Icing can create a significant problem within a cloud or where visible moisture exists, especially at levels where the outside temperature is approximately -10 °C. Under these conditions, supercooled water droplets (water existing in a liquid state 0 °C and below) can rapidly freeze upon contact with wings and other glider surfaces. Early precipitation below a cloud base from a developing storm may be difficult to see. At times, precipitation can even be falling through an updraft feeding the cloud. Snow, graupel (a snow/frozen water combination), or ice pellets falling from the forming storm above can stick to the leading edge of the wing and degrade performance. Ice can form on the canopy and interfere with the pilot's forward vision.

Low Visibility

Poor visibility due to precipitation or low ceilings as air below a thunderstorm cools creates another concern for the pilot. Even light or moderate precipitation can reduce visibility dramatically. Often, under a precipitating Cb, confusion distinguishing between the precipitation and cloud can occur.

Lightning

Lightning strikes are completely unpredictable. Lightning in a thunderstorm may occur within one cloud, jump from one cloud to another if a nearby storm exists, or jump from cloud to ground. Some strikes emanate from the side of the Cb and travel horizontally for miles before turning abruptly toward the ground. Inflight damage to gliders has included burnt control cables and blown-off canopies. In some cases, strikes have caused little more than mild shock and cosmetic damage. At the other extreme, a composite training glider in Great Britain suffered a strike that caused complete destruction of one wing; fortunately, both pilots parachuted to safety. In that case, the glider was two or three miles from the thunderstorm. Pilots and ground personnel should avoid ground launching, especially with a metal cable, with a thunderstorm in the vicinity.

Tornadoes

Severe thunderstorms can sometimes spawn tornadoes a few hundred to a few thousand feet across with winds that can exceed 200 mph. Tornadoes that do not reach the ground are called funnel clouds. Pilots should consider any tornado watch or warning information and remain well clear of funnel clouds and tornadoes.

Weather for Slope Soaring

Wind deflects horizontally, vertically, or some combination of the two when it encounters topography. Slope or ridge soaring relies on updrafts produced by the mechanical lifting of air as it encounters the upwind slope of a hill, ridge, or mountain.

Individual or isolated hills tend not to produce slope lift because the wind deflects around the hill, rather than over it. A somewhat broader hill with a windward face of approximately a mile, might produce some slope lift over a small area. The best ridges for slope soaring span at least a few miles. While ridges only 100 or 200 feet high can produce slope lift, the pilot might not find sufficient rising air to climb above the top of the ridge.

Slope lift can extend to a maximum of two or three times the ridge height. [*Figure 9-16*] Generally, the higher the ridge extends above the adjacent valley, the higher the glider pilot can climb. The pilot should maintain a safe maneuvering altitude and sufficient altitude to land if necessary. Pilots should consider flight between 500 to 1,000 feet above the adjacent valley as a minimum safe height.

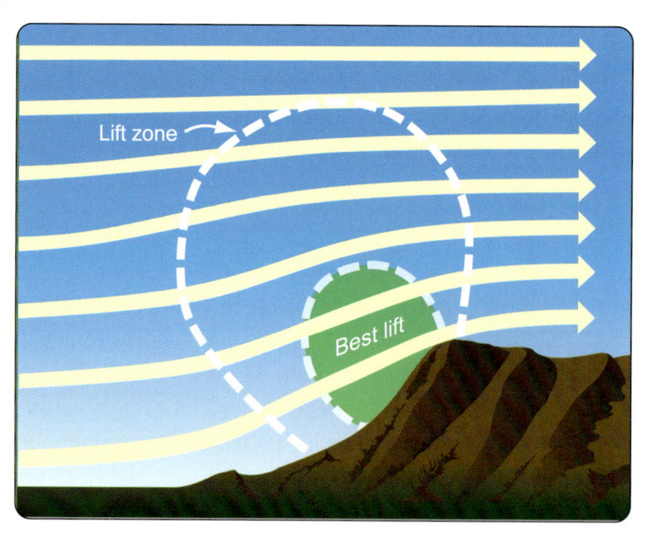

Figure 9-16. *Lift zone for slope soaring.*

Depending on the slope, windspeed of 10 to 15 knots blowing nearly perpendicular to the ridge produces usable updrafts. However, wind directions up to 30° or 40° from perpendicular may still produce slope lift. High ridges may have little or no wind along the lower slopes, but the upper parts of the ridge may be in winds strong enough to produce slope lift, creating a vertical wind shear.

The area of best lift varies with height. Below the ridge crest, the best slope lift is found within a few hundred feet of the ridge, depending on the slope and wind strength. Very steep ridges require extra speed and caution since eddies and turbulence can form on the upwind side. Above the ridge crest, the best lift usually occurs further upwind from the ridge as the glider gains altitude. [*Figure 9-17*]

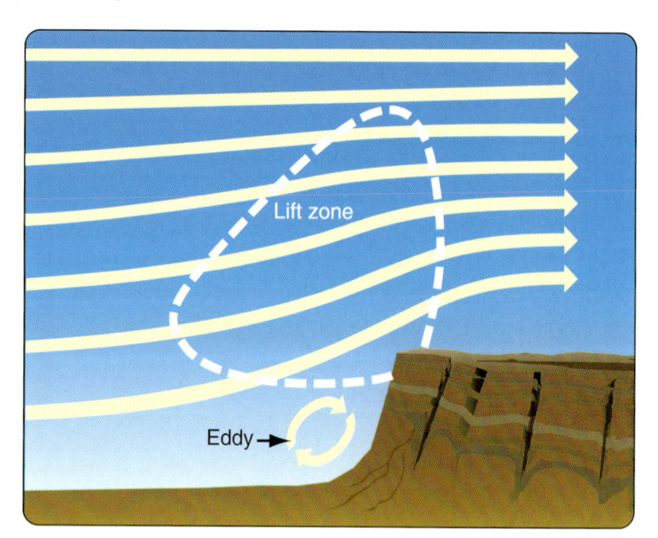

Figure 9-17. *Slope lift and eddy with near-vertical slope.*

An ideal ridge has a slope on the order of 1 to 4. For each four feet of horizontal travel, the terrain increases vertically by 1 foot. Shallower slopes do not create a vertical wind component strong enough to compensate for the glider's sink rate. Very steep, almost vertical slopes create lift, but may also produce turbulent eddies along the lower slope or anywhere close to the ridge itself. In such cases, only the upper part of the slope may produce updrafts, although steeper slopes allow a quick escape to the adjacent valley.

Slope lift in stable air can be very smooth, enabling safe soaring close to the terrain. In unstable air, thermals may flow up the slope. Depending on thermal strength and windspeed, the thermal may rise well above the ridge top, or it may drift into the lee downdraft and break apart. Downdrafts on the sides of thermals can easily cancel the slope lift, and require extra speed and caution, especially below the ridge crest near the terrain. The combination of unstable air and strong winds can make slope soaring unpleasant or even dangerous for an inexperienced glider pilot.

Upwind terrain can block the low-level wind flow of an otherwise promising ridge. Additionally, any waves produced by an upwind ridge or mountain can enhance or destroy ridge lift downstream, depending on the interference pattern of the waves. Locally, the downdraft from a thermal just upwind of a ridge can cancel slope lift for a short distance. The pilot should assume any slope lift could vanish and have plans for that eventuality.

While the flow deflects upward on the windward side of a ridge, it deflects downward on the lee side. [Figure 9- 18] This downdraft can reach 2,000 fpm or more near a steep ridge in strong winds (panel A). Flat-topped ridges offer little refuge since sink and turbulence can combine to make flight above the flat challenging or impossible (panel B). Finally, an uneven upwind slope with ledges or "steps" requires extra caution since small-scale eddies, turbulence, or sink can form (panel C). If crossing ridges in windy conditions, the pilot should plan for heavy sink on the lee side and stick to a set of personal minimum altitudes for crossing.

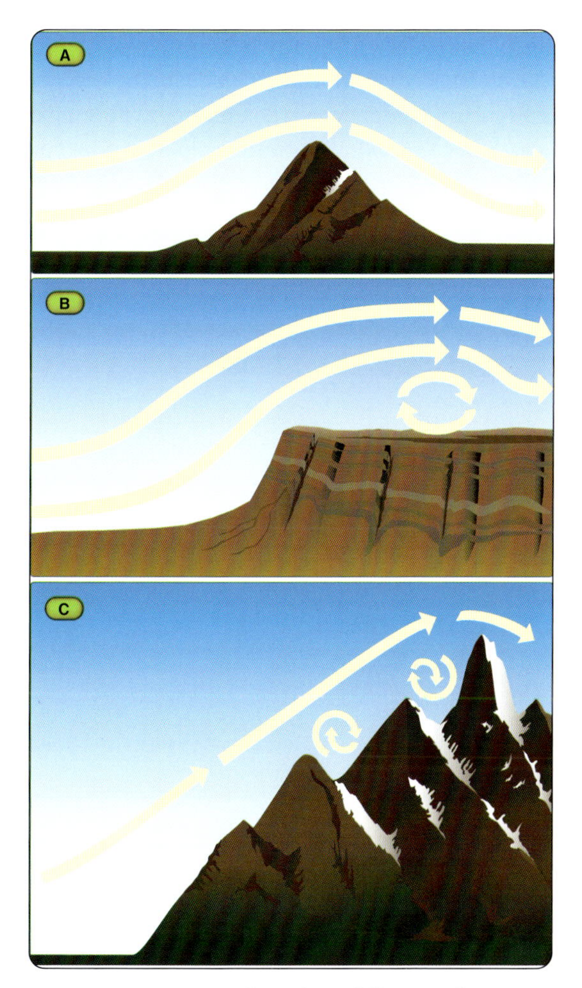

Figure 9-18. *Airflow along different ridges.*

Three-dimensional effects are important as well. For instance, a ridge with cusps or bowls may produce better lift on the upwind-facing edge of a bowl if the wind is at an angle from the ridge. However, the pilot may encounter sink on the lee side of a bowl edge. [*Figure 9-19*]

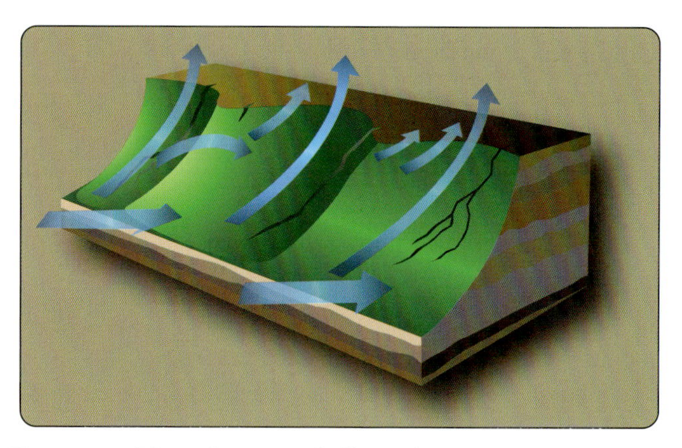

Figure 9-19. *Three-dimensional effects of oblique winds and bowls.*

Moist air rising from contact with a slope that cools sufficiently may form a so-called cap cloud. [*Figure 9-20*] The cloud may form above the ridge, and if the air moistens over time, the cloud slowly lowers onto the ridge and down the upwind slope, limiting the usable height of the slope lift. Under certain conditions, a morning cap cloud may rise as the day warms, then slowly lower again as the day cools. Since the updraft forms the cloud, a pilot could climb into the cap cloud and lose outside visual references.

Figure 9-20. *Cap cloud.*

Mountain Waves

Wind blowing over mountains or ridges can produce waves, the most powerful of which have lifted gliders to above 49,000 feet in the United States. With strong and widespread winds aloft and stable atmosphere, mountain waves can extend downwind along the entire length of a mountain range. Pilots refer to these waves as mountain waves, lee waves, mountain lee waves, or standing waves. Pilots have achieved multi-leg flights of over 2,000 kilometers in mountain waves. Note that for brief periods in some parts of the world, mountain waves reach into the stratosphere and receive a boost from the polar vortex. Pilots in an experimental pressurized glider reached altitudes above 76,000 feet in 2018 using this phenomenon in Argentina.

Water flowing in a stream or small river illustrates mountain wave formation. A submerged rock causes ripples (waves) in the water downstream, which slowly dampen out. In the case of mountain waves, the airflow over the mountain displaces a parcel of air from its equilibrium level. Since the atmosphere contains variations in the stability profile, wind blowing over a mountain does not always produce downstream waves. Anyone seeking more information regarding mountain wave formation than presented in this chapter should consult the Aviation Weather Handbook (FAA-H-8083-28).

Mountain waves differ fundamentally from slope lift. Slope soaring occurs on the upwind side of a ridge or mountain, while mountain wave soaring occurs on the downwind side. Mountain waves can tilt upward with height, and at times near the top of the wave, the glider pilot may be almost directly over the mountain or ridge that produced the wave.

Isolated small hills or conical mountains do not form classic lee waves. In some cases, they do form waves emanating at an angle to the wind flow like water waves created by the wake of a ship. A single peak may require only a mile or two in the dimension perpendicular to the wind for high-amplitude lee waves to form, though wave lift produced this way occurs in a relatively small area.

Mechanism for Wave Formation

Stable air can support wave formation when a disturbance causes vertical motion of the air and wind causes horizontal displacement. As illustrated in *Figure 9-21* at dashed line 1, the dry unsaturated parcel (depicted by the red dot) sits at rest at its equilibrium level. After upward displacement of the parcel (dashed line 2), the lifted parcel, now cooler than the surrounding air, accelerates downward toward its equilibrium level. It overshoots the level due to momentum and keeps going down. Dashed line 3 shows that the parcel, now warmer than the surrounding air and at a lower altitude than at dashed line 1, moves upward again. The process continues with the motion eventually damping out. The number of oscillations depends on the initial parcel displacement and the stability of the air. In the lower part of the figure, wind has been added, illustrating the wave pattern the parcel makes as it oscillates vertically. If there were no wind, a vertically displaced parcel would just oscillate up and down, while damping at one spot over the ground.

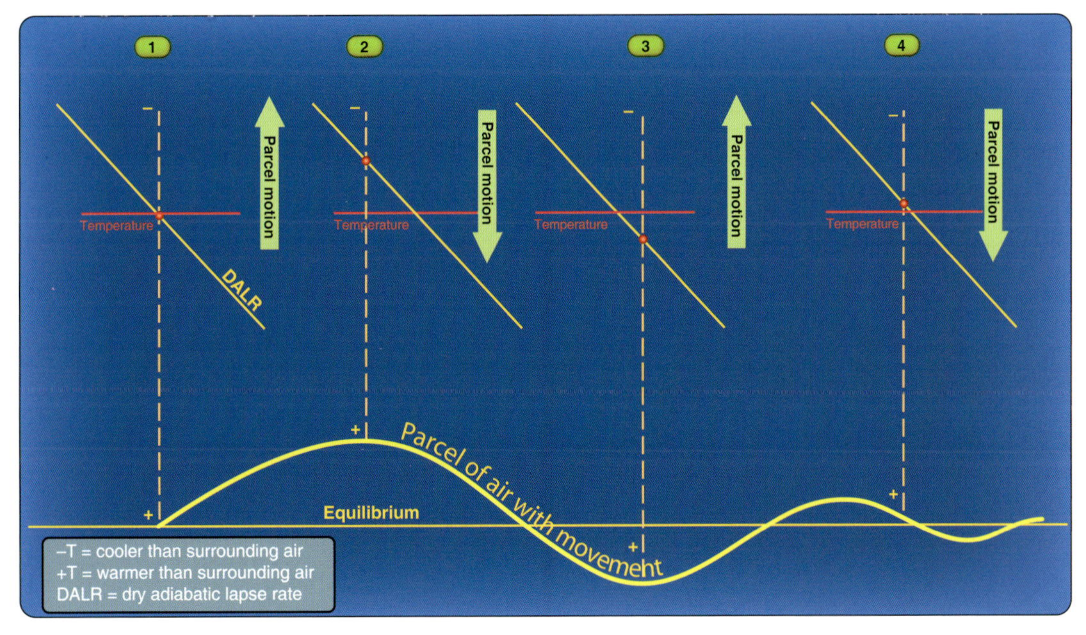

Figure 9-21. *Parcel displaced vertically and oscillating around its equilibrium level.*

The lower part of *Figure 9-21* also illustrates two features of any wave. The wavelength is the horizontal distance between two adjacent wave crests. Typical mountain wavelengths vary considerably, between 2 and 20 miles. The amplitude is half the vertical distance between the trough and crest of the wave.

Figure 9-22 illustrates a two-dimensional conceptual model of a mountain with wind and temperature profiles. Note the increase in windspeed (blowing from left to right) with altitude and a stable layer near mountaintop height with less stable air above and below. As the air flows over the mountain, it descends the lee slope (below its equilibrium level in stable air), which sets up a series of oscillations downstream. While the wave exhibits smooth flow, a low-level turbulent zone exists below, with an embedded rotor circulation under each crest. Turbulence, especially within the individual rotors, while usually moderate to severe, can occasionally become extreme.

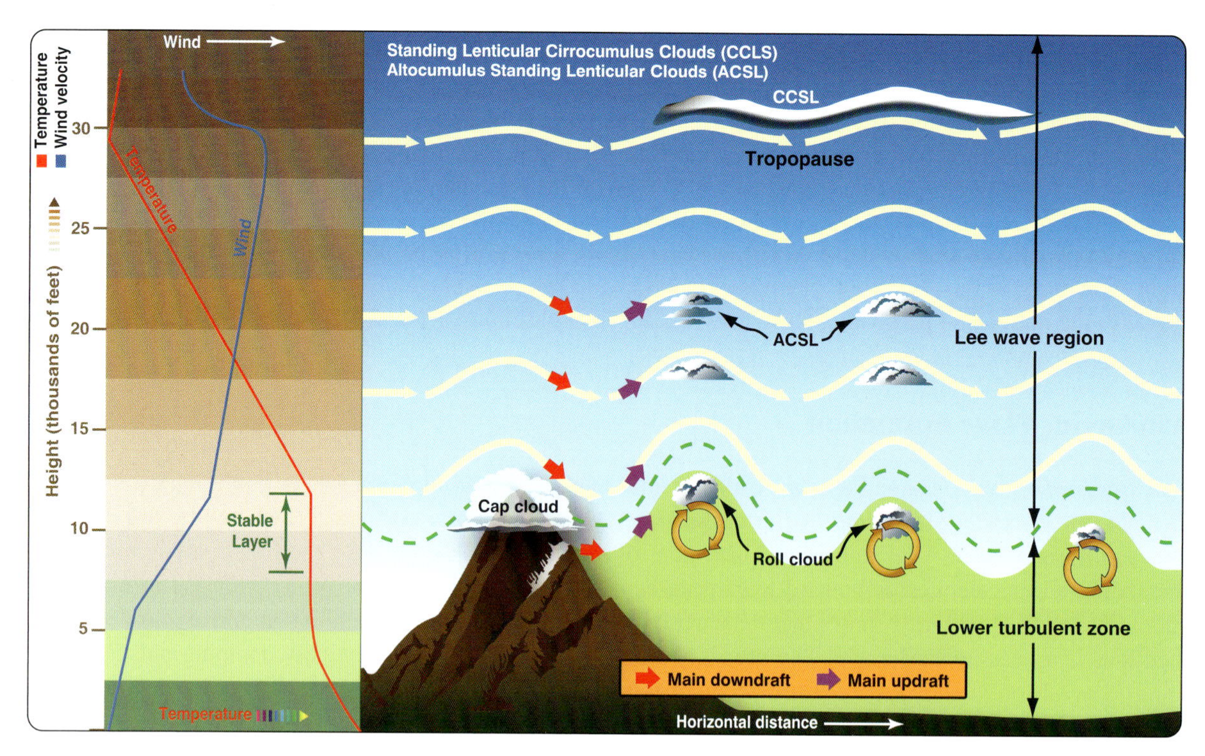

Figure 9-22. *Mountain lee wave system.*

This simple conceptual model has many possible variations. For instance, the topography could have many complex three-dimensional features, such as large ridges, or spurs at right angles to the main range. Variations can occur when a North-South range curves to become oriented Northeast-Southwest. In addition, numerous variations of the wind and stability profiles can occur. In some conditions, the second or third wave crests increases in amplitude, and the pilot can fly higher if flying that portion of the wave.

Low-level turbulence can range from unpleasant to dangerous. While difficult to predict, the intensity of rotor turbulence increases with the amplitude of lee waves and can vary based on location and conditions. At times, rotor turbulence is uniformly rough everywhere below the smooth wave flow. At other times, turbulence becomes severe under wave crests. On occasion, moderate or severe turbulence exists only within a small-scale rotor under the wave crest. Typically, the worst turbulence occurs on the leading edge of the primary rotor.

Figure 9-22 above indicates cloud types associated with a mountain wave system. A cap cloud flowing over the mountain tends to dissipate as the air forced down the mountain slope warms and dries. The first (or primary) wave crest features a roll or rotor cloud with one or more lenticulars forming above. Wave harmonics further downstream may also create lenticulars or rotor clouds. If the wave reaches high enough altitudes, lenticulars may also form at cirrus levels.

The entire mountain wave system can form in completely dry conditions without clouds, and the presence of clouds depends on the amount of moisture at various levels. If only lower-level moisture exists, only a cap cloud and rotor clouds might occur with no lenticulars above, as in *Figure 9-23* (panel A). On other days, only mid-level or upper-level lenticulars appear with no rotor clouds beneath them. When low- and mid-levels contain enough moisture, a deep rotor cloud may form, with lenticulars right on top of the rotor cloud, with no clear air between the two cloud types.

Figure 9-23. *Cloud cover associated with variations in moisture.*

In wet climates, the moist air moving horizontally, can completely close the gap between the cap cloud and primary rotor. This closure could strand a glider on top of the clouds [*Figure 9-23* (Panel B)]. Pilots should consider this possibility when above rotor clouds in moist conditions.

Wave amplitude depends partly on topography. Relatively low ridges of 1,000 feet or less vertical relief can produce lee waves, and uniform height of the mountaintops along a range produces better organized waves. The shape of the lee slope also affects wave production. Very shallow lee slopes can produce waves of sufficient amplitude to support a glider. Different wind and stability profiles favor different topography profiles, and one mountain height, width, and lee slope may produce usable waves only in certain weather conditions. Hence, experience with a particular soaring site can assist pilot prediction of wave conditions.

The weather requirements for wave soaring include sufficient wind and a proper stability profile. Windspeed should be at least 15 to 20 knots at mountaintop level with increasing winds above. The wind direction should be within about 30° perpendicular to the ridge or mountain range. Air stability at the DALR near the mountaintop would not likely produce lee waves even with adequate winds. A well-defined inversion at or near the mountaintop with less stable air above would increase the likelihood of wave production.

Weak lee waves may form without much increase in windspeed as altitude increases, but an actual decrease in windspeed with height usually caps the wave at that level. When winds decrease dramatically with height, for instance, from 30 to 10 knots over two or three thousand feet, turbulence is common at the top of the wave. On some occasions, the flow at mountain level may be sufficient for wave flight, but then begins to decrease with altitude just above the mountain, leading to a phenomenon called "rotor streaming." In this case, the air downstream of the mountain breaks up and becomes turbulent, with no lee waves above.

Lee waves experience diurnal effects, especially in the spring, summer, and fall. Height of the topography also influences diurnal effects. For smaller topography, as morning leads to afternoon and the air becomes unstable to heights above the wave-producing topography, lee waves tend to disappear. On occasion, the lee wave still exists but the pilot needs more height to reach the smooth wave lift. Toward evening as thermals again die down and the air stabilizes, lee waves may again form. During the cooler season, when the air remains stable all day, lee waves are often present all day, provided the winds aloft continue. Large-mountain dissipation of lee waves during daytime appears less significant. For instance, during the 1950s Sierra Wave Project (searchable online), the wave amplitude reached a maximum in mid to late afternoon, with convective heating at a maximum. Rotor turbulence also increased dramatically at that time.

Topography upwind of the wave-producing range can also affect propagation, as illustrated in *Figure 9-24*. In the first case [*Figure 9-24* Panel A], referred to as destructive interference, the wavelength of the wave from the first range is out of phase with the distance between the ranges. Lee waves do not form downwind of the second range, despite favorable winds and stability aloft. In the second case [*Figure 9-24* (Panel B)], referred to as constructive interference, the ranges are in phase, and the lee wave from the second range has a larger amplitude than it might otherwise.

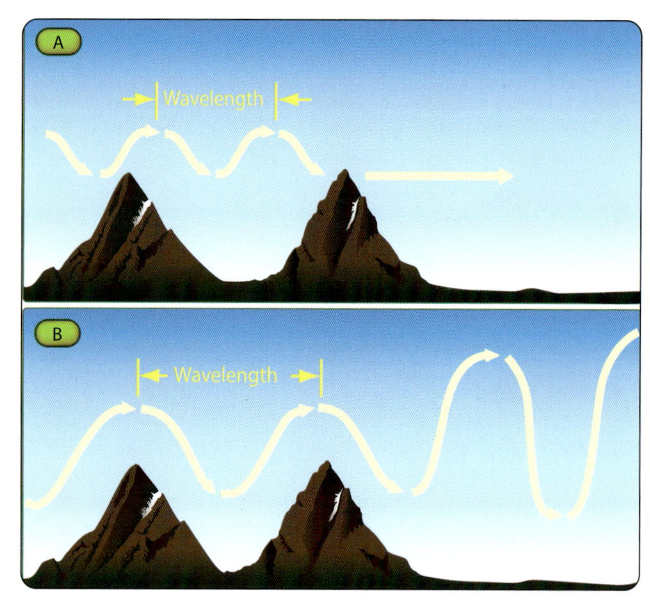

Figure 9-24. *Destructive and constructive interference.*

Wave flight requires planning and appropriate equipment. Additionally, flight in Class A airspace requires Federal Aviation Administration (FAA) notification. The Soaring Society of America (SSA) offers soaring pilots Lennie Awards for completing and documenting a wave flight. [*Figure 9-25*]

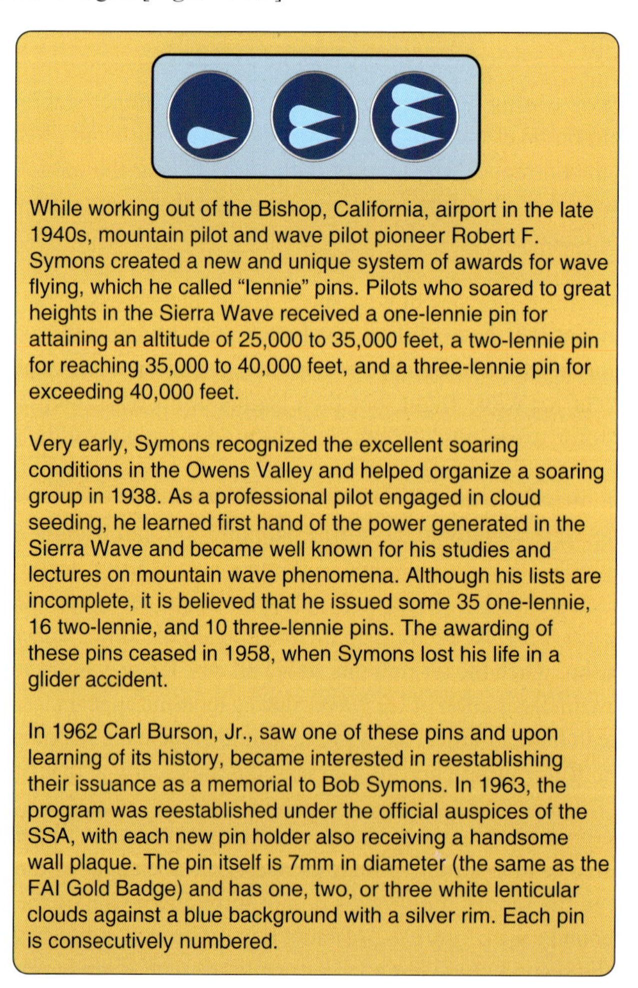

While working out of the Bishop, California, airport in the late 1940s, mountain pilot and wave pilot pioneer Robert F. Symons created a new and unique system of awards for wave flying, which he called "lennie" pins. Pilots who soared to great heights in the Sierra Wave received a one-lennie pin for attaining an altitude of 25,000 to 35,000 feet, a two-lennie pin for reaching 35,000 to 40,000 feet, and a three-lennie pin for exceeding 40,000 feet.

Very early, Symons recognized the excellent soaring conditions in the Owens Valley and helped organize a soaring group in 1938. As a professional pilot engaged in cloud seeding, he learned first hand of the power generated in the Sierra Wave and became well known for his studies and lectures on mountain wave phenomena. Although his lists are incomplete, it is believed that he issued some 35 one-lennie, 16 two-lennie, and 10 three-lennie pins. The awarding of these pins ceased in 1958, when Symons lost his life in a glider accident.

In 1962 Carl Burson, Jr., saw one of these pins and upon learning of its history, became interested in reestablishing their issuance as a memorial to Bob Symons. In 1963, the program was reestablished under the official auspices of the SSA, with each new pin holder also receiving a handsome wall plaque. The pin itself is 7mm in diameter (the same as the FAI Gold Badge) and has one, two, or three white lenticular clouds against a blue background with a silver rim. Each pin is consecutively numbered.

Figure 9-25. *Lennie Awards are given for completing and documenting a wave flight.*

Convergence Lift

Convergence lift occurs when two opposing air masses meet, and air moves upward in response to the opposing winds. Air does not need to meet head on to go up. Wherever air piles up in this fashion, it leads to convergence and rising air.

One type of convergence commonly found near coastal areas results from a sea-breeze. Inland areas heat during the day, while the air over the adjacent water maintains about the same temperature. Inland heating leads to lower pressure, drawing in cooler air. Sometimes, as the cooler air moves inland, it behaves like a miniature shallow cold front, and lift forms along a convergence line. At other times, cooler air acts as a focus for a line of thermals. When unstable inland air exists, a sea-breeze can cause frontal effects and act as a focus for a line of thunderstorms. Additionally, since the air on the coast side of the sea-breeze front consists of cool air, passage of the front can end thermal soaring for the day.

Air over water often has a higher dewpoint than drier inland air. As shown in *Figure 9-26*, a curtain cloud sometimes forms, which marks the area of strongest lift. Since colder and warmer air mixes in the convergence zone, pilots should expect turbulence in any thermals associated with a sea breeze front.

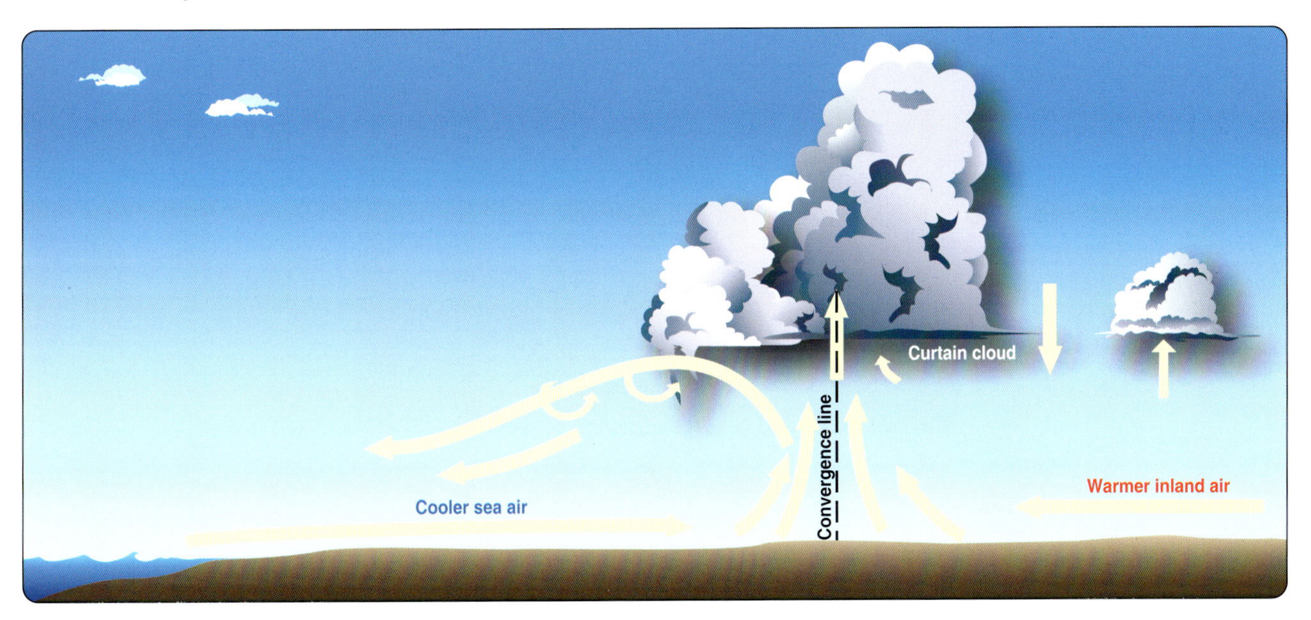

Figure 9-26. *Sea-breeze front.*

Several factors influence the sea-breeze front characteristics (e.g., turbulence, strength, speed of inland penetration, the degree of inland heating, and the land/sea temperature difference). For instance, a small land/sea temperature difference at sunrise with overcast cirrus clouds can reduce ground heating and impede or weaken sea-breeze front formation. Another factor is the synoptic wind flow (general flow of wind for a particular place and time). A weak synoptic onshore flow may allow quicker inland penetration of the sea-breeze front, while a strong onshore flow may remove conditions that allow development of a sea-breeze front. A moderate offshore flow generally prevents any inland penetration of a sea-breeze front.

In a well-defined sea-breeze front marked by a curtain cloud, the pilot can fly straight along the line in steady lift. A poorly defined, weaker convergence line often produces more lift than sink. The pilot should fly slower in lift and faster in sink.

Convergence can also occur along and around mountains or ridges. In *Figure 9-27* Panel A, flow deflected around a ridgeline meets as a convergence line on the lee side of the ridge. The line may be marked by cumulus or a boundary with a sharp visibility contrast. The latter occurs if the air coming around one end of the ridge flows past a polluted urban area, such as in the Lake Elsinore soaring area in southern California. In very complex terrain, with ridges or ranges oriented at different angles to one another, or with passes between high peaks, small-scale convergence zones can be found in adjacent valleys, depending on wind strength and direction. *Figure 9-27* Panel B illustrates a smaller-scale convergence line flowing around a single hill or peak and forming a line of lift stretching downwind from the peak.

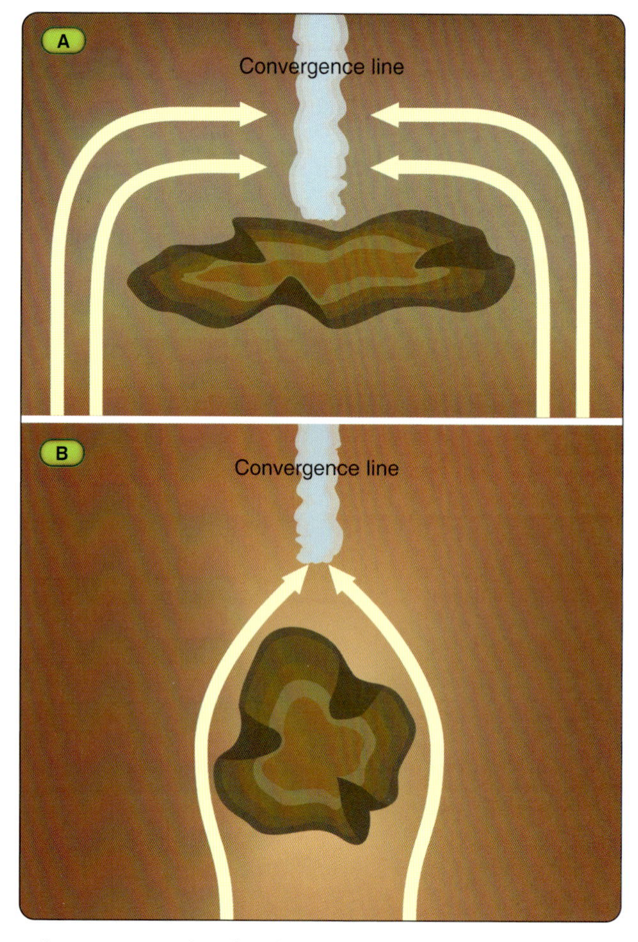

Figure 9-27. *Convergence induced by flow around topography (as viewed from above).*

Convergence can sometimes occur along the top of a ridgeline or mountain range. In *Figure 9-28*, drier synoptic- scale wind flows up the left side of the mountain, while a moist valley breeze flows up the right side of the slope. The two flows meet at the mountain top and form lift along the entire range. If clouds are present, the air from the moist side condenses first, often forming one cloud with a well-defined step, marking the convergence zone. For this scenario, the better lift conditions occur on the Western dry side rather than the East side where clouds are more likely to form.

Figure 9-28. *Mountaintop convergence.*

As a final example, when daytime heating abates in mountainous terrain, a cool katabatic wind or drainage wind may flow down mountain slopes. The flow down the slope converges with air in the adjacent valley to form an area of weak lift. The convergence, often too weak to create lift to support glider flight, may act as a trigger for the last thermal of the day. [*Figure 9-29*]

Figure 9-29. *Convergence induced by flow around topography.*

Obtaining Weather Information

For visual flight rules (VFR) flights, federal regulations require pilots to review weather reports and forecasts if they plan to depart the airport vicinity. Even for a local flight, glider pilots should know the current and forecast weather to avoid flight in hazardous conditions.

For additional details regarding available weather services and products, glider pilots should refer to the current version of the Pilot's Handbook of Aeronautical Knowledge, FAA-H-8083-25, Chapter 12, Aviation Weather Services and the Aviation Weather Handbook, FAA-H-8083-28. These sources include a wealth of information. A pilot familiar with relevant and available weather reports and forecasts can analyze weather hazards and mitigate any associated risks. Pilots who understand weather can experience safe flight in a variety of conditions.

Preflight Weather Briefing

The FAA delivers flight services to pilots in the CONUS, Alaska, Hawaii, and Puerto Rico. Services are provided by phone at 1-800-WX-BRIEF, on the internet through the Flight Service Pilot Web Portal, and in person (Alaska only) at 17 Flight Service Stations (FSS). Services include, but are not limited to: preflight weather briefings, flight planning, and in-flight advisory services.

Weather briefers do not actually predict the weather; they simply translate and interpret weather reports and forecasts within the vicinity of the airport, route of flight, or the destination airport if the flight is a cross-country. A pilot may request one of four types of briefings: standard, abbreviated, soaring [*Figure 9-30*], or outlook.

Soaring Forecast

National Weather Service Denver/Boulder, Colorado
645 AM MDT Wednesday August 25, 2010

This forecast is for Wednesday August 25, 2010:

If the trigger temperature of 77.3 F/25.2 C is reached...then
Thermal Soaring Index....................... Excellent
Maximum rate of lift........................ 911 ft/min (4.6 m/s)
Maximum height of thermals.................. 16119 ft MSL (10834 ft AGL)

Forecast maximum temperature................... 89.0 F/32.1 C
Time of trigger temperature.................... 1100 MDT
Time of overdevelopment........................ None
Middle/high clouds during soaring window....... None
Surface winds during soaring window............ 20 mph or less
Height of the -3 thermal index................. 10937 ft MSL (5652 ft AGL)
Thermal soaring outlook for Thursday 08/26..... Excellent

Wave Soaring Index............................. Poor
Wave Soaring Index trend (to 1800 MDT)......... No change
Height of stable layer (12-18K ft MSL)......... None
Weak PVA/NVA (through 1800 MDT)................ Neither
Potential height of wave....................... 14392 ft MSL (9107 ft AGL)
Wave soaring outlook for Thursday 08/26........ Poor

Remarks...

Sunrise/Sunset.................... 06:20:55 / 19:42:44 MDT
Total possible sunshine........... 13 hr 21 min 49 sec (801 min 49 sec)
Altitude of sun at 13:01:25 MDT... 60.82 degrees

Upper air data from rawinsonde observation taken on 08/25/2010 at 0600 MDT

Freezing level................. 15581 ft MSL (10296 ft AGL)
Additional freezing level....... 54494 ft MSL (49209 ft AGL)
Convective condensation level... 13902 ft MSL (8617 ft AGL)
Lifted condensation level....... 14927 ft MSL (9641 ft AGL)
Lifted index................... -3.4
K index........................ +9.7

* * * * * * Numerical weather prediction model forecast data valid * * * * * *

08/25/2010 at 0900 MDT | 08/25/2010 at 1200 MDT
|
K index... +4.0 | K index... -0.7

This product is issued twice per day, once by approximately 0630 MST/0730 MDT (1330 UTC) and again by approximately 1830 MST/1930 MDT (0130 UTC). It is notcontinuously monitored nor updated after its initial issuance.

The information contained herein is based on rawinsonde observation and/or numerical weather prediction model data taken near the old Stapleton Airport site in Denver, Colorado at

North Latitude: 39 deg 46 min 5.016 sec
West Longitude: 104 deg 52 min 9.984 sec
Elevation: 5285 feet (1611 meters)

and may not be representative of other areas along the Front Range of the Colorado Rocky Mountains. Note that some elevations in numerical weather prediction models differ from actual station elevations, which can lead to data which appear to be below ground. Erroneous data such as these should not be used.

The content and format of this report as well as the issuance times are subject to change without prior notice. Comments and suggestions are welcome and should be directed to one of the addresses or phone numbers shown at the bottom of this page. To expedite a response to comments, be sure to mention your interest in the soaring forecast.

DEFINITIONS:

Convective Condensation Level - The height to which an air parcel possessing the average saturation mixing ratio in the lowest 4000 feet of the airmass, if heated sufficiently from below, will rise dry adiabatically until it just becomes saturated. It estimates the base of cumulus clouds that are produced by surface heating only.

Convection Temperature (ConvectionT) - The surface temperature required to make the airmass dry adiabatic up to the given level. It can be considered a "trigger temperature" for that level.

Freezing Level - The height where the temperature is zero degrees Celsius.

Height of Stable Layer - The height (between 12,000 and 18,000 feet above mean sea level) where the smallest lapse rate exists. The location and existence of this feature is important in the generation of mountain waves.

K Index - A measure of stability which combines the temperature difference between approximately 5,000 and 18,000 feet above the surface, the amount of moisture at approximately 5,000 feet above the surface, and a measure of the dryness at approximately 10,000 feet above the surface. Larger positive numbers indicate more instability and a greater likelihood of thunderstorm development. One interpretation of K index values regarding soaring in the western United States is given in WMO Technical Note 158 and is reproduced in the following table:

below -10	no or weak thermals
-10 to 5	dry thermals or 1/8 cumulus with moderate thermals
5 to 15	good soaring conditions
15 to 20	good soaring conditions with occasional showers
20 to 30	excellent soaring conditions, but increasing probability of showers and thunderstorms
above 30	more than 60 percent probability of thunderstorms

Lapse Rate - The change with height of the temperature. Negative values indicate inversions.

Lifted Condensation Level - The height to which an air parcel possessing the average dewpoint in the lowest 4000 feet of the airmass and the forecast maximum temperature must be lifted dry adiabatically to attain saturation.

Lifted Index - The difference between the environmental temperature at a level approximately 18,000 feet above the surface and the temperature of an air parcel lifted dry adiabatically from the surface to its lifted condensation level and then pseudoadiabatically thereafter to this same level. The parcel`s initial temperature is the forecast maximum temperature and its dewpoint is the average dewpoint in the lowest 4000 feet of the airmass. Negative values are indicative of instability with positive values showing stable conditions.

Lift Rate - An experimental estimate of the strength of thermals. It is computed the same way as the maximum rate of lift but uses the actual level rather than the maximum height of thermals in the calculation. Also, none of the empirical adjustments based on cloudiness and K-index are applied to these calculations.

Maximum Height of Thermals - The height where the dry adiabat through the forecast maximum temperature intersects the environmental temperature.

Figure 9-30. *Soaring forecast.*

Weather-Related Information

Pilots can also find weather-related information on the Internet, including sites directed toward aviation. Pilots should verify the timeliness and source of the weather information provided by any Internet sites to ensure the information is up to date and accurate. For example, a source of accurate information includes the NWS website.

Interpreting Weather Charts, Reports, & Forecasts

Knowing how to interpret and understand weather information requires knowledge and practice. Weather charts and reports record observed atmospheric conditions at certain locations at specific times. The NWS collects data from automated sources or from trained observers using electronic instruments, computers, and personal observations as well as radiosonde observations or model generated soundings to produce the weather products pilots use to determine if a flight can be conducted safely. This same information can be used by soaring pilots to determine if sufficient lift for a planned flight may exist, where to find it, and how long the lift should last.

Chapter Summary

Pilots use several different types of lift to make extended glider flights. The lift comes from thermals, mountain ridges, wave formation, or convergence zones. The stability of the atmosphere affects several of these sources. Each source of lift has its own set of hazards and associated risks. For example, weather leading to thermals can also lead to thunderstorms. Pilots looking for weather conditions that favor extended flight should also know how to identify conditions and hazards that prompt them not to fly. Pilots should not fly near thunderstorms, in low visibility, or in high winds that affect the safety of takeoffs and landings. Pilots can get general and specific information from a variety of sources and should make certain that any source of weather information used before flight is accurate. The FAA has information and documents regarding weather products and services on its website.

Chapter 10: Soaring Techniques

Introduction

Many glider pilots enjoy searching for lift to gain altitude. Staying aloft often involves finding and staying within the strongest part of any updrafts. This chapter covers some basic soaring techniques that assist with that task.

In the early 1920s, soaring pilots discovered how to remain aloft using updrafts deflected by the hillsides they used for launch. Soon afterward, they discovered thermals in the valleys adjacent to the hills. In the 1930s, the discovery of mountain waves, which were not yet well understood by meteorologists, allowed glider pilots to make the first high-altitude flights. Since thermals occur over both flat terrain and hilly country, they remain the most-used type of lift for glider flights today.

As a note, glider pilots refer to rising air as lift, which differs from the lift generated by the wings. This chapter refers to lift as the rising air within an updraft and sink as the descending air in downdrafts.

Thermal Soaring

Successful thermalling requires several steps: locating the thermal, entering the thermal, centering within the thermal, and exiting the thermal.

When locating thermals, glider pilots look for nearby lift indicators. In sufficiently moist air and if and thermals rise high enough, cumulus clouds (Cu) (pronounced "cue") form. Glider pilots look for Cu in the developing stage, while the cloud builds from the thermal underneath it. The base of a developing Cu appears sharp and well defined. Dissipating Cu have a fuzzy appearance and little lift or sink underneath. [*Figure 10-1*]

Figure 10-1. *Photographs of (A) mature cumulus, which can produce good lift, and (B) dissipating cumulus.*

The lifetime of Cu often varies during a given day as Cu that develop early in the day may change into well-formed and longer-lived clouds. A promising Cu in the distance may also start dissipating, and glider pilots refer to such Cu as rapid or quick cycling, which means the Cu forms, matures, and dissipates in a short time.

Sometimes Cu cover much of the sky, which makes seeing the cloud tops difficult; however, the cloud bases also indicate the presence and strength of thermals. Generally, a dark area under the cloud base indicates a deeper cloud and a higher likelihood of a thermal underneath that spot. Since several thermals can feed one cloud, darker areas under the cloud may indicate a stronger thermal. At times, an otherwise flat cloud base under an individual Cu has wisps or tendrils of cloud hanging down, which indicate a particularly active area and the presence of warm rising air. Note the importance of distinguishing features under Cu that differentiate potential lift from virga. Virga consists of precipitation in the form of rain, snow, or ice crystals that descend from the cloud base and evaporate before striking the ground. Virga may signal that the Cu has reached towering cumulus or thunderstorm status. [Figure 10-2]

Figure 10-2. *Photographs of (A) towering cumulus, (B) cumulonimbus (Cb), and (C) virga.*

Lift near a cloud base often increases dramatically in a concave region under an otherwise flat cloud created by especially warm air. When trying to leave the strong lift in the concave area under the cloud, glider pilots can find themselves climbing rapidly with cloud all around. Adherence by a glider to the minimum distance below a cloud as listed in 14 of the Code of Federal Regulations (14 CFR) part 91, section 91.155, normally prevents an accidental climb into a cloud. In addition, maintaining the minimum separation gives a glider pilot the opportunity to avoid an aircraft operating under instrument flight rules in, near, or emerging from a cloud.

As any thermal rises from the surface and reaches the convective condensation level (CCL), a cloud begins to form. At first, only a few wisps form. The initial wisps of Cu in an otherwise blue (cloudless) sky indicate where an active thermal has begun building a cloud that will grow to a familiar cauliflower shape. When crossing a blue hole (a region anywhere from a few miles to several dozen miles of cloud-free sky in an otherwise Cu-filled sky), getting to an initial wisp of Cu can provide lift from the thermal underneath. On some days and depending on the moisture in the air, these wisps undergo no further development and provide the only indication of thermals.

Lack of Cu does not necessarily mean lack of thermals. If the air aloft is cool enough and the surface temperature warms sufficiently, thermals form even without enough moisture for cumulus formation. These dry, or "blue thermals" can be just as strong as their Cu-topped counterparts. Glider pilots sometimes find these thermals by chance while gliding. However, other indicators may exist and can make the search for thermals less random.

Other Indicators of Thermals

Another circling glider may indicate the presence of a thermal. Circling birds may also indicate thermal activity. Thermals tend to transport various aerosols, such as dust, upward with them. When a thermal reaches an inversion, it disturbs the stable air above it, spreads out horizontally, and deposits some of the aerosols at that level. Depending on the sun angle (and the pilot's type of sunglasses), haze domes may indicate dry thermals. If the air contains enough moisture, haze domes often form just before the first wisps of Cu.

Glider pilots talk about a house thermal or thermals that seem to form over and over in the same spot or in the same area. Glider pilots new to a soaring location should ask the local pilots about favored spots—doing so might make additional tows unnecessary. While some thermals may arise from consistent sources, no one can guarantee that a thermal currently exists where one existed before. In addition, if a thermal has recently formed, it takes time for the sun to reheat the area before another thermal might form at the same location.

On days without airborne indicators to mark thermals, pilots can look for clues on the ground. For example, in drier climates, dust devils often mark thermals triggering from the ground. At certain times of the day in hilly or mountainous terrain, the sun's radiation may strike a particular slope at or near a right angle. [*Figure 10-3*] A sun facing slope receives more energy per unit of area and can warm the surrounding air more effectively. Darker ground or surface features heat quicker than grass covered fields. Huge black asphalt parking lots can produce strong thermals. A large tilled black soil field can be a good source of lift if the pilot can find the associated narrow plume of rising air. Thermals often form where hills exist since slopes tend to be drier and heat better than surrounding lowlands. Finally, cooler air usually pools in low-lying areas overnight and takes longer to warm during the morning. Pilots might avoid searching for thermals from those areas early in the day.

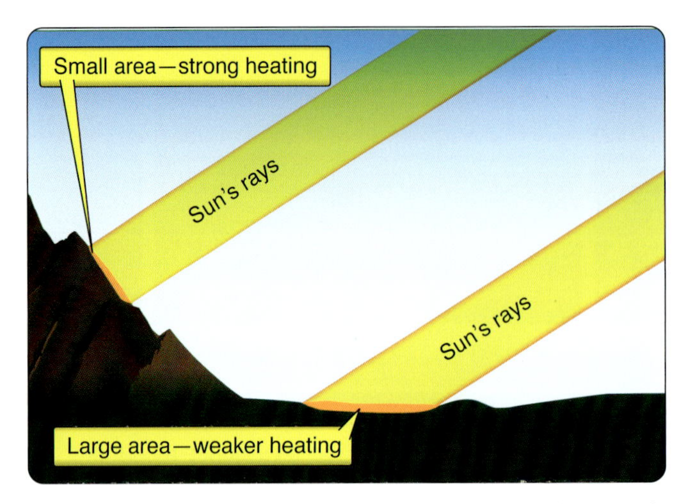

Figure 10-3. *Sun's rays are concentrated in a smaller area on a hillside than on adjacent flat ground.*

Other subtle ground markers exist. An open field surrounded by shade usually results in a space where the air suddenly warms. A town surrounded by green fields results in a rise in temperature since the town surfaces absorb more heat than the surrounding farmland. Likewise, a yellowish harvested field gets warmer than an adjacent wet field with green vegetation. Wet areas tend to use the sun's radiation to evaporate the moisture rather than heat the ground. A field with a rocky outcrop might produce better thermals since rocks can't hold moisture. Rocky outcrops along a snowy slope heat much more efficiently than surrounding snowfields. While searching for smaller features works at lower altitudes, pilots can use their knowledge of ground heating when at higher altitudes by avoiding areas such as a valley with many lakes.

Wind

Wind influences thermal structure and location. Strong shear can break thermals apart and effectively cap their height. Strong winds at the surface and aloft often break up thermals, making them turbulent and difficult or impossible to use. On the other hand, as discussed in Chapter 9, Glider Flight and Weather, moderately strong winds without too much wind shear sometimes organize thermals into long streets, which can provide lift when they lie along a cross-country course line. [*Figure 10-4*]

Figure 10-4. *Photograph of cloud streets.*

If suspecting the presence of waves associated with a thermal above the cloud, the pilot can climb in the thermal near a cloud base and then head toward the upwind side of the Cu. Often, only weak lift exists in smooth air upwind of the cloud. Once above cloud base and upwind of the Cu, the pilot should find climb rates of a few hundred fpm in the thermal wave. The glider can climb flying back and forth upwind of an individual Cu, or by flying along cloud streets if they exist. When thermal waves exist without clouds, the pilot should climb to the top of the thermal and penetrate upwind in search of smooth, weak lift. Without visual clues, thermal waves prove difficult to work. Sometimes the pilot stumbles upon a thermal wave by chance.

In light wind conditions, pilots should use a slanted search pattern. For instance, in Cu-filled skies, glider pilots need to search upwind of the cloud to find a thermal. How far upwind depends on the strength of the wind, thermal strength on that day, and distance below cloud base (the lower the glider, the further upwind the glider needs to be). The pilot should consider the fact that windspeed does not always increase at a constant rate with height and the possibility that wind direction can change dramatically with height. [*Figure 10-5*] Pilots can estimate wind direction and speed at a cloud base by watching cloud shadows on the ground.

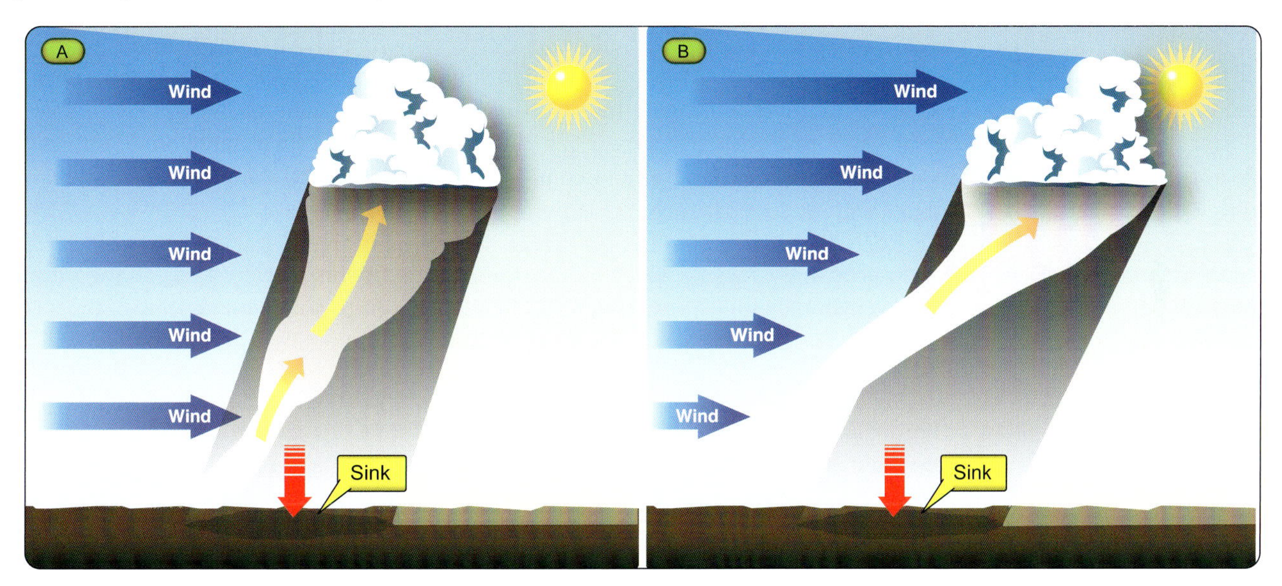

Figure 10-5. *Thermal tilt in shear that (A) does not change with height, and that (B) increases with height.*

Pilots can find where thermals appear in relation to clouds on a given day and use those encounters to determine how to search that day. If approaching Cu from the downwind side, heavy sink may occur near the cloud. From that position, the pilot should head for the darkest, best-defined part of the cloud base, then continue directly into the wind. Depending on the distance below cloud base, the pilot should find the lift forming the cloud about the time the glider passes upwind of the cloud. If approaching the cloud from a crosswind direction (for instance, heading north with westerly winds), the pilot can also estimate the thermal location from previous encounters that day. If only encountering reduced sink, lift may exist nearby, and a short leg upwind or downwind may locate the thermal.

Thermals also drift with the wind on cloudless days, and similar techniques can locate thermals using airborne or ground-based markers. For instance, if heading toward a circling glider but at a thousand feet lower, the pilot can estimate how much the thermal tilts in the wind and head for the most likely spot upwind of the circling glider. When in need of a thermal, pilots might consider searching on a line stretching upwind to downwind once under or over the circling glider. This may or may not work; if the thermal is a bubble rather than a column, the pilot may miss the bubble. The pilot in sink should limit the search if not finding lift. Searching for a thermal in sink near one spot, rather than leaving and searching for a new thermal can consume a lot of altitude.

Cool, stable air can also drift with the wind. Pilots should avoid areas downwind of known stable air, such as large lakes or large irrigated regions. [*Figure 10-6*] On a day with Cu, a big blue hole in an otherwise Cu-filled sky indicates a stable area. If the area is broad enough, a detour upwind of the stabilizing feature might conserve a lot of altitude.

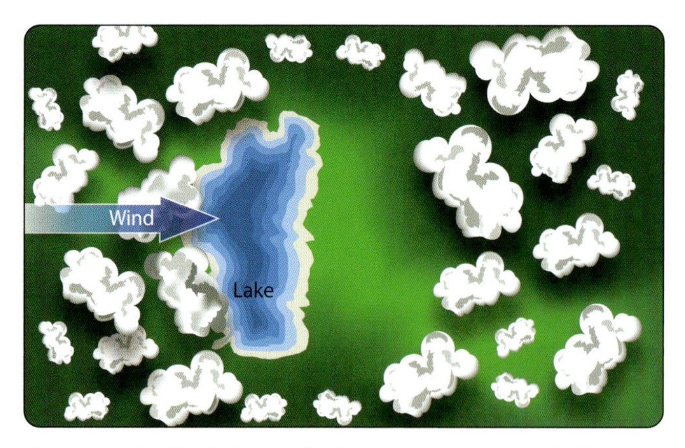

Figure 10-6. *Blue hole in a field of cumulus downwind of a lake.*

The Big Picture

With a sky full of Cu and if gliding in the upper part of the strongest lift, the pilot should focus on the Cu, and make choices based on the best clouds. At lower altitudes glider pilots may find it difficult to associate thermals with clouds above. If that happens, the pilot should use the Cu to find areas that appear generally active and then focus more on ground-based indicators, like dust devils, a hillside with sunshine on it, or a circling bird. When down low, the pilot can accept weak climbs. Often the thermal cycles again and rewards patience.

When searching for lift using the best speed to fly, L/D~MAX speed plus corrections for sink and any wind, allows a glider pilot to cover the most ground for a given altitude loss. See Chapter 5, Glider Performance, to review shifting of the glider polar for winds and sink.

Entering a Thermal

Oddly enough, increased sink often indicates a nearby thermal. Next, the pilot feels a subtle or obvious positive G- force, depending on the thermal strength. The "seat-of-the-pants" indication of lift occurs faster than shown by any variometer, which lags slightly. The pilot should reduce speed to between L/D and minimum sink and note the trend of the variometer needle (should be an upswing) or the audio variometer going from the drone to more rapid beeping. At the right time in the anticipated lift and in a perfect scenario, the pilot rolls into a coordinated turn at just the right bank angle and speed to center within the thermal.

Before going further, what vital step was left out of the above scenario? VISUALLY CLEAR THE AIRSPACE BEFORE TURNING! Pilots sometimes forget that basic primary step before any turn as the variometer attracts a lot of pilot visual attention upon entering lift. Low-altitude flight and a glider without an audio variometer increase the likelihood of this omission.

To help decide which way to turn, the pilot determines which wing seems to lift when entering the thermal. For instance, if the glider gently banks right when entering the thermal, the pilot should CLEAR LEFT, then turn left. A glider on its own tends to fly away from thermals. [*Figure 10-7*] Pilots who maintain a light touch on the controls can sense proximity to a thermal and avoid letting thermals continue to bank the glider away from the thermal. If no thermal-induced banking occurs, the pilot decides arbitrarily which way to turn. Note that new soaring pilots often turn in a favored direction. This could cause pilot inability to fly reasonable circles in the other direction. If this happens, the pilot should make a conscious effort to thermal in each direction half the time to improve proficiency and as preparation for thermalling in traffic.

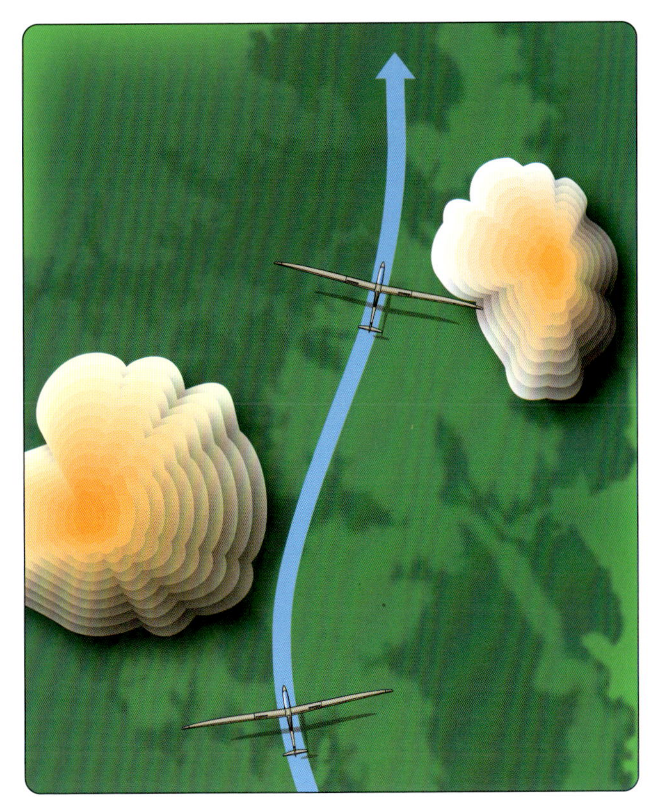

Figure 10-7. *Effect of glider being allowed to bank on its own when encountering thermals.*

As a glider encounters lift on one side, the wing in the stronger lift will start to rise, and some gliders will tend to slip laterally as indicated by the yaw string. The glider pilot should apply aileron pressure to bring the climbing wing down not just to level the wings, but to bank further to begin the turn into the lifting columns of air. If the rising air on one wing results in a sideslip, a turn toward the head of the yaw string should also bring the glider into the rising air.

Inside a Thermal

Optimum climb occurs when the pilot uses proper bank angle and speed after entering a thermal. Under ideal conditions, pilots use the shallowest possible bank angle at minimum sink speed. However, thermal size and associated turbulence usually do not favor this combination.

Bank Angle

The glider's sink rate increases as the bank angle increases, and the sink rate begins to increase more rapidly beyond about a 45° bank angle. A 40° compared to a 30° bank angle may allow for circling in the stronger lift near the center of the thermal. Maximizing thermal lift takes practice. The optimized bank angle depends on the size and strength of the thermal. Normally, the pilot does not use bank angles exceeding 50°, but exceptions exist. A pilot might use a bank of 60° to stay in the best lift. Thermals tend to be smaller at lower levels and expand in size as they rise higher. Therefore, a pilot generally uses a steeper bank angle to remain in lift at lower altitudes.

Thermalling illustrates the importance of understanding and training in steep turns (as previously discussed in Chapter 7: Launch & Recovery Procedures & Flight Maneuvers.) A steep turn during thermalling causes the outer wing to travel faster than the inner wing and results in an overbanking tendency. While the pilot holds aileron pressure against the turn to balance the lift between the wings, this pressure causes additional drag on the lowered wing and could result in a skid and lead to a spin entry. Applying slight top rudder pressure and slipping gently during a steep turn can compensate for the increased drag of the inside wing and result in better overall climb performance. Glider pilots should consult their flight instructors regarding the proper technique for steep turns during thermalling.

Speed

Flying at minimum sink speed for the G-load on the glider optimizes the climb in a well-formed thermal and light turbulence. Flying in thermals either above or below this speed will degrade glider performance. As discussed in Chapter 3, decreasing the airspeed in a turn will decrease the radius of the turn. The examples in this paragraph assume a minimum sink speed of 60 mph. At a 30° bank angle, decreasing speed from 60 to 40 mph decreases the radius of the circle by almost 250 feet. While this reduced radius may place the glider closer to strong lift near the thermal center, glider performance will suffer from the increased rate of descent while flying at an airspeed below the minimum sink speed. An increased bank angle to achieve a smaller diameter circle may provide better optimization. For example, while maintaining a minimum sink speed of 60 mph, increasing the bank angle from 30 to 45 degrees will decrease the turn radius by over 175 feet. Increasing the bank angle from 30 to 60 degrees will decrease the turn radius by over 275 feet. While some gliders can safely fly several knots below minimum sink speed to reduce the turn radius, the increased sink rate at lower speeds may offset any gain achieved.

Pilots should avoid thermalling speeds well below minimum sink speed due to the increased risk of a stall and the lack of controllability. Distractions while thermalling may include studying the cloud above or the ground below, quickly changing bank angles without remaining coordinated, thermal turbulence, or abrupt maneuvers to avoid other gliders in the thermal. The pilot may allow the airspeed to decay to the stall speed or may increase the bank angle thereby increasing the load factor and the stall speed. This increases the risk of an inadvertent stall. While thermalling, any stall recovery should occur instinctively and without delay. Without prompt corrective action a spin entry could occur, depending on the stall characteristics of the glider or if in turbulent thermals. Additionally, at airspeeds well below minimum sink speed, a lack of airflow over the wings and control surfaces may make the glider less controllable during stall recovery or during any abrupt maneuvers. Glider pilots should carefully monitor speed and nose attitude at lower altitudes. Using sufficient speed ensures that the pilot, and not the thermal turbulence, controls the glider.

Centering

Pilot opinions differ regarding how long to wait before rolling into a thermal after encountering lift. Some pilots advocate flying straight until the lift has peaked. Then, they start turning back into stronger lift. If using this technique, pilots should not wait too long after the first indication of decreasing lift. Other pilots favor rolling into the thermal before lift peaks, thus avoiding the possibility of losing the thermal. However, turning into the lift too quickly causes the glider to fly back out into sink. The choice depends on personal preference and the conditions on a given day.

Often upon entering a thermal, the glider experiences lift for part of the circle and sink for the other part. The pilot should determine where the best lift exists and move the glider into it for the most consistent climb. If the pilot turns in the wrong direction and almost immediately encounters sink, a 270° correction may correct the error. [*Figure 10-8*] The pilot should complete a 270° turn, straighten out for a few seconds, and if encountering lift again, turn back into it in the same direction. The pilot should avoid reversing the direction of turn since that procedure takes more time, covers more distance, and may lead away from the lift completely. [*Figure 10-9*]

Figure 10-8. *The 270° centering correction.*

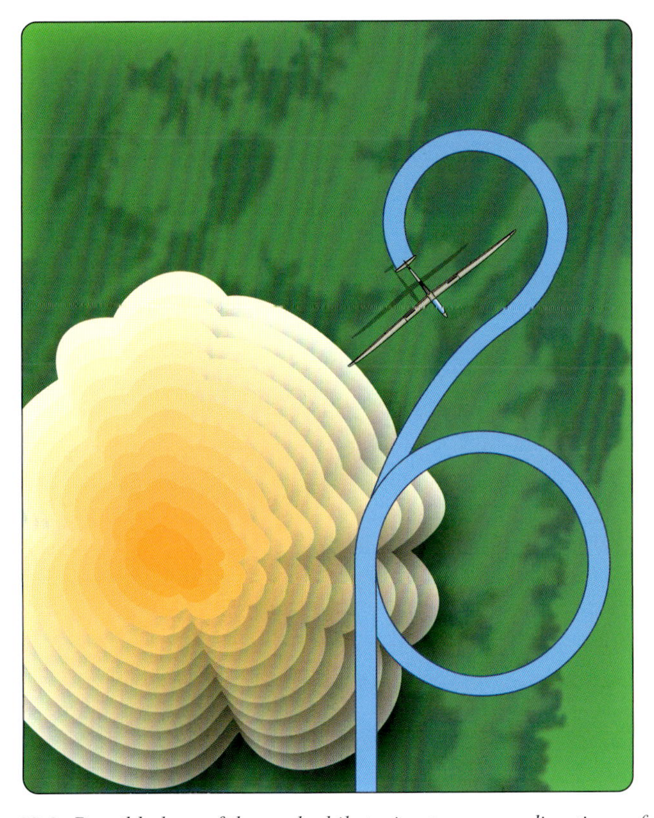

Figure 10-9. *Possible loss of thermal while trying to reverse directions of circle.*

If stronger lift exists on one side of the thermal, the pilot can use one of several centering techniques. One method involves noting the current position with the best lift: for instance, toward the Northeast or toward some feature on the ground that the pilot can see under the high wing when at the current point of best lift. [*Figure 10-10*] On the next turn, the pilot can adjust the circle by either straightening or shallowing the turn toward the stronger lift. The pilot should anticipate the turn and begin rolling out about 30° before the heading toward the strongest part. This allows rolling back toward the strongest part of the thermal rather than flying through it and turning away from the thermal center. How long a glider remains in a shallow bank or straight depends on the size of the thermal. Since gusts within the thermal can cause airspeed indicator variations, the pilot should pay attention to the nose attitude.

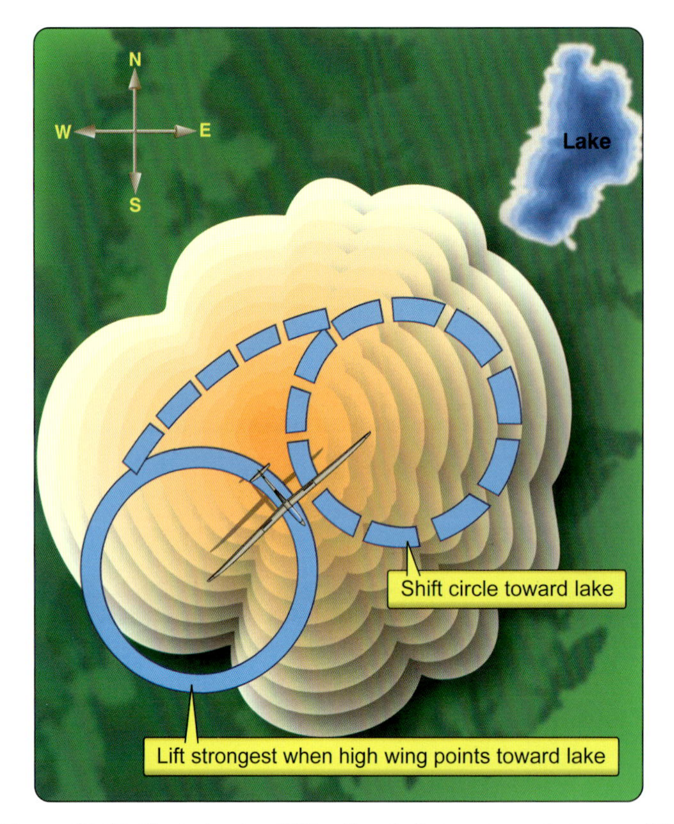

Figure 10-10. *Centering by shifting the circle turn toward stronger lift.*

A pilot can use other techniques as depicted in the three scenes of *Figure 10-11* and as described below. The pilot should:

1. Shallow the turn slightly (5° or 10°) when encountering weaker lift, then when encountering stronger lift again as indicated by increase in positive G-force or the variometer, the pilot resumes the original bank angle. If shallowing the turn too much, the glider may fly completely away from the lift.

2. Straighten or shallow the turn for a few seconds 60° after encountering the weakest lift or worst sink indicated by the variometer, then resume the original bank angle. This accounts for the lag in the variometer since the actual worst sink occurred a couple of seconds earlier than indicated.

3. Straighten or shallow the turn for a few seconds when the stronger seat-of-the-pants surge is felt. Then, the pilot should resume the original bank and verify the result with the variometer trend.

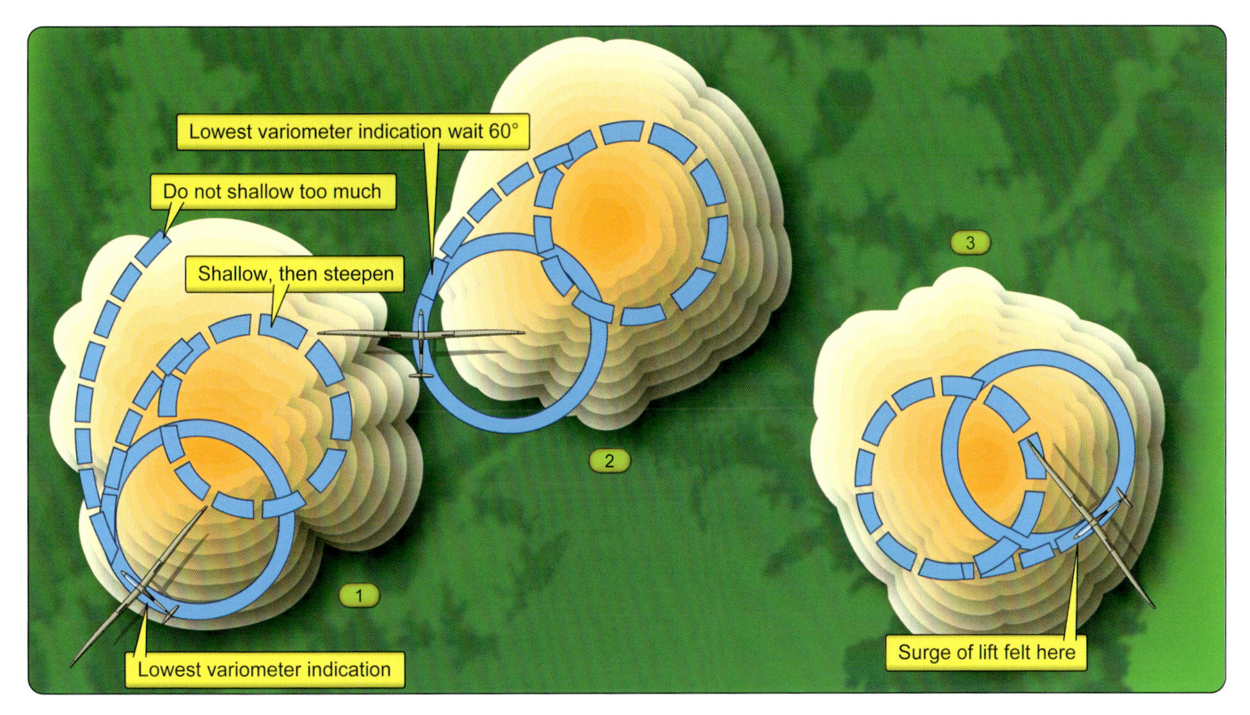

Figure 10-11. *Other centering corrections.*

New glider pilots should develop proficiency centering using one of the above methods first, and then experiment with other methods. As an additional note, thermals often deviate markedly from the conceptual model of concentric gradients with lift increasing toward the center. For instance, it sometimes feels as if two (or more) nearby thermal centers exist, which can make centering difficult. Glider pilots should continually adjust, and recenter to maintain the best rate of climb.

Other gliders can help pilots center a thermal as well. If a nearby glider seems to be climbing better, the pilot can adjust the turn to fly within the same circle. Similarly, if noting a bird soaring close by, a turn toward the soaring bird may lead to a better climb. Soaring birds have a much tighter turning radius than a glider, so they can stay in the strongest center of the lift while the glider pilot circles around them.

Collision Avoidance

Collision avoidance takes priority over aerodynamic efficiency when sharing a thermal with other gliders. The first glider in a thermal establishes the direction of turn, and all other gliders that join the thermal should turn in the same direction. Ideally, two gliders in a thermal at the same height or nearly so should position themselves across from each other so they can maintain visual separation. [*Figure 10-12*] A pilot should enter a thermal in a way that does not interfere with gliders already in the thermal. Pulling up to bleed off excess speed in the middle of a crowded thermal would create a hazard to other gliders. Safe technique involves bleeding off speed before reaching the thermal. Announcing the entry to the other glider(s) on the radio (if equipped) may enhance collision avoidance.

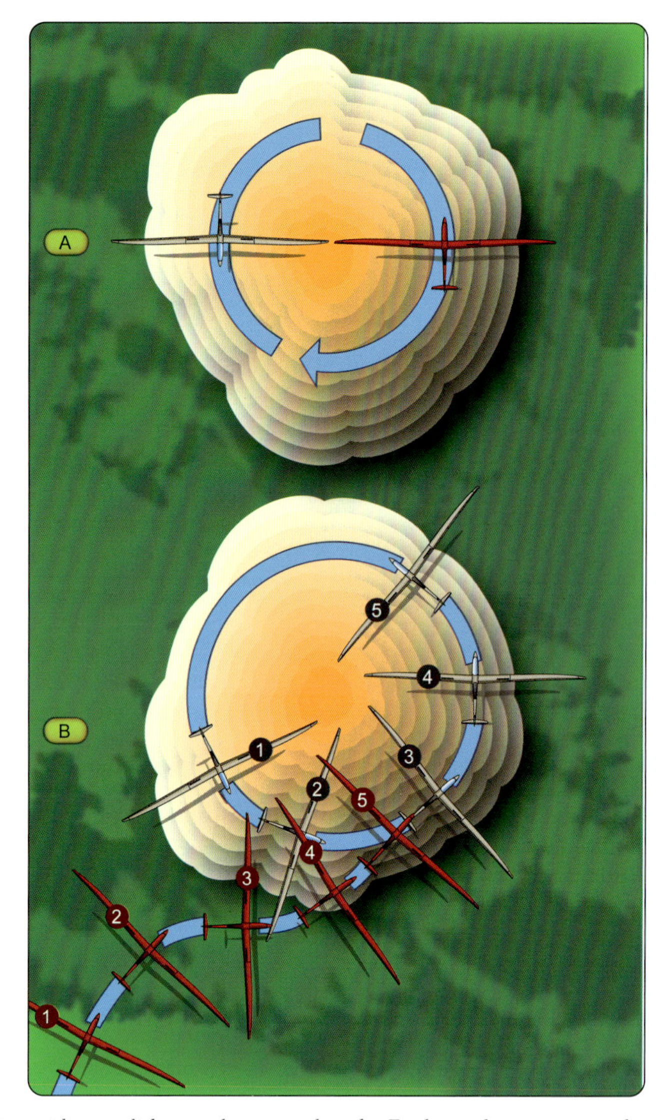

Figure 10-12. *Proper positioning with two gliders at the same altitude. Each number corresponds to the position of both gliders at a given time.*

Different types of gliders in the same thermal may have different minimum sink speeds, which makes it more difficult for each glider to remain directly across from the other. Each pilot should remain in visual contact with the other glider. Radio communication may help avoid a collision, but too much talking clogs the frequency and can impede the broadcast of another pilot's message. Pilots should not fly directly above or below another glider in a thermal since differences in performance, or even minor changes in speed, can lead to unexpected altitude changes. If a pilot loses sight of another glider in a thermal and cannot verify the other glider's position via a radio call, the pilot should leave the thermal. After 10 or 20 seconds, the pilot can come back around to rejoin the thermal in a better position to see the other traffic. Unsafe thermalling practices make a mid-air collision more likely, which could result in fatalities.

Exiting a Thermal

When exiting a thermal, appropriate methods can preserve some altitude. The pilot should increase speed as needed to penetrate any sink often found on the edge of the thermal and leave the thermal in a manner that does not hinder or endanger other gliders. While circling, the pilot scans the full 360° of sky. This allows the pilot to continually check for other traffic in the vicinity and decide where to go for the next climb. Experienced pilots often decide where to go next while still in lift.

Managing Expectations

Glider pilots should adapt quickly to whatever the air has to offer at the time. While thermalling techniques become second nature with practice, pilots should expect to land early if unable to find sufficient lift as part of normal flying experience.

Ridge/Slope Soaring

Efficient slope soaring (also called ridge soaring or ridge running) involves flying in the updraft along the upwind side of a ridge. Although the concept seems simple, slope soaring presents significant hazards and places demands on the glider and the pilot. Thorough preflight planning and route planning include ridge selection based on the current winds.

Surface winds of 15–20 knots perpendicular to the ridge optimize ridge soaring. Wind flow within 45° of the perpendicular line also provides adequate lift. Winds less than 10 knots might produce adequate ridge soaring depending on the terrain, but with 10 knots of wind or less, pilots should avoid flying low over any ridge due to the possibility of encountering sink. Local ridge pilots know about of these conditions. [*Figure 10-13*]

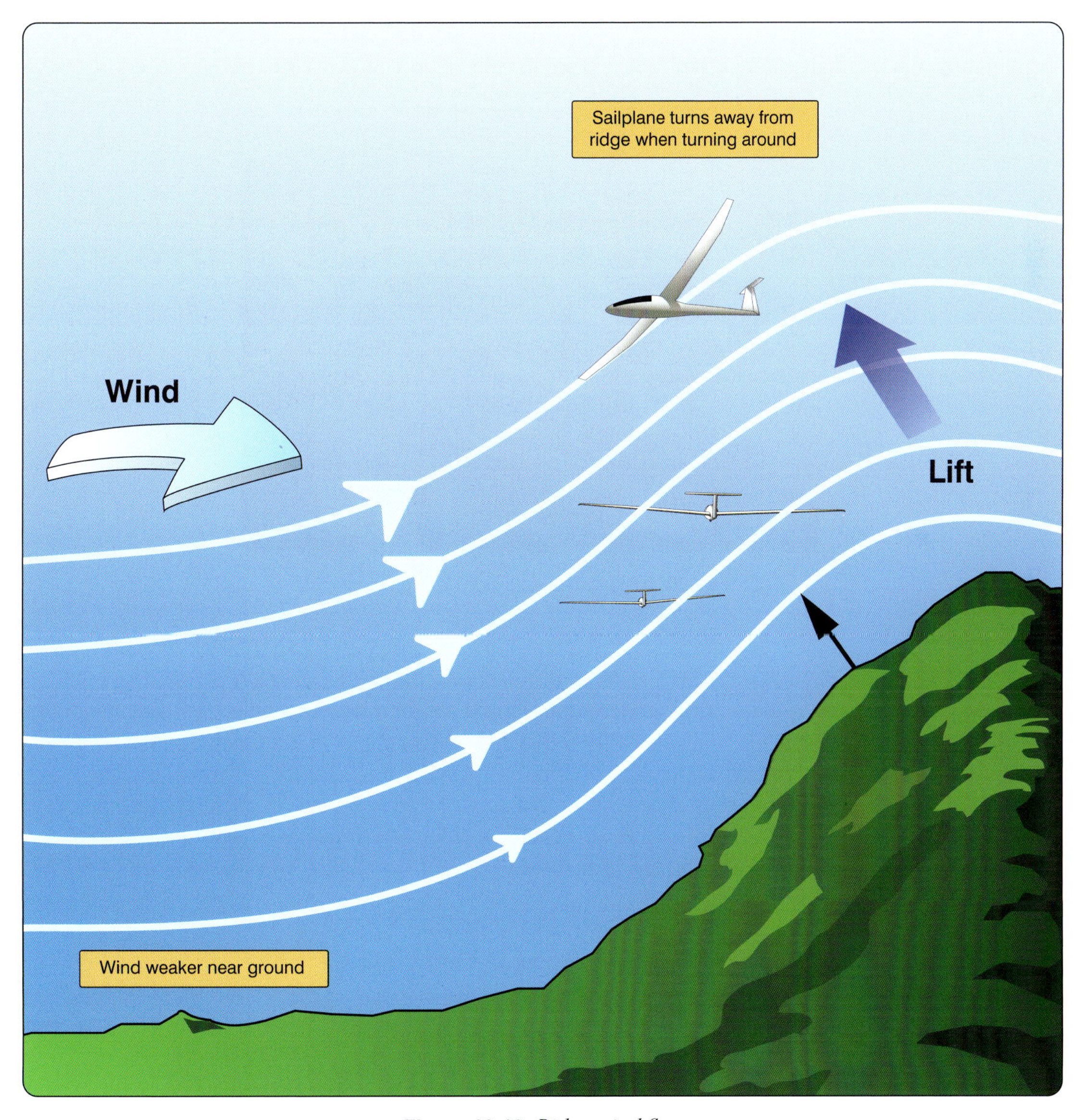

Figure 10-13. *Ridge wind flow.*

Airflow follows the hill or ridge shape. The pilot can imagine a flow of water around the ridge instead of air. However, air can compress and may develop local variations and eddies. [*Figure 10-14*]

Figure 10-14. *Airflow generally follows the hill's shape.*

The more complicated the ridge, the more erratic and hazardous the airflow may become. [*Figure 10-15*]

Figure 10-15. *Irregular profiles may create a hazard.*

Traps

Since traps or dangers exist during ridge soaring, glider pilots should obtain instruction when first learning to ridge soar or slope soar. Pilots should approach the upwind side of the ridge at a 45° angle, so that a quick egress away from the ridge can occur in the absence of lift.

NOTE: When approaching the ridge from downwind, approach the ridge at a diagonal. If excess sink is encountered, this method allows a quick turn away from the ridge. [*Figure 10-16*]

Figure 10-16. *Approach ridges diagonally.*

If gliding above the ridge, an appropriate crab angle prevents drifting over the top into the lee-side downdraft. For the new glider pilot, crabbing along the ridge may seem strange, and the pilot might resort to uncoordinated control input to point the nose along the ridge. This could result in an inefficient and dangerous skid toward the ridge.

Thermal sink can turn the glider upside down, a phenomenon known as upset. A thermal may appear anywhere. When it appears from the opposite side of the ridge, it has strong energy. When flying in complex conditions (winds and thermals), fly with extra speed for positive control of the glider. DO NOT fly on the ridge crest or below the ridge on the downwind side. [*Figure 10-17*]

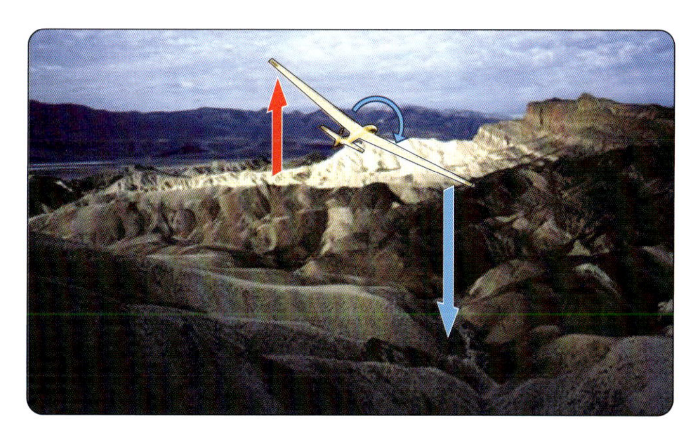

Figure 10-17. *Thermal sink can roll the craft toward the mountain.*

Whenever flying downwind, groundspeed and the radius of any turn will increase. [*Figure 10-18*]

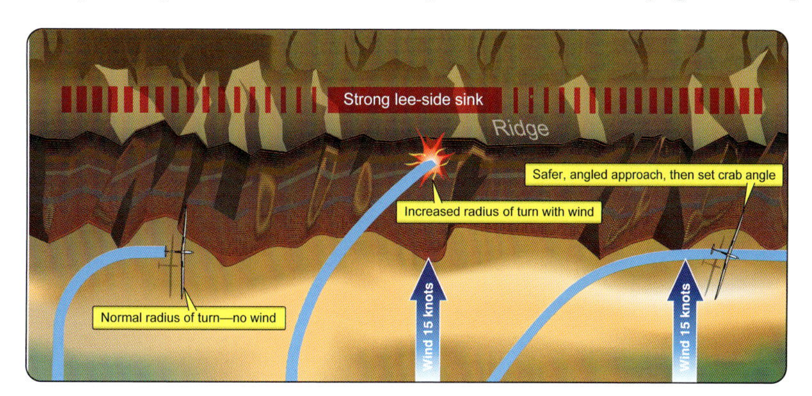

Figure 10-18. *Flying with a wind increases the turn radius over the ground, so approach the ridge at a shallow angle.*

In theory, to obtain the best climb, the pilot should fly at minimum sink speed. Since minimum sink speed gives less margin above stall speed, flying at this speed near terrain may create danger. Maneuverability at minimum sink speed may not provide for adequate control near terrain, especially if gusty wind or thermals complicate the ridge lift. When gliding at or below ridge-top height, the pilot should fly faster than minimum sink speed—how much faster depends on the glider, terrain, and turbulence. When the glider climbs to at least several hundred feet above the ridge and when shifting upwind away from it within the best lift zone, the pilot can reduce speed.

NOTE: When flying close to the ridge, use extra speed for safety—extra speed gives the glider more positive flight control input and enables the glider to fly through areas of sink quickly. [*Figure 10-19*]

Figure 10-19. *When flying close to a ridge, pilots use extra speed for more control and to pass quickly through sink.*

Procedures for Safe Flying

Several procedures enhance safe slope soaring and allow many gliders to use the same ridge simultaneously. The following paragraphs and *Figure 10-20* explain the procedures.

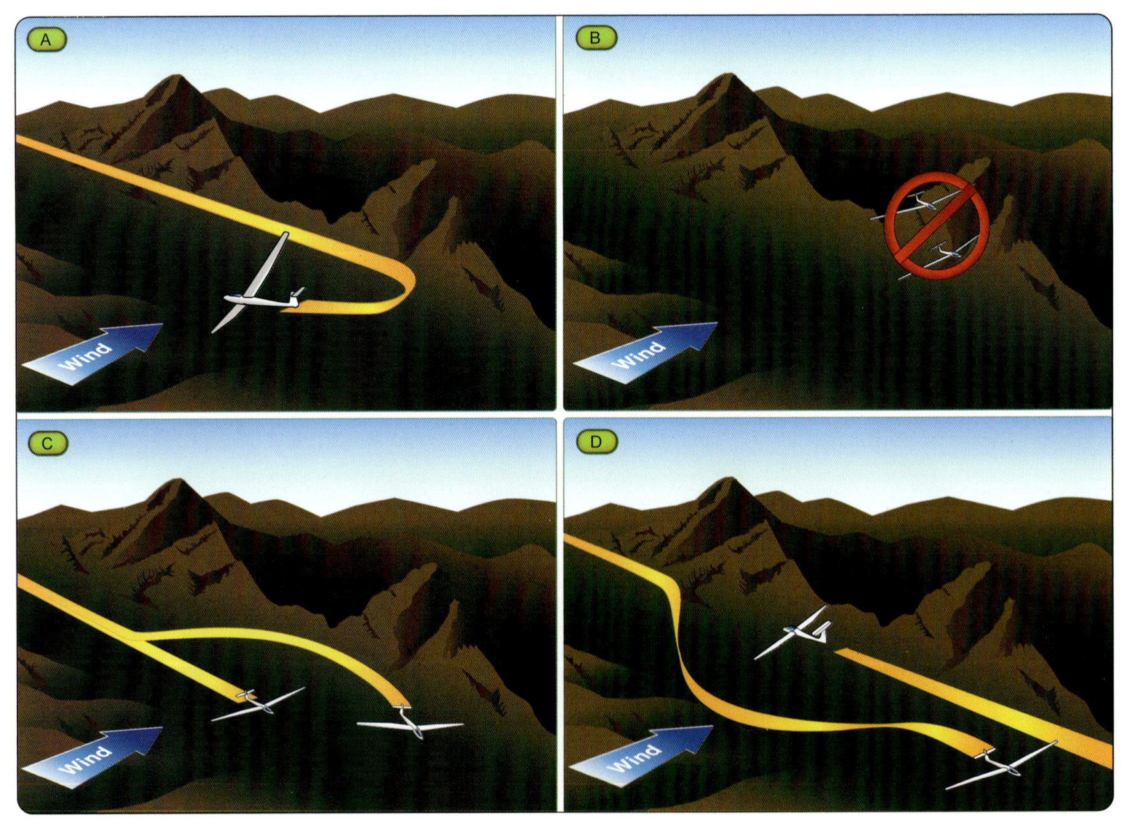

Figure 10-20. *Ridge rules.*

The pilot should:

- Make all turns away from the ridge. [*Figure 10-20A*] A turn toward the ridge creates unnecessary risk, even if positioned well away from the ridge. The groundspeed on the downwind portion of the turn may increase dramatically and lead to a collision with the ridge. Even if above the ridge, a downwind turn may take the glider over the ridge crest and into heavy sink.

- Not fly directly above or below another glider. [*Figure 10-20B*] Pilots in gliders in close vertical proximity might not see the other glider. A slight change in climb rate between the gliders can lead to a collision.

- Pass another glider on the ridge side, anticipating that the other pilot might turn away from the ridge. [*Figure 10- 20C*] If the space available to pass appears inadequate or if the overtaking glider encounters sink, turbulence, etc., it may maneuver away from the ridge. The passing glider should either turn back in the other direction (away from the ridge if spacing permits) or fly upwind away from the ridge and rejoin the slope lift as traffic allows. If using a radio, the pilot passing can try to contact the pilot of the other glider and coordinate. Procedures may differ outside of the United States.

- Understand that Title 14 of the Code of Federal Regulations (14 CFR) requires both aircraft approaching head-on to give way to the right. If a glider with the ridge to the right does not have room to move in that direction, the glider with its left side to the ridge should give way as needed. [*Figure 10-20D*] In general, gliders approaching head-on are difficult to see; therefore, when piloting the glider with its right side to the ridge, the pilot should ensure the approaching glider yields early on. While ridge soaring, pilots should enhance their vigilance. The use of a radio may provide additional safety. Pilots should look at 14 CFR part 91, section 91.111, Operating near other aircraft, and section 91.113, Right-of-way rules: Except water operations.

Bowls & Spurs

With wind blowing at an angle to the ridge, bowls or spurs, formations with recessed or protruding rock formations extending from the main ridge, can create better lift on the upwind side and sink on the downwind side. If at or near the height of the ridge, the pilot can detour around the spur to avoid the sink, then drift back into the bowl to take advantage of the better lift. [*Figure 10-21*] After passing such a spur, the pilot should consider any other traffic and not make abrupt turns toward the ridge. If soaring hundreds of feet above a spur, the pilot may consider flying over it and increase speed in any sink. This requires caution since a thermal in the upwind bowl, or even an imperceptible increase in the wind, can cause greater than anticipated sink on the downwind side. The pilot should always have an escape route available.

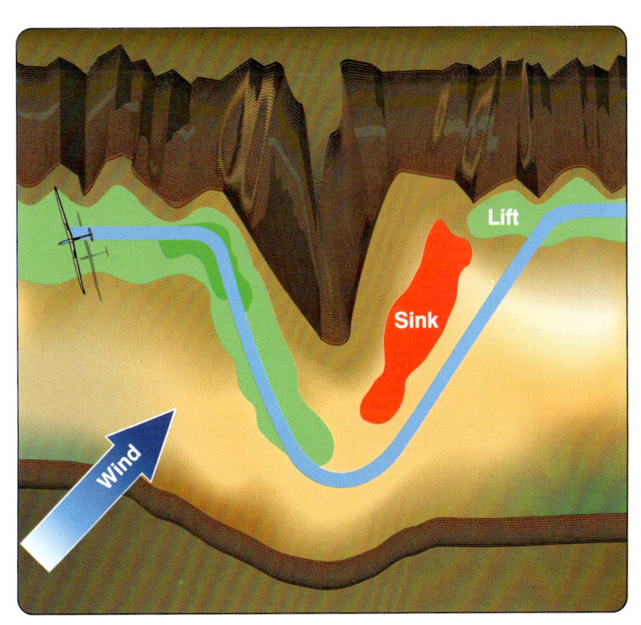

Figure 10-21. *Avoid sink on the downwind side of spurs by detouring around them.*

Slope Lift & Thermalling

Combinations of thermals with slope lift can occur, and slope soaring can provide lift when thermals temporarily shut down. A pilot encountering a thermal along the ridge can make a series of S-turns with each turn into the wind. The glider can drift back to the thermal after each turn if needed. The glider pilot should never continue the turn toward the ridge. Speed helps to control any encounter with strong sink that can occur on the sides of the thermal. The maneuver takes practice, but when done properly, the pilot can make a rapid climb in the thermal to a point well above the ridge crest, where thermalling 360° turns can begin. Even when well above the ridge, the pilot should use caution and ensure the glider does not drift into the lee-side sink during a slow climb. Before trying an S-turn, the pilot should ensure it would not interfere with other traffic along the ridge. [*Figure 10-22*]

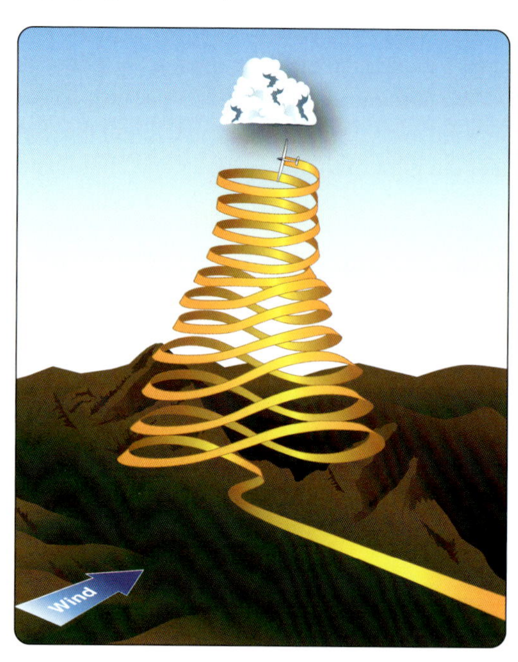

Figure 10-22. *One technique for catching a thermal from ridge lift.*

A pilot can use a second technique for catching thermals when slope soaring by heading upwind away from the ridge. This works best when Cu mark potential thermals, and aids timing. If not finding a thermal, the pilot should cut the search short while still high enough to dash back downwind to the safety of the upwind slope lift. [*Figure 10-23*]

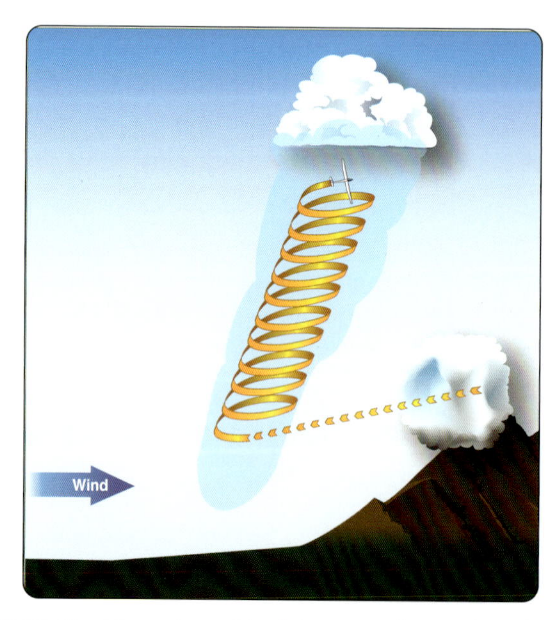

Figure 10-23. *Catching a thermal by flying upwind away from the slope lift.*

Obstructions

A risk of collision exists when flying at extremely low altitudes along a ridge (tree top level). Obstructions include wires, cables, and power lines, all of which the pilot might not see. The pilot should ensure completion of an adequate reconnaissance when flying at these altitudes. Aeronautical charts show high-tension towers that have many wires between them, and soaring pilots familiar with the area can also provide useful information regarding local ridge obstructions.

Tips & Techniques

Observe the ridges slope for collectors or dividers of the wind flow and determine which of the slopes could gather wind flow. [*Figure 10-24*] Due to the changes in wind direction and or sun angle, wind flow can change in a few minutes.

Figure 10-24. *Analyze sloping ground for collectors and dividers of wind.*

- Collectors include mountain/ridge bowls and ends of canyons that can offer extreme areas of lift when the wind blows into them. Remember to have a way out.

- Dividers are ridges parallel with the wind and separate airflow. A collector downwind from a divider may receive more airflow, making better lift possible.

The downwind side of any ridge or hill produces turbulence and sink. Larger or higher ridges and greater wind velocities create wider areas of turbulence. During these conditions, the pilot should remember to keep the seat and shoulder harness tight. Sink calls for speed, and turbulence and speed stress the glider airframe and reduce pilot comfort. Pilots should not exceed glider limitations set forth in the GFM/POH including the design speed for maximum gust intensity (Vb). [*Figure 10-25*]

Figure 10-25. *Expect turbulence and sink on the downwind side of any hill.*

Pilots should assume airflow starts down at the ridge crest. While steep ridges with narrow ridge tops can collect thermal action from both sides of the crest and the best lift may exist directly above the crest, the pilot should not plan to use that lift and always remain upwind of the crest. [*Figure 10-26*]

Figure 10-26. *The crest is where rising air starts back down.*

Deeply shaded areas along the ridge often enhance sinking air. If strong lift exists on the sunny side of the ridge, then strong sink usually exists on the shady or dark side of the ridge. This often happens with low sun angles in the late afternoon. [*Figure 10-27*]

Figure 10-27. *Deep shade can enhance sink.*

A thermal source occurs where drainage areas along the ridge meet. For example, an area with numerous ridges or peaks that slope down into a valley. Canyons or large bowls hold areas of warm air.

Cloud streets may form above the ridge. A glider pilot can try to climb in a thermal to reach these streets. Pilots can use fast cruising speeds under these streets but should stay upwind of the ridge when using the air circulation marked by the streets.

Since glider pilots tend to seek the same known good conditions for lift, thermals and areas of ridge lift can attract significant traffic. Pilots should remain alert for conflict with other aircraft including low flying airplanes and helicopters that use the same area. Sharing a common radio frequency for traffic calls may enhance safety, but nothing replaces a good visual scan for other aircraft.

Density altitude and true airspeed increase as the glider climbs. At 10,000 feet mean sea level (MSL), a pilot needs approximately 40 to 45 percent more room to maneuver. The air is less dense by 2 percent per one thousand feet of altitude gained. [*Figure 10-28*]

Figure 10-28. *At 10,000' MSL, 44 percent more room is needed to complete a turn due to density altitude.*

Wave Soaring

Almost all high-altitude glider flights use mountain lee waves as the primary source of lift. As covered in Chapter 9, Glider Flight and Weather, lee wave systems can contain separate rotor turbulence and smooth wave flow. The use of lee waves for cross-country soaring has enabled flights exceeding 1,500 miles, with average speeds of over 100 mph. [*Figure 10-29*]

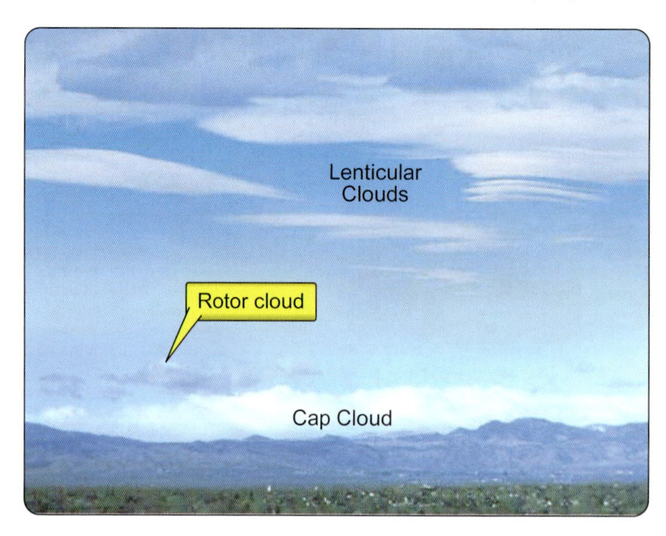

Figure 10-29. *Rotor and cap clouds with lenticulars above.*

Preflight Preparation

Special areas within the continental United States allow glider operations in Class A airspace above 18,000 feet MSL under visual flight rules (VFR). Air Traffic Control (ATC) may open a "wave window" to a specific altitude at specified times. Each wave window has its own set of procedures agreed to by ATC through a Letter of Agreement. Glider pilots should understand the special provisions in the letter of agreement before flying within a wave window.

A flight above 18,000 feet MSL requires extensive preflight preparation. Pilots planning wave flights to lower altitudes can reduce the list of preparation items accordingly.

14 CFR part 91, section 91.211 requires that crewmembers use supplemental oxygen for flight of more than 30 minutes above cabin pressure altitudes of 12,500 feet MSL up to and including 14,000 feet MSL. While above cabin pressure altitudes of 14,000 feet MSL, required crewmembers must use supplemental oxygen. Pilots should preflight the oxygen system, understand signs of hypoxia, know their reactions to high altitude, and consider using oxygen at altitudes well below 12,500 feet MSL.

When flying at high altitudes, the outside air chills the glider interior. Sunlight can help warm portions of the pilot's upper body, but the pilot's lower extremities and feet normally get cold. Pilots planning for a long flight in cold air should wear thermal underwear, warm socks, and shoes and have gloves easily accessible during the flight. Clothing with rechargeable electric heating components can provide hours of warmth.

True airspeed (TAS) becomes a consideration at higher altitudes. To avoid flutter, some glider manuals reduce never-exceed speed (V_{NE}) as a function of altitude. For instance, the Pilot's Operating Handbook (POH) for one common two-seat glider, lists a V_{NE} at sea level of 135 knots. However, at 19,000 feet MSL, it lowers to only 109 knots. Pilots should study the glider's POH carefully for any indicated airspeed limitations.

Some flights might not contact the wave. Sink on the downside of a lee wave can reach 2,000 fpm or more. In addition, missing the wave often means a trip back through the turbulent rotor. To reduce the workload and stress, pilots should calculate minimum return altitudes from several locations before flight. In addition, pilots should plan for worst-case scenarios and consider available off-field landing options if the planned minimum return altitude proves inadequate.

The pilot should begin with a normal preflight of the glider. In addition, the pilot should check the lubricant used on control fittings. Some lubricants get stiff when cold. The pilot should ensure the glider is totally devoid of excess water before flying into freezing temperatures. This includes checking the bottom of the fuselage where water can freeze around rudder and elevator cables and checking spoilers or dive brakes for water from rain or from melting snow cleared from the wing. Water in the spoilers or drive brakes at altitude can freeze and make them difficult or impossible to open. Checking the spoilers or dive brakes occasionally during a high climb helps avoid this problem.

The pilot should check for a freshly charged battery since cold temperatures can reduce battery effectiveness and affect the avionics. The preflight should include checking the radio and accessory equipment, such as the microphone in the oxygen mask. Other specific items to check depend on the systems in the glider.

A briefing with the tow pilot should occur before a wave tow. Prior to flight, the pilots should discuss routes, minimum altitudes, rotor avoidance (if possible), anticipated tow altitude, and other potential situations.

A pilot wearing a bail-out parachute on wave flights should check its proper fitting and use. When wearing a parachute, the seat can suddenly seem cramped. Once seated in the glider, the pilot should check for full, free rudder movement since larger than normal footwear can affect rudder control. In addition, given the bulky cold-weather clothing, the pilot should check canopy clearance. The pilot's head can break a canopy in rotor turbulence, so seat and shoulder belts should be tightly secured even if difficult to achieve with the extra clothing. Proper placement of the oxygen mask should make it easily accessible within a few seconds since the climb in the wave can be very rapid. Securing everything else before takeoff prevents disorder while encountering the rotor.

Getting into the Wave

Access into the wave occurs in two ways: soaring into it or being towed directly into it. Three main wave entries while soaring include thermalling into the wave, climbing the rotor, and transitioning into the wave from slope soaring.

At times, an unstable layer lower than the mountaintop has a strong, stable layer cap, which may support lee waves. A line of cumulus clouds downwind of and aligned parallel to the ridge or mountain range suggests the presence of these waves. With these conditions present, the pilot can avoid the rotor area and thermal into the wave. Whether lee waves are suspected or not, the air near the thermal top may become turbulent. At this point, the pilot should attempt a penetration upwind into the smooth wave. [*Figure 10-30*]

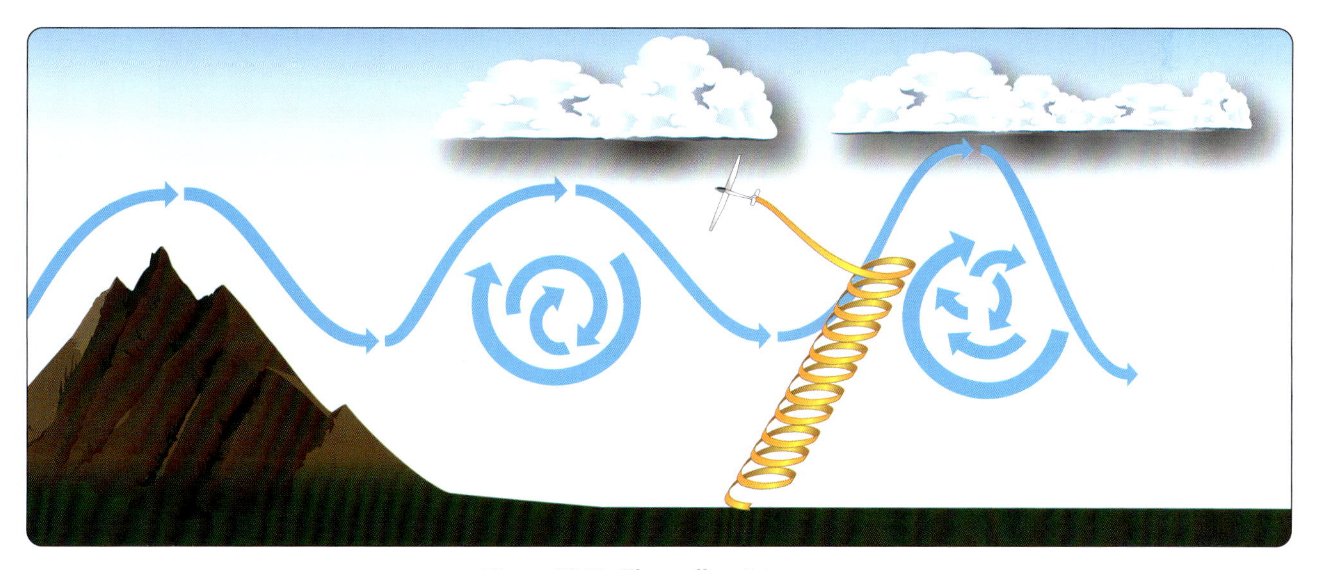

Figure 10-30. *Thermalling into wave.*

Depending on the topography near the soaring site, it may be possible to transition from slope lift into a lee wave created by upwind topography where multiple ridges exist. In this case, the pilot climbs as high as possible in slope lift, then penetrates upwind into the lee wave. With the lee waves in phase with the topography, the pilot can often climb from slope to wave lift without the rotor. At times, the glider pilot may not realize wave has been encountered until finding lift steadily increasing as the glider climbs. Climbing in slope lift and then turning downwind to encounter possible lee waves produced downwind of the ridge should generally not occur. Even with a tailwind, the lee-side sink can put the glider on the ground before contact with the wave.

Another possibility involves a tow into the upside of the rotor and a climb using the rotor to reach the wave. The technique relies on finding the rising part of the rotor. Since rotor lift usually remains stationary over the ground, this may involve flying a "figure-8" in the rotor lift to avoid flying downwind. The pilot can also fly several circles with an occasional straight leg or fly straight into the wind for several seconds until the lift diminishes followed by circling to reposition within the lift. Which choice works best depends on the size of the lift and the strength of the wind. Since the rotor may contain regions of rapidly changing and turbulent lift and sink, staying in it as well as simple airspeed and bank control may prove difficult. Inexperienced pilots should avoid using a tow to the upside of a rotor to reach the wave.

Towing into the wave occurs by towing ahead of the rotor or through the rotor. Complete avoidance of the rotor by towing around it will generally increase the time on tow, but the reduced turbulence increases the tow pilot's willingness to perform future wave tows. [*Figure 10-31*] If the launch site sits near one end of the wave-producing ridge or mountain range, a tow around the rotor and then directly into the wave lift becomes more feasible.

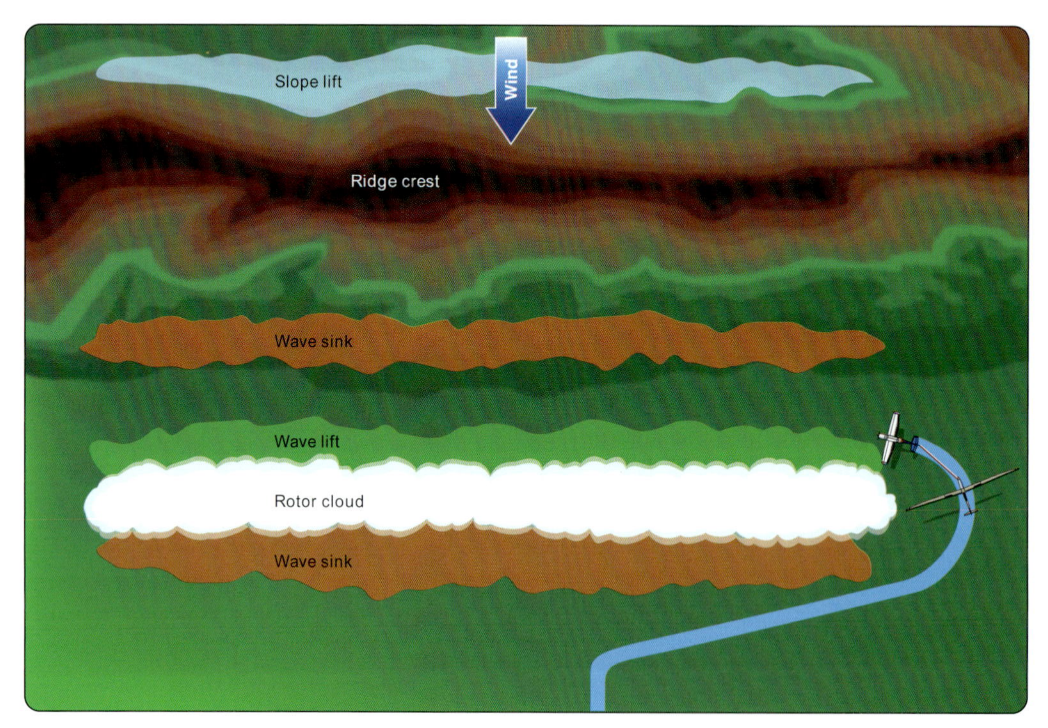

Figure 10-31. *A tow around the rotor directly into the wave avoids turbulence.*

Often, a tow directly through the rotor provides the only route to the wave. The tow will usually encounter moderate to severe rotor turbulence. The nature of rotor turbulence differs from a turbulent thermal. The rotor subjects the aircraft to sharp, chaotic horizontal and vertical gusts along with rapid accelerations and decelerations. At times, the rotor can become so rough that even experienced pilots will elect to remain on the ground. Any pilot without experience flying through rotors should obtain instruction before attempting a tow through a rotor.

During a tow through a rotor, the glider often gets out of position, and the glider pilot should attempt to maintain position horizontally and vertically. Turbulence too violent to handle may require an immediate release. Slack-producing situations occur commonly due to a rapid deceleration of the towplane. The glider pilot should recognize the onset of slack line and correct accordingly. The glider pilot should maintain the high-tow position because any tow position lower than normal runs the risk of the slack line coming back over the glider. On the other hand, the glider should fly no higher than normal to avoid a forced release should the towplane suddenly drop. Gusts may also cause an excessive bank of the glider, and it may take a moment to roll back to a level attitude. The pilot may need to use full aileron and rudder deflection for a few seconds.

The trend of the variometer often indicates the progress through the rotor. General downswings get replaced by general upswings, usually along with increasing turbulence. The penetration into the smooth wave lift can occur in a matter of seconds or it can occur gradually. The glider pilot should note any lenticulars above as a position upwind of the clouds helps confirm contact with the wave. If in doubt, the tow may continue for a few moments longer to confirm wave contact. Once confident of the wave lift, the glider pilot makes the release. If on a crosswind heading, the glider should release and fly straight or with a crab angle. If flying directly into the wind, the glider should turn a few degrees to establish a crosswind crab angle. The pilot should avoid drifting downwind and immediately losing the wave. After release, the towplane should descend and/or turn away and create separation from the glider. The glider and tow pilot should brief any nonstandard release procedures before takeoff. [*Figure 10-32* and *Figure 10-33*]

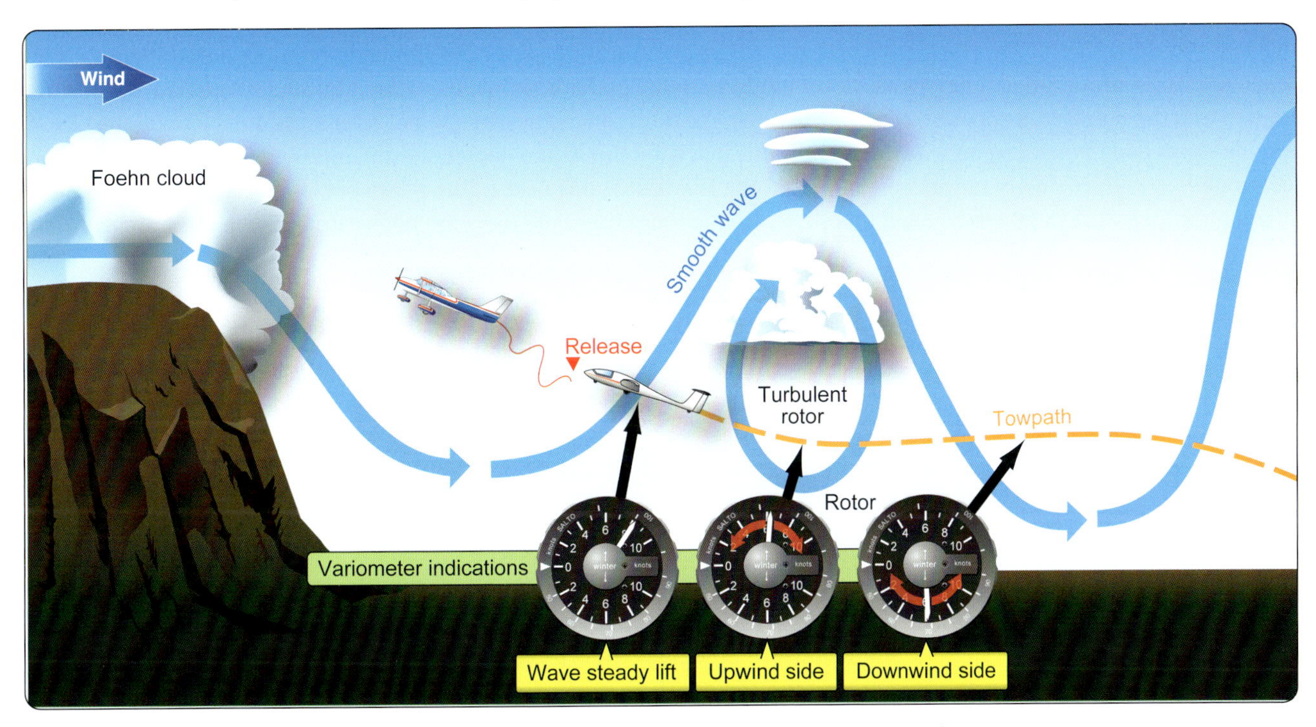

Figure 10-32. *Variometer indications during the penetration into the wave.*

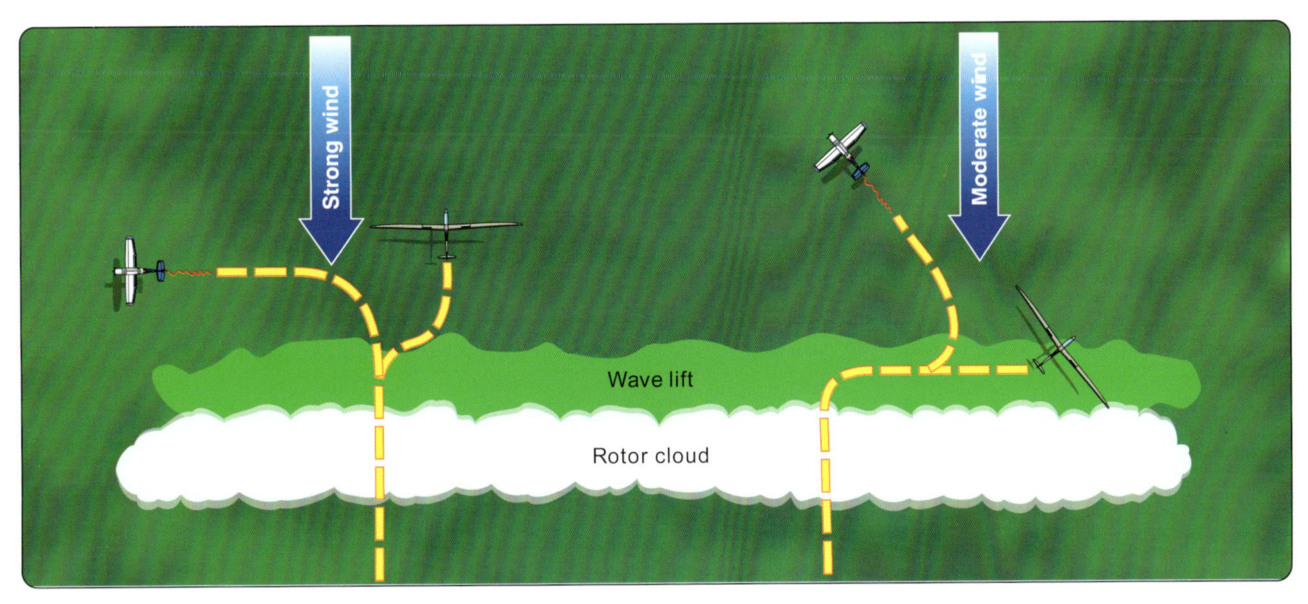

Figure 10-33. *Possible release and separation on a wave tow.*

Flying in the Wave

After wave contact, the best technique for utilizing the lift depends on the extent of the lift and the strength of the wind. In weak lift, the pilot should stay with the initial slow climb as better lift should develop as the climb continues. At other times, the variometer may peg at 1,000 fpm directly after release from tow.

In strong winds (40 knots or more), the pilot should find the strongest portion of the wave, point into the wind, and adjust speed so that the glider remains in the strong lift. The best lift usually occurs along the upwind side of the rotor cloud or just upwind of any lenticulars. In the best-case scenario, the required speed matches the glider's minimum sink speed. In quite strong winds, the pilot flies faster than minimum sink to maintain position in the best lift. Under those conditions, flying slower would allow the glider to drift downwind (fly backward over the ground) and into the downside of the wave. Once on the downside, getting back to the frontside requires penetrating a strong headwind. With strong lift, stronger winds aloft might push the glider downwind, so the pilot should monitor the position relative to rotor clouds or lenticulars. If no clouds exist, the pilot can use nearby ground references and increase speed with altitude as needed to maintain position in the best lift. In a wind not strong enough for the glider to remain stationary over the ground, the glider slowly moves upwind out of the best lift. If this occurs, the pilot should turn slightly from a direct upwind heading, drift slowly downwind into better lift, and turn back into the wind before drifting too far. [*Figure 10-34*]

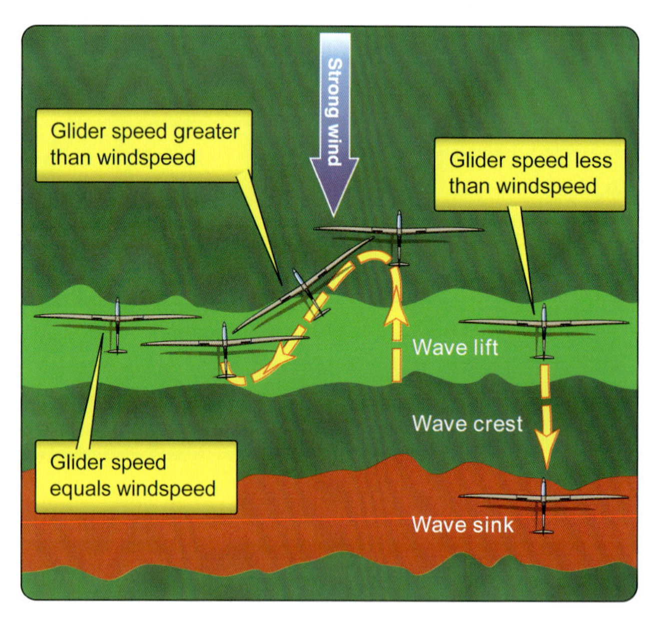

Figure 10-34. *Managing wave position with speed.*

Often, the wave lift moves over the ground since small changes in windspeed or stability can alter the wavelength of the lee wave within a few minutes. If lift begins to decrease while climbing in the wave, one of these things has occurred: the glider approached the top of the wave, the glider moved out of the best lift, or the wavelength of the lee wave has changed. In any case, the pilot can explore the area for better lift by searching upwind first. Searching upwind allows the pilot to drift downwind back into the rising part of the wave if not finding better lift upwind. Searching downwind first can make it difficult or impossible to contact the lift again if encountering sink on the downwind side of the wave. In addition, the pilot might exceed the glider's maneuvering speed or redline for rough air as gliding from the downwind to upwind could put the glider back in the rotor. [*Figure 10-35*]

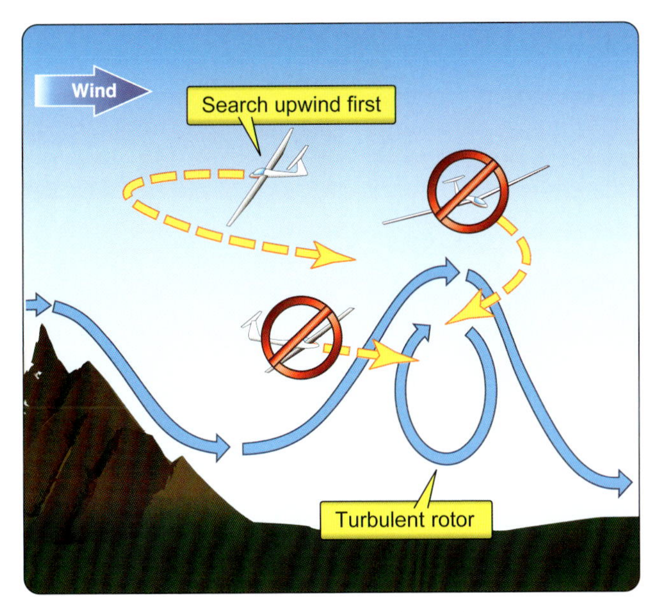

Figure 10-35. *Search upwind first to avoid sink behind the wave crest or the rotor.*

In moderate winds (20 to 40 knots) and if the wave extends along the ridge or mountain range for a few miles, the pilot can fly back and forth along the wave lift while crabbing into the wind. The pilot can use the rotor cloud or lenticular as a reference. All turns should occur into the wind to avoid moving to the downside of the wave or back into the rotor. When making an upwind turn to change course 180°, the pilot changes heading less than 180°, with the reduction in turn based on the strength of the wind. The pilot notes the crab angle needed to stay in lift on the first leg and can use that same amount of into-the-wind crab angle initially after completing the next upwind turn. With no cloud, ground references allow the pilot to establish and maintain the proper crab angle. While climbing higher into sufficiently strong winds, the pilot may transition from crabbing back and forth to a stationary upwind heading. [*Figure 10-36*]

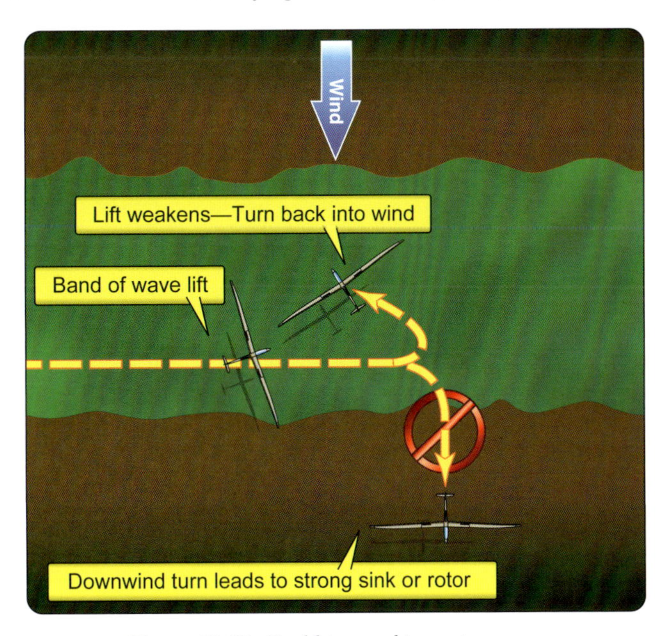

Figure 10-36. *Crabbing and turns in a wave*

Weaker winds (15 to 20 knots) may call for different techniques. Lee waves from smaller ridges can form in relatively weak winds of approximately 15 knots, and wave lift from larger mountains rapidly decreases when climbing to a height where winds aloft diminish. In a small area that still provides lift near the wave top, the pilot can fly shorter figure 8 patterns to reach the maximum altitude. The pilot can also fly an oval-shaped pattern straight into the wind in lift and fly a quick 360° turn to reposition and as it diminishes. If a consistent climb is not possible, the pilot can fly a series of circles with an occasional leg into the wind to avoid drifting too far downwind. In a sufficiently large lift area, the pilot can use a technique like that used in moderate winds. [*Figure 10-37*]

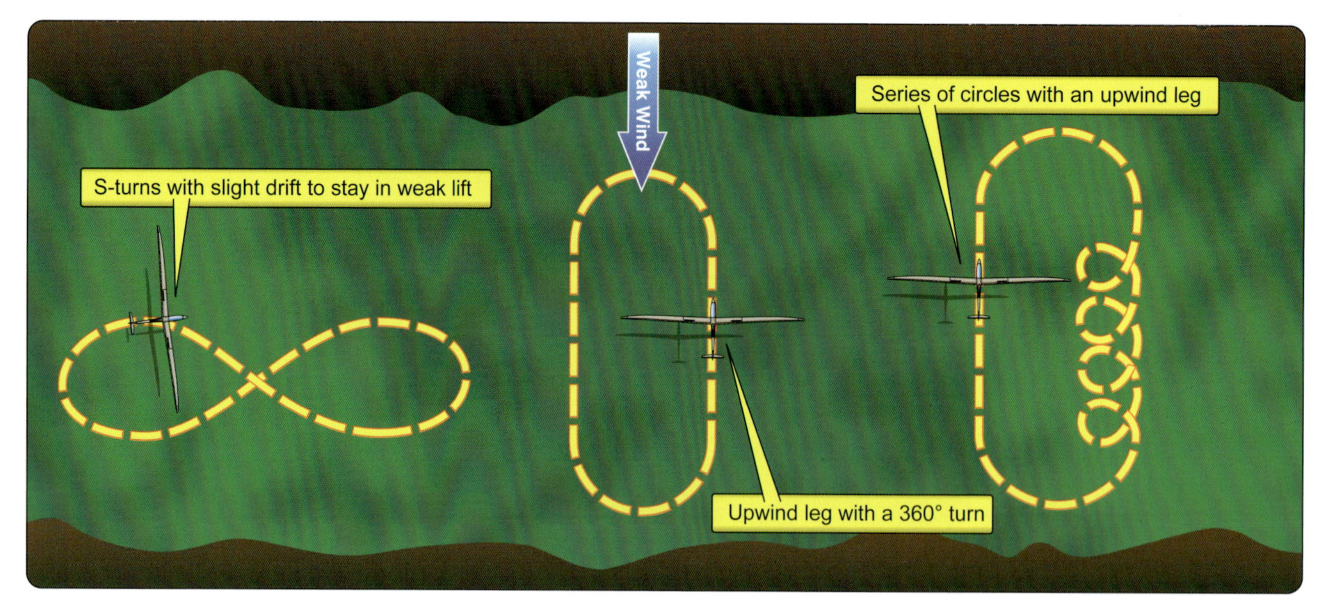

Figure 10-37. *Techniques for working lift near the top of the wave in weak winds.*

The discussion thus far assumed a climb in the primary wave. The pilot can also climb using any secondary or tertiary lee wave and then penetrate the next wave upwind. The success of this strategy depends on wind strength, clouds, the intensity of sink downwind of wave crests, and the performance of the glider. Depending on the height attained in the secondary or tertiary lee wave, a trip through the rotor of the next wave upwind could occur. Pilots should exercise caution if penetrating upwind at high speed. The transition into the downwind side of the rotor can be as abrupt as on the upwind side, so the pilot should reduce speed at the first hint of turbulence. In any case, the glider could lose a significant amount of altitude while penetrating upwind through the sinking side of the next upwind wave. [*Figure 10-38*]

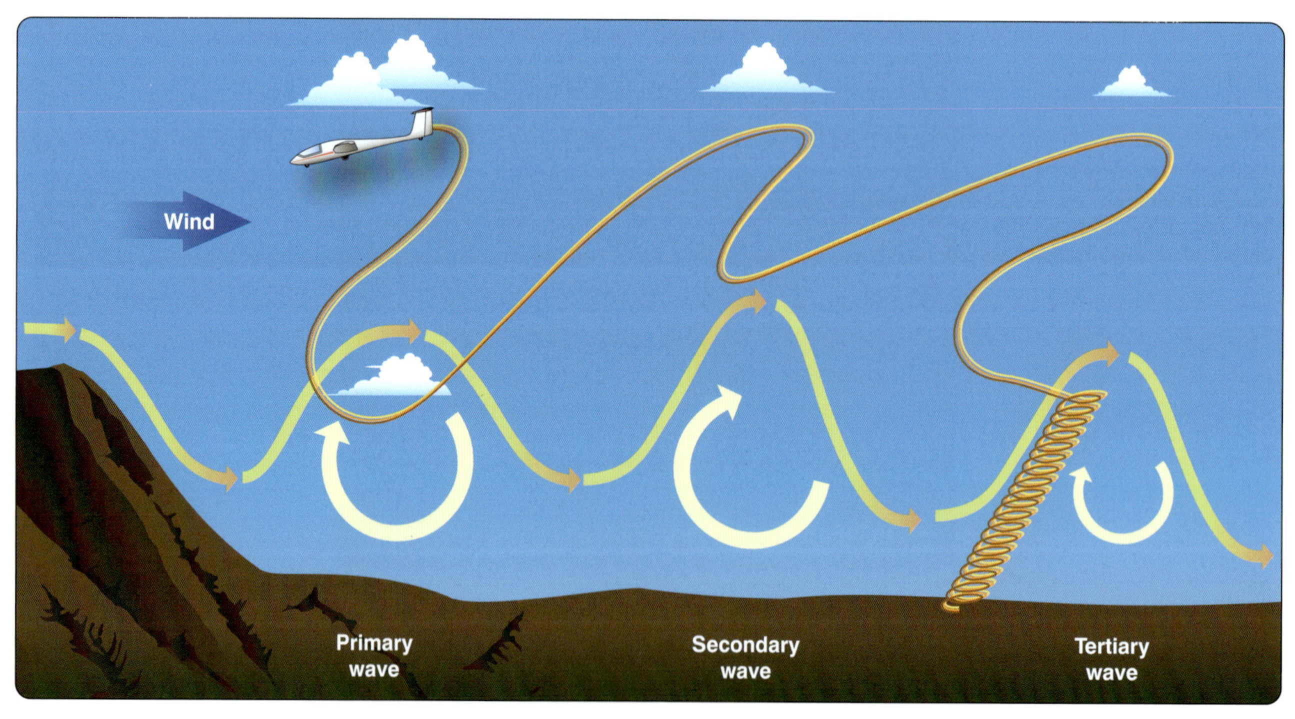

Figure 10-38. *Possible flightpath while transitioning from the tertiary into the secondary and then into the primary.*

The sink downwind of the wave crest can assist a pilot who decides to make a quick descent as sink can easily attain twice the strength of the lift encountered on the upwind side of the wave crest. Eventual descent into downwind rotor might also occur. An inadequate space between a rotor cloud and overlying lenticulars can prevent a safe downwind transition that might then occur with reduced visibility. In this case, the pilot can make a crosswind detour if a short ridge or mountain range produces the wave. If clouds negate a downwind or crosswind departure from the wave, a descent on the upwind side of the wave crest can occur. Spoilers or dive brakes may be used to descend through the updraft, followed by a transition under the rotor cloud and through the rotor. The pilot should control speed during flight through the rotor. In addition, lift on the upwind side of the rotor may make it difficult to stay out of the rotor cloud. This type of descent requires caution and emphasizes the importance of an exit strategy before climbing too high in the wave. Pilots should remember that conditions and clouds can evolve rapidly during the climb.

Some of the dangers and precautions associated with wave soaring include:

- Symptoms of hypoxia—check the oxygen system, and immediately begin a descent to lower altitudes that do not require supplemental oxygen. Do not delay!
- Extreme cold—descend before becoming uncomfortably cold.
- Severe or extreme rotor turbulence—exercise caution on tow and when transitioning from smooth wave flow (lift or sink) to rotor. Rotors near the landing area can cause strong shifting surface winds of 20 or 30 knots. Wind shifts up to 180° sometimes occur in less than a minute at the surface under rotors.
- Restricted vision—warm, moist exhaled air may cause frost formation on the canopy and restrict vision. Opening air vents may alleviate the problem or delay frost formation. The pilot can use heated panels or descend before frost becomes a hazard.
- Entrapment above clouds—wet waves associated with a great deal of cloud formation may close gaps beneath the glider and the pilot should descend in visual conditions before becoming trapped. If trapped above clouds, the pilot could attempt a benign spiral through the cloud as an emergency maneuver only if previously explored and stable for the glider in visual conditions.
- Inadvertent night flight—at sunset, bright sunshine still exists at 25,000 feet while the ground below gets quite dark. Know the time of actual sunset. Even at an average 1,000 fpm descent, it takes 20 minutes to lose 20,000 feet.

Caution: Flights under a rotor cloud can encounter high sink rates and pilots should approach those areas with caution.

Soaring Convergence Zones

Pilots can most easily spot a convergence zone in the presence of cumulus clouds. They may appear as a single well-defined straight or curved cloud street. The edge of a field of cumulus can mark convergence between a relatively moist or unstable mesoscale air mass from a drier or more stable one. Often, the cumulus along convergence lines have a base lower on one side.

With no cloud present, pilots can sometimes spot a convergence zone by a difference in visibility across it, which may be subtle or distinct. Even without any clues in the sky, conditions on the ground can indicate a convergence zone. Pilots can look for wind differences on lakes a few miles apart. A lake showing a wind direction different from the ambient flow for the day may indicate conditions that can create a convergence zone. Wind direction shown by blowing smoke can also indicate convergent conditions. A few dust devils, or a short line of them, may indicate the presence of ordinary thermals versus those triggered by convergence. Spotting these subtle clues takes practice and good observational skills and explains why a few pilots can continue soaring while other cannot.

The best soaring technique for this type of lift depends on the nature of the convergence zone itself. For instance, curtain clouds mark a well-defined, sea-breeze front, and the pilot can fly straight along the line in steady lift. A weaker convergence line often produces more lift than sink. An even weaker convergence line may simply serve as the focus for more frequent thermals, and the pilot can use normal thermalling techniques such as flying slower in lift and faster in sink along the convergence line. Some combination of straight legs along the line with an occasional stop to thermal might provide sufficient lift to stay aloft.

Convergence zone lift can at times become turbulent, especially if air mixes from different sources, such as along a sea-breeze front. The general roughness could indicate a convergence line. When narrow, rough, and strong thermals exist within the convergence line, the pilot can work these areas like any other difficult thermals by using steeper bank angles and more speed for maneuverability.

Combined Sources of Updrafts

Lift sources categorize into four types: thermal, slope, wave, and convergence. Often, more than one type of lift exists at the same time, such as thermals with slope lift, thermals leading to an existing wave, convergence zones enhancing thermals, thermal waves, and wave and slope lift. In mountainous terrain, all four lift types may exist on a single day. The glider pilot needs to remain mentally nimble to take advantage of various types and locations of rising air during the flight.

Rising air might not always come from these four lift categories. Sources of lift that do not fit one of the four lift types discussed probably exist. For instance, a few reports suggest pilots soared in travelling waves from an unknown source. At some soaring sites, debate exists over the classification of the type of lift. This should not create a problem if the pilot can work the lift as needed, get safely back on the ground, and ponder the source of lift after the flight.

Chapter Summary

Pilots should understand the source of lift they intend to use for a given flight. They should also understand how to maximize that potential lift and avoid spending extra time in sink. Learning to take full advantage of each kind of lift takes patience and experience. In the interest of safety, pilots should follow certain rules that apply to different soaring regimes. These include positioning the glider so the pilot can see other gliders in a thermal, yielding the right-of-way as needed when approaching another glider during ridge soaring, avoiding downwind turns that could bring the glider in contact with ridge terrain, respecting the potential for rotor turbulence, and knowing how to control the glider if encountering extreme turbulence during wave soaring. Since flying a glider involves dynamic conditions, a pilot should always have a place to land in mind should a landing become necessary.

Chapter 11: Cross-Country Soaring

Introduction

A cross-country flight occurs when the pilot flies the glider beyond gliding distance from the local soaring site. Cross-country flying requires more preparation and decision-making than a local flight. The pilot should determine if the glider, equipment, and the pilot can maintain safely given the known and expected environmental conditions along the route of flight.

Flight Preparation & Planning

If planning to use thermals, the pilot should consider the availability and strength of thermals, if they will remain active, landing possibilities, and which airports along the course have a runway compatible with prevailing wind conditions. The pilot should also consider what effect winds will have on gliding distance and the best speed to fly in sink between thermals. While the main part of this chapter describes flying cross-country using thermals, increased preparation for cross-country flights also involves other sources of lift, and this chapter provides a brief description of cross-country soaring using ridge or wave lift.

Getting Ready for Cross-Country Glider Flights

Adequate soaring skills indicate pilot readiness for cross-country soaring. Until the pilot has flown several flights more than 2 hours and can locate and utilize thermals consistently, the pilot should focus on improving those skills before attempting cross-country flights.

Any cross-country flight could end in an off-field landing, so pilots planning to fly on cross-country flights should also perfect their short-field landing skills. Pilots can practice these landings on local flights by setting up a simulated off-field landing area at an airport or glider port. The first few simulated landings should utilize the services of an instructor, and the pilot should practice several landings without using the altimeter. Pilots should try to avoid interfering with the normal flow of traffic during simulated off-field landings.

In addition, the pilot should know what airspace exits along the planned route including different classes (B, C, D, E, G), restricted areas, prohibited areas, and military operations areas. The pilot should understand implications of any airways along the flightpath. Once comfortable with the sectional during ground study, pilots can locate landmarks and features within a few miles of the soaring site from the air.

While short-field training can use a known area, site selection training involves selecting a landing site from the air. Using a self-launching glider or other powered aircraft to practice landing area selection allows simulated approaches to different selected areas in a condensed time frame.

Glider pilots can use an electronic or paper Sectional Aeronautical Chart to determine position along a route of flight during cross-country flights. These charts publish with updates every 56 days and contain general information, such as topography, cities, major and minor roads and highways, lakes, and other features that may stand out from the air, such as a ranch in an otherwise featureless prairie. In addition, sectionals show the location of public, and some private, airports, airways, restricted and warning areas, and boundaries and vertical limits of different classes of airspace. Charted information on airports includes field elevation, orientation and length of all paved runways, runway lighting, and radio frequencies in use. Each sectional features a comprehensive legend. A detailed description of the sectional chart is found in FAA-H-8083-25, the Pilot's Handbook of Aeronautical Knowledge. *Figure 11-1* shows a sample sectional chart.

Figure 11-1. *Excerpt from a Sectional Aeronautical Chart.*

Pilots should spend a significant amount of time studying sectional charts on the ground. This includes flying some "virtual" cross-country flights in various directions from the local soaring site. In addition to studying the terrain (hills, mountains, large lakes) that may affect the soaring along the route, the pilot should study the various lines and symbols and answer the following questions:

- What airports are available on course?
- Do any have a control tower?
- What do all the numbers and symbols for each airport mean?
- What do other chart symbols mean?

GPS navigation and moving map displays enhance awareness of position during flight or after an off-field landing, and pilots should take advantage of this technology. However, pilots should practice verifying position with a sectional chart while in the air in case the GPS fails.

Any cross-country flight may end with a landing away from the home soaring site, so pilots and crews should prepare for that occurrence prior to flight. Sometimes an aerotow retrieval can take place if the flight terminates at an airport; however, trailer retrievals occur more often. Both the trailer and tow vehicle should be in good operating condition before the pilot departs on a cross-country flight, and the pilot and retrieval crew should discuss communication options prior to flight.

The pilot should obtain a standard briefing and a soaring forecast from an approved weather source. The briefing should include general weather information for the planned route, as well as any NOTAMs, AIRMETs, or SIGMETs, winds aloft, approaching front, or areas of likely thunderstorm activity. Depending on the weather outlook, inexperienced pilots may find it useful to discuss options with more experienced pilots at their soaring site.

The pilot should contact a briefer or NWS-derived source for a current soaring forecast. As briefly discussed in Chapter 9, Glider Flight and Weather, certain NWS Weather Forecast Offices (WFO) issue soaring forecasts. These automated forecasts primarily derive from the radiosonde observation or model-generated soundings. The content and format of a soaring forecast vary depending on the NWS WFO providing the forecast and the needs of the local soaring community. A soaring forecast issues once per day without continuous monitoring or updating after initial issuance. The content and format of a soaring forecast as well as the issuance times are subject to change without prior notice. The following sample soaring forecast came from the WFO in Salt Lake City, Utah [*Figure 11-2*]:

Soaring Forecast
National Weather Service Denver/Boulder, Colorado
645 AM MDT Wednesday August 25, 2010

This forecast is for Wednesday August 25, 2010:

If the trigger temperature of 77.3 F/25.2 C is reached...then
Thermal Soaring Index...................... Excellent
Maximum rate of lift........................ 911 ft/min (4.6 m/s)
Maximum height of thermals................. 16119 ft MSL (10834 ft AGL)

Forecast maximum temperature.................. 89.0 F/32.1 C
Time of trigger temperature.................... 1100 MDT
Time of overdevelopment........................ None
Middle/high clouds during soaring window....... None
Surface winds during soaring window........... 20 mph or less
Height of the -3 thermal index................. 10937 ft MSL (5652 ft AGL)
Thermal soaring outlook for Thursday 08/26..... Excellent

Wave Soaring Index............................ Poor
Wave Soaring Index trend (to 1800 MDT)......... No change
Height of stable layer (12-18K ft MSL)......... None
Weak PVA/NVA (through 1800 MDT)............... Neither
Potential height of wave....................... 14392 ft MSL (9107 ft AGL)
Wave soaring outlook for Thursday 08/26........ Poor

Remarks...

Sunrise/Sunset.................... 06:20:55 / 19:42:44 MDT
Total possible sunshine........... 13 hr 21 min 49 sec (801 min 49 sec)
Altitude of sun at 13:01:25 MDT... 60.82 degrees

Upper air data from rawinsonde observation taken on 08/25/2010 at 0600 MDT

Freezing level................. 15581 ft MSL (10296 ft AGL)
Additional freezing level....... 54494 ft MSL (49209 ft AGL)
Convective condensation level... 13902 ft MSL (8617 ft AGL)
Lifted condensation level....... 14927 ft MSL (9641 ft AGL)
Lifted index.................... -3.4
K index......................... +9.7

* * * * * * Numerical weather prediction model forecast data valid * * * * * *

08/25/2010 at 0900 MDT | 08/25/2010 at 1200 MDT
|
K index... +4.0 | K index... -0.7

This product is issued twice per day, once by approximately 0630 MST/0730 MDT (1330 UTC) and again by approximately 1830 MST/1930 MDT (0130 UTC). It is notcontinuously monitored nor updated after its initial issuance.

The information contained herein is based on rawinsonde observation and/or numerical weather prediction model data taken near the old Stapleton Airport site in Denver, Colorado at

North Latitude: 39 deg 46 min 5.016 sec
West Longitude: 104 deg 52 min 9.984 sec
Elevation: 5285 feet (1611 meters)

and may not be representative of other areas along the Front Range of the Colorado Rocky Mountains. Note that some elevations in numerical weather prediction models differ from actual station elevations, which can lead to data which appear to be below ground. Erroneous data such as these should not be used.

The content and format of this report as well as the issuance times are subject to change without prior notice. Comments and suggestions are welcome and should be directed to one of the addresses or phone numbers shown at the bottom of this page. To expedite a response to comments, be sure to mention your interest in the soaring forecast.

DEFINITIONS:

Convective Condensation Level - The height to which an air parcel possessing the average saturation mixing ratio in the lowest 4000 feet of the airmass, if heated sufficiently from below, will rise dry adiabatically until it just becomes saturated. It estimates the base of cumulus clouds that are produced by surface heating only.

Convection Temperature (ConvectionT) - The surface temperature required to make the airmass dry adiabatic up to the given level. It can be considered a "trigger temperature" for that level.

Freezing Level - The height where the temperature is zero degrees Celsius.

Height of Stable Layer - The height (between 12,000 and 18,000 feet above mean sea level) where the smallest lapse rate exists. The location and existence of this feature is important in the generation of mountain waves.

K Index - A measure of stability which combines the temperature difference between approximately 5,000 and 18,000 feet above the surface, the amount of moisture at approximately 5,000 feet above the surface, and a measure of the dryness at approximately 10,000 feet above the surface. Larger positive numbers indicate more instability and a greater likelihood of thunderstorm development. One interpretation of K index values regarding soaring in the western United States is given in WMO Technical Note 158 and is reproduced in the following table:

below -10	no or weak thermals
-10 to 5	dry thermals or 1/8 cumulus with moderate thermals
5 to 15	good soaring conditions
15 to 20	good soaring conditions with occasional showers
20 to 30	excellent soaring conditions, but increasing probability of showers and thunderstorms
above 30	more than 60 percent probability of thunderstorms

Lapse Rate - The change with height of the temperature. Negative values indicate inversions.

Lifted Condensation Level - The height to which an air parcel possessing the average dewpoint in the lowest 4000 feet of the airmass and the forecast maximum temperature must be lifted dry adiabatically to attain saturation.

Lifted Index - The difference between the environmental temperature at a level approximately 18,000 feet above the surface and the temperature of an air parcel lifted dry adiabatically from the surface to its lifted condensation level and then pseudoadiabatically thereafter to this same level. The parcel's initial temperature is the forecast maximum temperature and its dewpoint is the average dewpoint in the lowest 4000 feet of the airmass. Negative values are indicative of instability with positive values showing stable conditions.

Lift Rate - An experimental estimate of the strength of thermals. It is computed the same way as the maximum rate of lift but uses the actual level rather than the maximum height of thermals in the calculation. Also, none of the empirical adjustments based on cloudiness and K-index are applied to these calculations.

Maximum Height of Thermals - The height where the dry adiabat through the forecast maximum temperature intersects the environmental temperature.

Figure 11-2. *Sample Soaring Forecast.*

In the sample soaring forecast depicted above, the line for Height of the -3 Thermal Index represents the difference between the environmental temperature and the temperature at a particular level determined by following the dry adiabat through the forecast maximum temperature up to that level. Increasing magnitude of a negative value indicates stronger thermal lift. A value of -3 or below generally indicates thermal activity favorable for cross-country soaring. Soaring pilots should consult with the NWS WFO in their soaring area for more information about the soaring forecast.

Finalizing plans

Pilots may have specific goals for their upcoming cross-country flight and should plan based on the area and different weather scenarios. If planning for a closed-course 300 nautical mile (NM) flight, the pilot should consider several possible out-and-return or triangle courses ahead of time. On the day of departure, the pilot can select the best pre-planned option based on the weather outlook. Since numerous final details need attention on the morning of the flight, accounting for the items used during flight should take place the day before.

Lack of preparation can lead to delays, which may not leave enough of the soaring day to accomplish the planned flight. Even worse, poor planning leads to hasty last-minute preparation and a rush to launch, making it easy to miss critical safety items.

Inexperienced and experienced pilots alike should use checklists for various phases of the cross-country preparation to organize the details. When properly used, checklists can help avoid oversights, such as missed assembly items, sectionals left at home, barograph not turned on before takeoff, oxygen status, drinking water in the glider, etc. Checklists can cover the following:

- Items to take to the gliderport (food, water, battery, charts, barograph).
- Assembly in accordance with the Glider Flight Manual/Pilot's Operating Handbook (GFM/POH), including assembly check and any other items as needed.
- Positive control check.
- Prelaunch (charts, barograph, glide calculator, oxygen on).
- Briefings for tow pilot, ground crew, and retrieval crew.
- Pre-takeoff.

Being better organized before the flight leads to less stress during the flight and enhances flight safety.

Personal & Special Equipment

Many items not required for local soaring ensure pilot comfort on long cross-country flights. An adequate supply of drinking water prevents dehydration. Some pilots use a backpack drinking system with a readily accessible hose and bite valve. This system stows easily beside the pilot and allows frequent sips of water. Pilots should also have food onboard since cross-country flights can last up to 8 hours or more. Pilots should also have a relief system on longer flights.

Several items carried onboard can assist in case of an off-field landing. (For more details, see Chapter 8, Abnormal and Emergency Procedures.) For example, the land-out kit should include a system for securing the glider. The remaining contents of a land-out kit depend on the population density and climate of the soaring area. A safe landing site may occur many miles from the nearest road, and the land-out kit should include extra water and food for this contingency. Walking shoes could prove valuable should the pilot need to hike to a structure some distance away. A mobile phone proves useful for landings in areas with cell coverage. Some pilots elect to carry an Emergency Position Indicating Radio Beacon (EPIRB) in case of mishap during a remote off-field landing.

Cross-country soaring requires some means of measuring distances to the next source of lift or the next suitable landing area. Pilots can use a GPS system for measuring distances or a mechanical system such as a plotter and paper chart. [*Figure 11-3*]

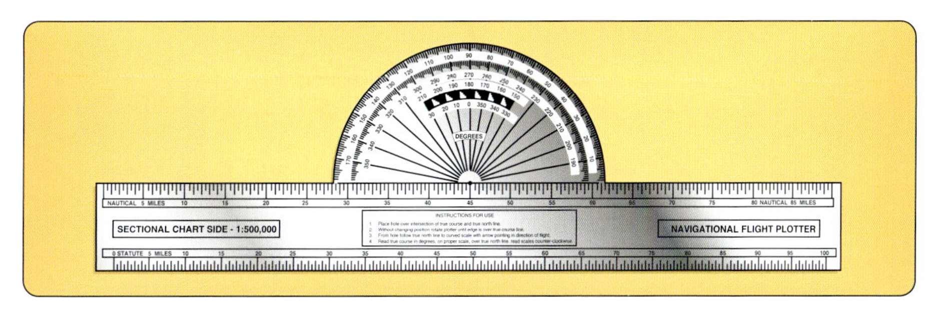

Figure 11-3. *Navigational plotter.*

Glide calculations take headwinds or tailwinds into account, as well as speeds to fly through varying sink rates as discussed in chapter 5, Glider Performance. Tools range widely in their level of sophistication, but all account for the performance polar for the glider. The simplest glide aid derived from the polar consists of a table showing altitudes required for distance versus wind. Another option consists of a circular glide calculator as shown in *Figure 11-4*. The settings in *Figure 11-4* indicate that a glide of 18 miles in an estimated 10 knot headwind takes 3,600 feet. Note that this only gives the altitude required to make the glide. High-performance gliders often have glide/navigation computers that automatically compute the glide ratio (L/D).

Figure 11-4. *Circular glider calculator.*

Another method allows a pilot to compute effective L/D by utilizing a standard formula. Glide ratio, with respect to the air (GRA) or L/D, remains constant at a given airspeed. For example, the pilot might know the glide ratio, and lift over drag (L/D) being 30 to 1 (expressed as 30:1) for a speed of 50 knots in a specific glider. At 50 knots airspeed with an L/D of 30:1, a 10-knot tailwind results in an effective L/D of 36:1. [*Figure 11-5*]

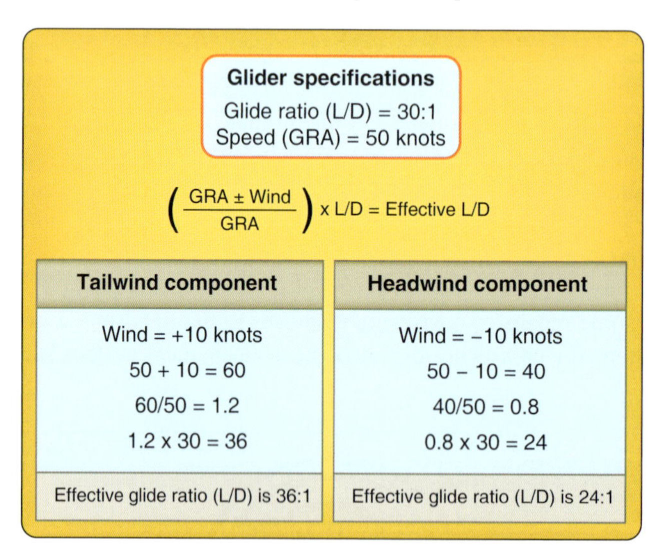

Figure 11-5. *Glide calculation examples for a headwind and a tailwind.*

In addition to a glide calculator, a flight computer or MacCready ring on the variometer gives the pilot the appropriate speed to fly for different sink rates. Data such as windspeed and direction may be manually or automatically input depending on the age and capability of the flight computer. Accurately flying the correct speed in sinking air can extend the achieved glide considerably.

Many models of electronic glide calculators exist. Often coupled with an electronic variometer, they display the altitude necessary for distance and wind as input by the pilot. In addition, many electronic glide calculators feature speed-to-fly functions that indicate whether the pilot should fly faster or slower. Most electronic speed-to-fly directors include audio indications, so the pilot can remain visually focused outside the cockpit. The pilot should have manual backups for electronic glide calculators and speed-to-fly directors in case of a low battery or other electronic system failure.

Other equipment may verify soaring performance that allows a pilot to receive a Federation Aeronautique Internationale (FAI) badge or to record flights. These include turn-point cameras, barographs, and GPS flight recorders. For complete descriptions of these items, as well as badge or record rules, check the Soaring Society of America website for details.

Finally, pilots should consider using a notepad or kneeboard on which to make notes before and during the flight. Notes prior to flight could include weather information such as winds aloft forecasts or distance between turn points. In flight, noting takeoff and start time and time around any turn points helps gauge average speed around the course.

Navigation

Airplane pilots navigate by pilotage (flying by reference to ground landmarks) or dead reckoning (computing a heading from true airspeed and wind, and then estimating time needed to fly to a destination). Glider pilots generally use pilotage since they often deviate from a course line over a long distance and do not fly one speed for any length of time. At times, glider pilots might use a combination of the two methods.

A Sample Cross-Country Flight

For training purposes, a pilot could plan a triangle course starting at Portales Airport (PRZ), with turn points at Benger Airport (X54), and the town of Circle Back. The preflight preparation includes drawing the course lines for the three legs using an electronic system or chart. [*Figure 11-6*]

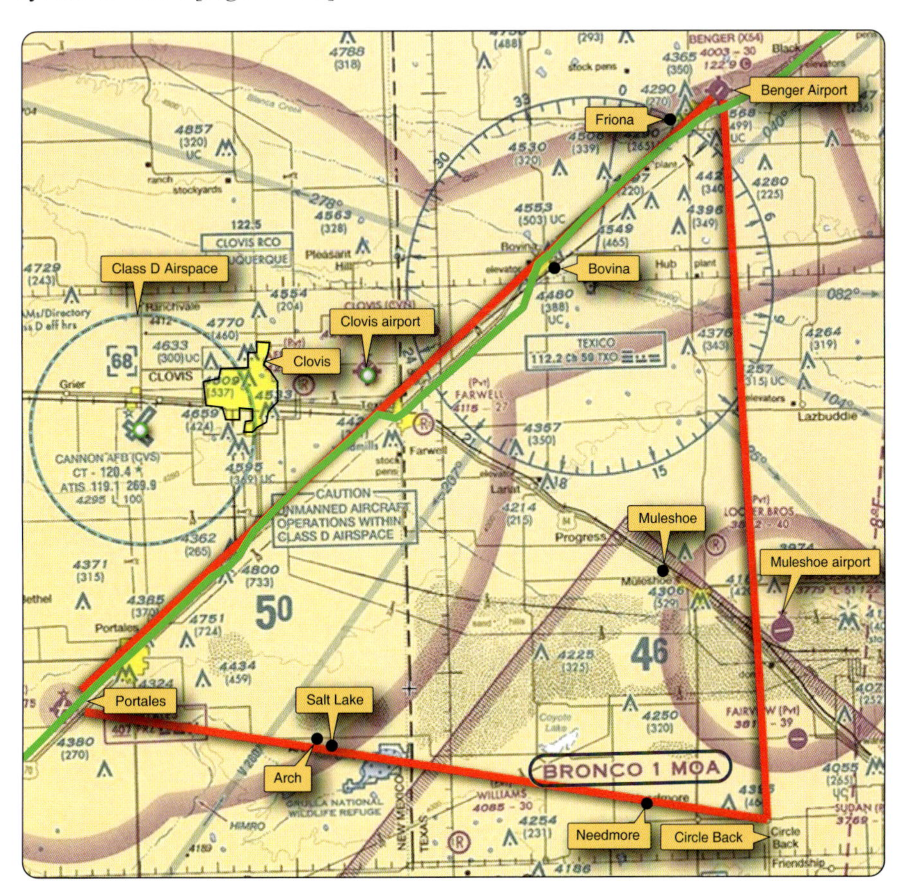

Figure 11-6. *Cross-country triangle drawn on the Albuquerque Sectional Chart.*

The forecast shows the expected base of the cumulus at 11,000 MSL, and the winds aloft indicate 320° at 10 knots at 9,000 MSL and 330° at 20 knots at 12,000 MSL during the flight. The pilot should take note of the winds aloft for reference during the flight. For instance, the first leg has an almost direct crosswind from the left; the second leg has a weaker crosswind component from the right; the final leg has an almost direct headwind. For this reason, the glider pilot may decide to fly the course in the opposite direction. In this example however, the pilot accepts a headwind on the final leg. The forecast shows the expected base of the cumulus at 11,000 MSL, and the winds aloft indicate 320° at 10 knots at 9,000 MSL and 330° at 20 knots at 12,000 MSL during the flight. The pilot should take note of the winds aloft for reference during the flight. For instance, the first leg has an almost direct crosswind from the left; the second leg has a weaker crosswind component from the right; the final leg has an almost direct headwind. For this reason, the glider pilot may decide to fly the course in the opposite direction. In this example however, the pilot accepts a headwind on the final leg.

During preflight preparation, the pilot should study the course line along each leg for expected landmarks and possible alternate landing sites. For instance, the first leg follows highway and parallels railroad tracks for several miles before the highway turns north just due south of Clovis. The town of Clovis should become obvious on the left. Note the Class D airspace around Cannon Air Force Base (CVS) just west of Clovis. If better soaring conditions exist north of course track and with the northwesterly wind, the glider might cross the path of aircraft on a long final approach to the northwest-southeast runway at the air base. Knowing the courses and the approximate heading for each leg helps keep the pilot from getting lost even when making deviations toward the best lift. During the flight, if the sky ahead shows several equally promising cumulus clouds, choosing the one closest to the course line keeps the flight distance to a minimum and makes the most sense.

The pilot should see the Clovis airport (CVN) next and check for traffic operating in and out of the airport. After Clovis, the towns of Bovina and Friona can serve as landmarks for the flight. The Texico (TXO) VOR, a VHF Omnidirectional Range station near Bovina serves as approach aid to the Clovis airport, and the pilot should remain alert for more powered aircraft traffic as a result.

The pilot can locate the first turn point easily because of good landmarks, including Benger Airport (X54). [*Figure 11-7*] The second leg has fewer landmarks. After about 25 miles, the town of Muleshoe and the Muleshoe airport (2T1) should appear. (See *Figure 11-6*). The town should appear on the right and the airport on the left of the intended course.

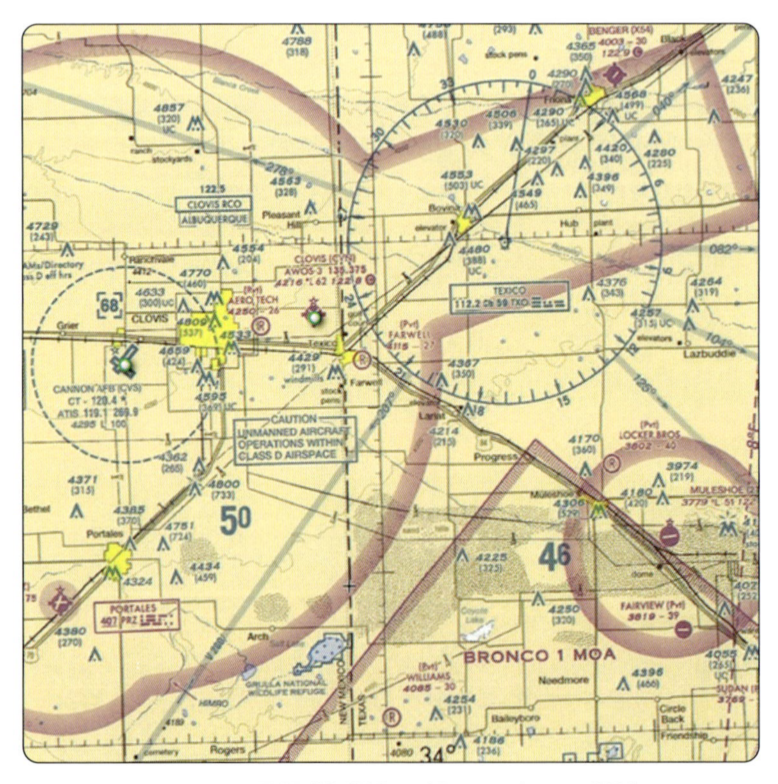

Figure 11-7. *TEXICO VOR and Benger Airport (X54).*

Next, the course enters the Bronco 1 Military Operations Area (MOA). The dimensions of the MOA appear on the sectional chart, and the pilot should determine the active times of this airspace. Approaching the second turn point, the pilot could confuse the towns of Circle Back and Needmore. [*Figure 11-6*] The relative position of an obstruction 466 feet above ground level (AGL) and a road that heads north out of Needmore provide clues for identification. Landmarks on the third leg [*Figure 11-8*] include power transmission lines, Salt Lake (possibly dry), the small town of Arch, and a major road coming south out of Portales.

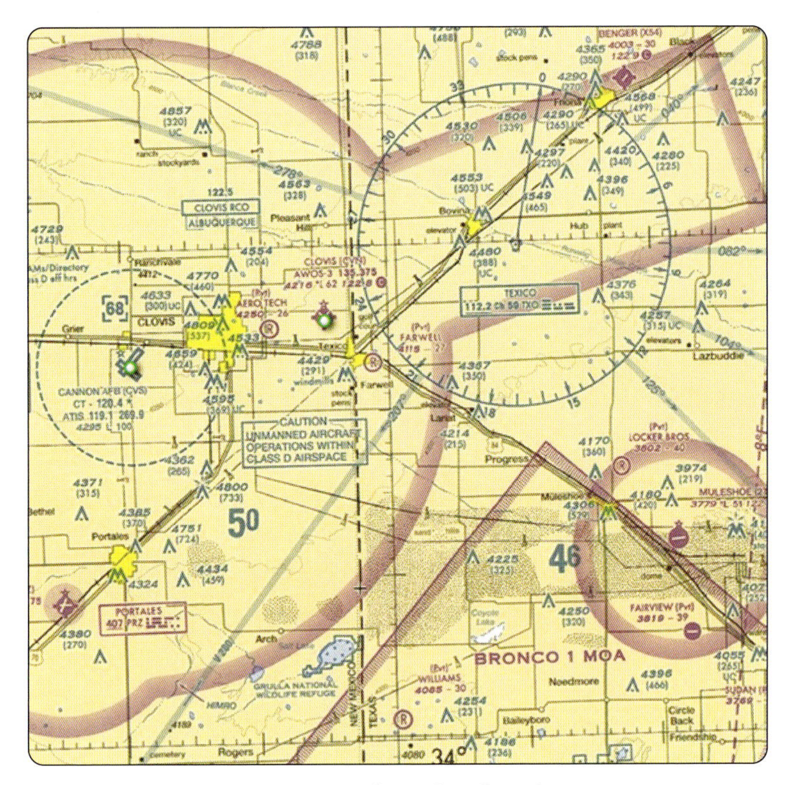

Figure 11-8. *Circle Back and Needmore.*

The flight can begin after a final check of the weather, a thorough preflight of the glider, and after stowing all the appropriate equipment. Once in the air and on course, the pilot can try to verify the winds aloft while using pilotage to remain as close to the course line as soaring conditions permit. If making course deviations, the pilot should remain aware of the location of the course line to the next turning point. For example, a Cu directly ahead may indicate lift, but one 30° off course indicates possibly even more lift. The pilot chooses whether to proceed toward the larger lift while accepting a longer distance to fly, or to accept lesser lift with a smaller off-course deviation.

Sometimes a pilot determines an approximate course while in the air. Assume a few miles before reaching the town of Muleshoe, on the second leg, the weather ahead has deteriorated—a shower developed at the final turn point (Circle Back). Rather than continuing, the pilot can cut the triangle short and return directly to Portales. The pilot determines that Portales is about 37 miles away on an estimated heading of about 240° true or about 231° magnetic after accounting for magnetic variation. The pilot adjusts for the northwesterly wind at almost 90° across the new course and uses a 10° or 20° crab to the right, allowing for some drift in thermal climbs. With practice, a pilot can manage an inflight course change with relative ease.

The sky towards Portales indicates favorable lift conditions. However, the area along the new course includes sand hills, an area that may not have good choices for off-field landings. The pilot decides to take time to gain altitude in thermals until beyond this point and until within gliding distance of suitable landing sites.

Navigation Using GPS

A GPS system makes navigation easier. A GPS unit displays distance and heading to a specified point, usually found by scrolling through an internal database of waypoints. Many GPS units also continuously calculate and display ground speed. Given TAS information, the GPS can calculate the headwind component.

When using a GPS unit, the pilot should continue focusing on flying the glider, finding lift, and scanning for traffic. Like any electronic instrument, a GPS unit can fail, and the pilot should have a backup system.

Cross-Country Techniques

For safe cross-country soaring, the pilot should always stay within glide range of a suitable landing area. The landing area may be an airport or other suitable spot to land out. If following this practice, even with high sink rates between thermals, the pilot should never need a thermal to obtain the range to a suitable landing area.

Before venturing beyond gliding distance from the home airport, the pilot can practice thermalling and cross- country techniques using small triangles or other short courses. *Figure 11-9* shows three examples. The length of each leg, typically between 5 and 10 miles, depends on the performance of the glider. The pilot does not need excellent soaring conditions to fly these short courses, but conditions should not make it difficult to stay aloft. On a good day, the pilot can fly the short-course pattern more than once and can also practice switching communication frequencies and listening to transmissions from aircraft or controllers at nearby airports. While progressing along each leg of the triangle, the pilot should frequently cross check the altitude needed to return to the home airport and abandon the course if needed. Setting a minimum altitude for arrival at the home site of 1,500 or 2,000 feet AGL adds a margin of safety. The pilot should make every landing at the conclusion of a soaring flight an accuracy landing to keep the pilot's attention focused until the conclusion of the flight.

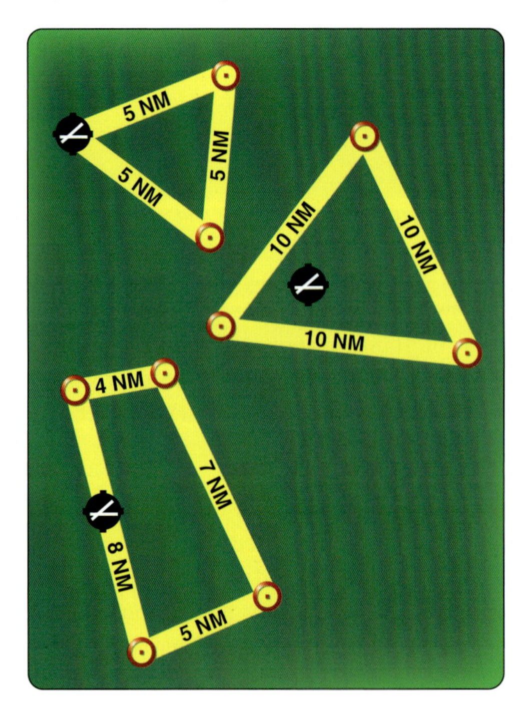

Figure 11-9. *Examples of practice cross-country courses.*

Some flight computers automatically calculate the winds aloft while other GPS systems estimate winds by calculating the drift after several thermal turns. When flying with GPS, the pilot can determine a headwind component from TAS by simple subtraction of groundspeed while maintaining a particular heading. Determining winds aloft without either system can prove difficult. A first estimate comes from winds aloft forecasts. Once aloft, the pilot can estimate windspeed at cloud level from the track of cumulus shadows over the ground. However, the winds at lower levels can differ from those at cloud level. On cloudless days, the pilot can estimate wind by noting drift while thermalling. If losing more height on glides than expected, the pilot should increase the headwind estimate.

A common flight planning technique involves drawing 5 and 10 nautical mile radius circles around alternate landing sites along the planned route. This helps the pilot visualize the altitude needed to safely reach an alternate site should thermal activity be insufficient to continue the cross-country. Alternatively, a pilot using a glide calculator or computerized tool can quickly determine the altitude needed to glide a specific distance. For instance, while on a cross-country flight and over a good landing spot, the next good landing site appears 12 miles distant into a 10-knot headwind. [*Figure 11-10*] A glide calculation shows that the glider will lose 3,200 feet during the glide, and the pilot should add at least 1,500 feet to allow for setting up for a landing, which makes the total altitude needed to make the glide as 4,700 feet. While not high enough to accomplish the 12-mile glide, the 3,800 current foot altitude allows the pilot to start along the course provided the pilot remains within gliding distance of the landing spot where the glide began. After two miles with no lift, the glider has descended to an altitude of almost 3,300 feet. While not high enough to glide the remaining 10 miles, the pilot can still glide back to the landing site two miles behind. After almost 4 miles, the pilot encounters a 4-knot thermal at about 2,700 feet and climbs to 4,300 feet.

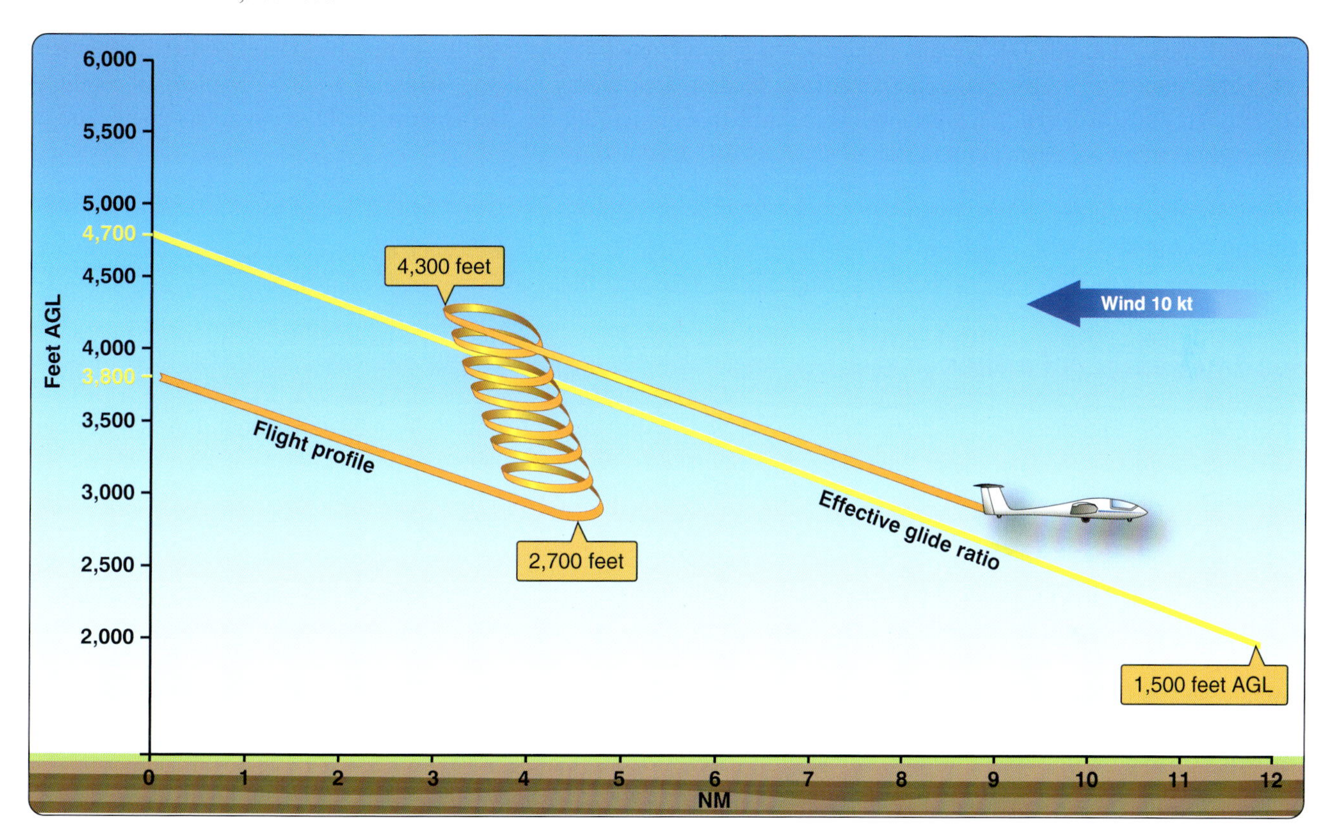

Figure 11-10. *Example of a flight profile during a cross-country course.*

The glide calculator or computer accounts for the glider's calm air rate of descent, which the pilot can adjust for headwinds and tailwinds. However, any vertical currents (sink) can drastically affect these calculations and distort the results. Areas of good thermal lift exist with areas of strong sink, and a glider pilot should read the sky to find the lift and avoid or pass through the sink as quickly as possible. A competent glider pilot understands the polar curves of the glider and the effects of different conditions of lift, sink, and winds.

During the climb in the example above, the downwind drift of the thermal moves the glider back approximately a half mile. The pilot would like to glide almost 9 miles to the next landing spot, and a check of the glide calculator indicates 2,400 feet needed plus 1,500 feet at the destination, for a total of 3,900 feet. The pilot has 400 feet above the altitude needed for the glide with a margin to plan the landing. In the previous example, had the thermal topped at 3,600 feet (instead of 4,300 feet) the pilot could continue on course in hopes of finding more lift before needing to turn downwind and back to the previous landing spot. Any cross-country soaring flight involves dozens of decisions and calculations such as this. In addition, a pilot should plan for increasing the altitude safety margin if conditions might cause a lower effective glide ratio. For example, other pilots reporting heavy sink along the intended course would alert a pilot to increase the safety margin.

On any soaring flight, a critical altitude exists where a decision must be made to cease attempts to work thermals and commit to a landing. Cross-country flights can have landings in unfamiliar places and feature additional pressures like those discussed in Chapter 8, Abnormal and Emergency Procedures. In the event not reaching a planned landing site, a reasonable procedure involves choosing a general area by 2,000 feet AGL, picking a landing site by 1,500 feet AGL, and committing to landing by 1,000 feet AGL. The exact altitude where the thought processes should shift from soaring to landing preparation depends on the terrain. In areas where numerous fields suitable for landing may exist, the pilot can delay field selection to a lower altitude. In areas with landing sites spaced by 30 miles or more apart, the pilot's focus on committing to a landing spot should begin at much higher altitudes above the ground.

Attempts to thermal in the pattern may lead to a stall or spin accident. Therefore, once committed in the pattern, the pilot should not try thermalling. When over a safe landing spot, the pilot should perform the prelanding checklist.

A common first cross-country flight consists of a 50-kilometer (32 statute miles) straight distance flight with a landing at another field. The pilot can fly this distance at a leisurely pace on an average soaring day. The pilot should review the planned course carefully, research all available landing areas along the way, arrive early, and complete all preflight preparations. Once airborne, the pilot should take time to get a feel for the day's thermals. If the day looks good enough and the glider gains adequate altitude, the pilot can set off on course.

Pilots gain cross-country skills through practice but should also continue to review theory while gaining that experience. A theory or technique that initially made little sense takes on a lot more significance after several cross-country flights. Postflight self-critique, which can occur at any time before the next flight, also improves pilot skills.

Soaring Faster & Farther

An average cross-country speed of 20 or 30 miles per hour (mph) seems adequate for a 32-mile flight, but that average speed does not accommodate longer flights. Flying at higher average cross-country speeds allows for increased soaring flight distance.

In the context of cross-country soaring, flying faster means achieving a faster average groundspeed. The secret to faster cross-country flight lies in spending less time climbing and more time gliding. This occurs when using better thermals and spending more time in lifting air and less time in sinking air. MacCready ring theory and/or speed-to-fly theory determines the optimum speeds between thermals. Proper use of the MacCready speed ring or equivalent electronic speed director displays the appropriate speed.

Height Bands

On most soaring days an altitude range called a height band describes where maximum thermal strength exists. Height bands give the optimum altitude range in which to climb and glide on a given day. For instance, a thermal may have 200 to 300 fpm lift between 3,000 and 5,000 feet AGL which then weakens before topping out at 6,000 feet AGL. In this case, a 2,000-foot height band exists between 3,000 feet and 5,000 feet AGL. Staying within the height band gives the best (fastest) climbs. On a long cross-country flight, the pilot should thermal while within the height band and avoid stopping for weak thermals unless needing additional altitude at that moment.

On another day, thermals may be strong from 1,000 feet to 6,000 feet AGL before weakening, which would suggest a height band 5,000 feet deep. In this case, however, depending on the thermal spacing, terrain, pilot experience level, and other factors, a height band would run from 2,000 feet or 3,000 feet up to 6,000 feet AGL. Pilots should avoid gliding to the lower bounds of strong thermals (1,000 feet AGL) since the thermal could dissipate and commit the pilot to a poorly planned landing. [*Figure 11-10*]

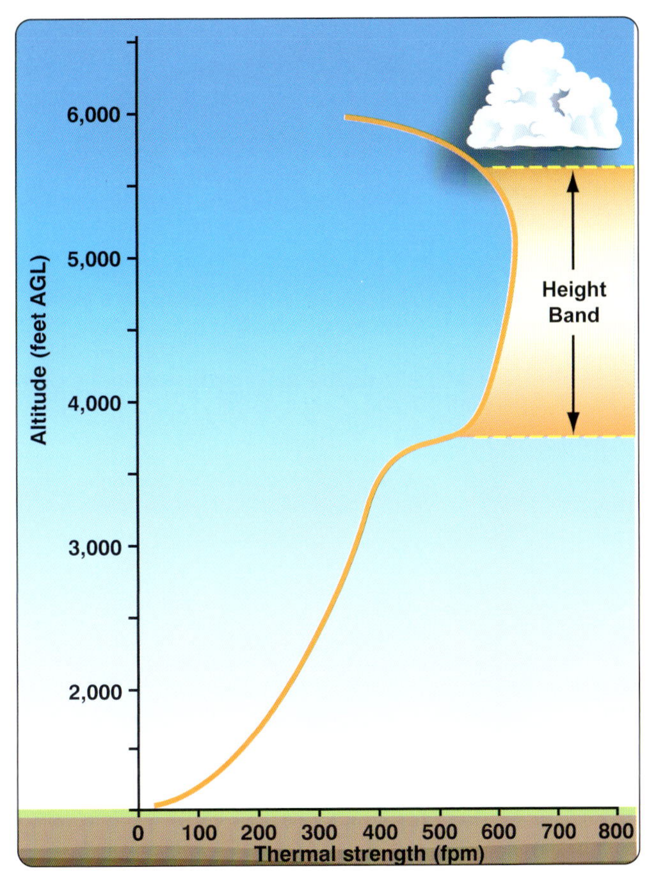

Figure 11-11. *Example of the height band.*

Note: METARs give cloud levels as AGL, while PIREPS report clouds using MSL. Graphical Area Forecasts (GFA) give clouds as MSL. Pilots should interpret the reported cloud heights with care. This interpretation takes on added significance when glider pilots travel to higher elevation airports and subtract field elevation from MSL reports to ensure cloud clearances.

Determining the top of the height band depends upon personal preference and experience, but a rule of thumb puts the top at an altitude where thermals drop off to 75 percent of the best achieved climb. A pilot who finds 400 fpm as the maximum thermal strength in the height band might leave when thermals decrease to 300 fpm for more than a turn or two. Pilots can also compare the current climb with average climb achieved to determine the height band top. If the climb falls to 75 percent or less of the average climb, the pilot should consider leaving the thermal. Many electronic variometers have an average function that displays average climb over specific time intervals. Another technique involves simply timing the altitude gained over 30 seconds or 1 minute.

Theoretically, the pilot can achieve the highest average speed on a cross-country flight when setting the MacCready ring, if available, for the rate of climb within the height band. To do this, the pilot rotates the ring so that the index mark lines up with the rate of climb (for instance, 400 fpm) rather than at zero (the setting used for maximum distance). [*Figure 11-12*] This setting optimizes the time distribution between climbing and gliding. If flying slower than the MacCready setting, the pilot consumes more time between thermals than can be saved during shorter time in strong thermals. If flying faster than the MacCready setting, the pilot loses too much altitude between thermals and uses more than the optimum amount of time to regain the altitude.

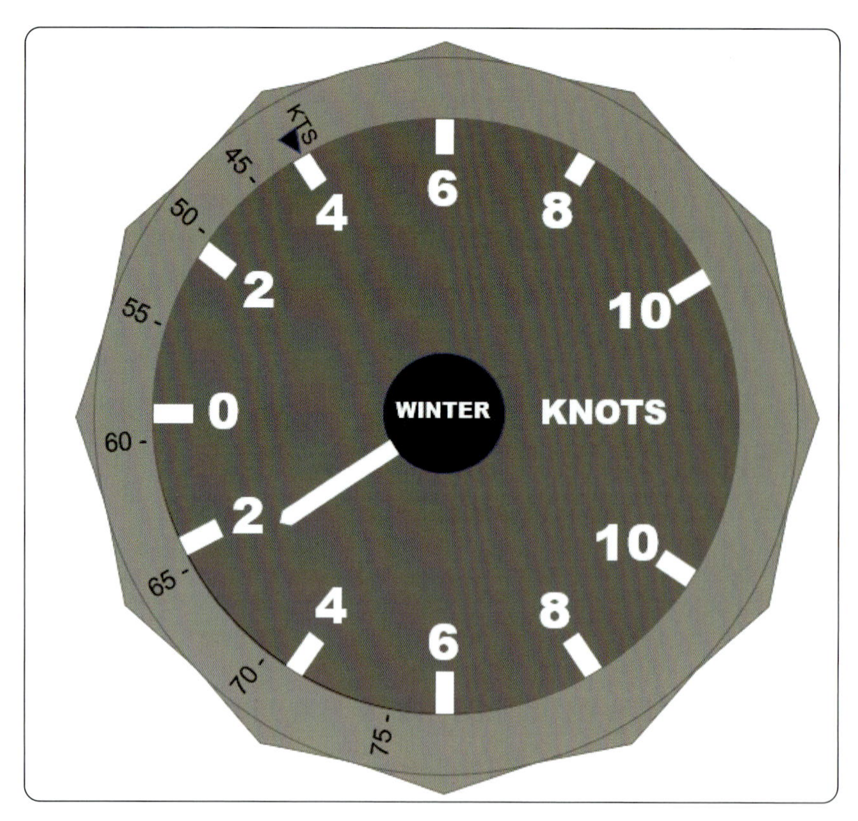

Figure 11-12. *A MacCready Ring set for an expected 4 knot climb in the next thermal. With current sink of a little over 2 knots, the MacCready ring suggests 65 knots as the speed to the next thermal.*

MacCready ring theory assumes that the next thermal has at least the same strength as that set on the ring, and the glider can reach the next thermal with sufficient altitude. The pilot should judge whether actual conditions support adherence to the numbers. Factors that may require departure from the MacCready ring theory include terrain (extra height needed to clear a ridge), distance to the next suitable landing spot, or deteriorating soaring conditions ahead. If the next thermal appears to be out of reach before dropping below the height band, the pilot should climb higher, glide more slowly, or both.

To illustrate the use of speed-to-fly theory, assume there are four gliders at the same height. The scenario includes three weak cumulus clouds each produced by 200-fpm thermals followed by a larger cumulus with 600 fpm thermals under it, illustrated in Figure 11-13.

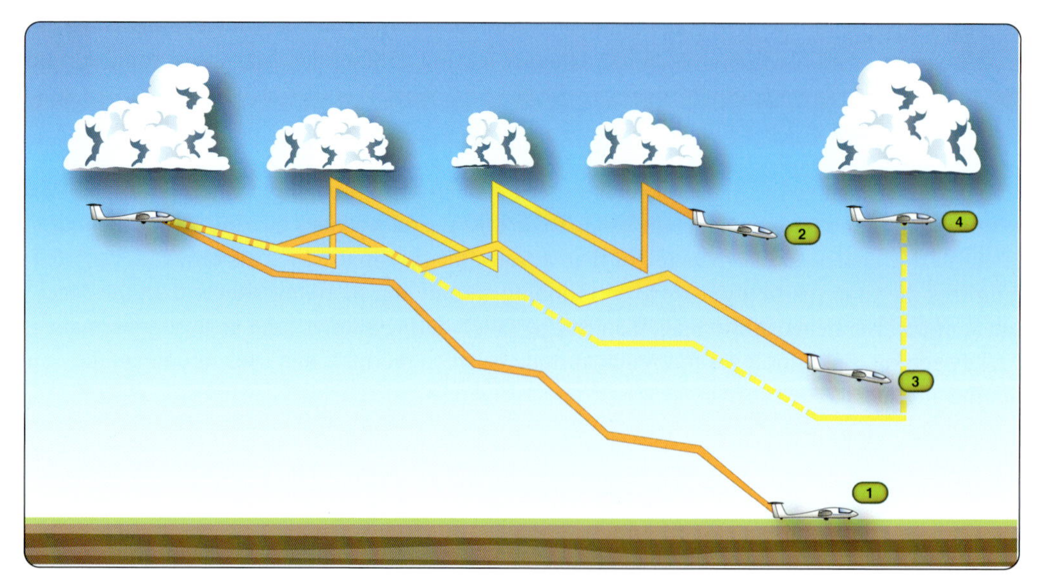

Figure 11-13. *Example of glides achieved for different MacCready ring settings.*

- Pilot 1 sets the ring to 6 knots for the anticipated strong climb under the large cumulus, but the aggressive approach has the glider on the ground before reaching the cloud.
- Pilot 2 sets the ring for 2 knots and climbs under each cloud until resetting the ring to 6 knots after climbing under the third weak cumulus, in accordance with strict speed-to-fly theory.
- Pilot 3 is conservative and sets the ring to zero for the maximum glide.
- Pilot 4 calculates the altitude needed to glide to the large cumulus using an intermediate setting of 3 knots and finds the glider can glide to the cloud and still be within the height band.

By the time pilot 4 climbs under the large cumulus, the pilot sits well ahead of pilots 2 and 3 and can relay retrieval instructions for pilot 1. This example illustrates the science and art of faster cross-country soaring. The science comes from speed-to-fly theory, while the art involves interpreting and modifying the theory for the actual conditions.

Tips & Techniques

The height band changes during the day. On a typical soaring day, thermal height and strength often increase rapidly during late morning, and then both the strength and height remain somewhat steady for several hours during the afternoon. The height band rises and broadens with overall thermal height. Sometimes a base of cumulus clouds limits the top of the height band. The cloud base may slowly increase by thousands of feet over several hours, during which time the height band also increases. Pilots should stop traveling and thermal when at or near the bottom of the height band. Pushing too hard for distance can lead to an early off-field landing or lead to lost time spent climbing inefficiently at lower altitudes. Thermals often "shut off" rapidly late in the day, and pilots should consider staying higher late in the day. [*Figure 11-14*]

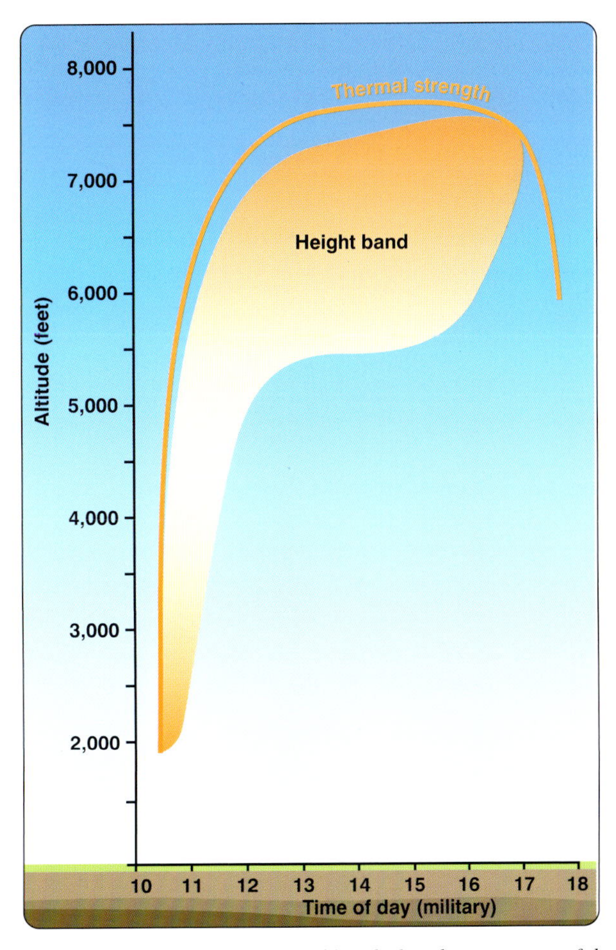

Figure 11-14. *Thermal height and height band versus time of day.*

Another way to increase cross-country speed involves a technique known as "dolphin flight," which covers surprising distances with little to no turning or circling. The idea involves diving to speed up while in sink and slowing down and climbing in lift while maintaining a straight course line. This technique utilizes closely spaced thermals, as occur along a cloud street. The speed to fly between lift areas depends on the appropriate MacCready setting.

As an example, assume two gliders start at the same point and fly under a cloud street with frequent thermals with weak sink between the thermals. Glider 1 uses "dolphin flight" by flying faster in the sink and slower in lift. Glider 2 conserves altitude and stays close to the cloud base by flying best L/D through weak sink. To get to the next cloud, the pilot of glider 2 flies faster in areas of lift. At the end of the cloud street, one good climb quickly puts glider 1 near the cloud base and well ahead of glider 2. [*Figure 11-15*] The best-speed-to-fly decreases time in sink and therefore decreases the overall amount of descent but produces the best forward progress. Being slower in sink increases time descending and slows forward progress. Being fast in lift decreases time in lift and altitude gained.

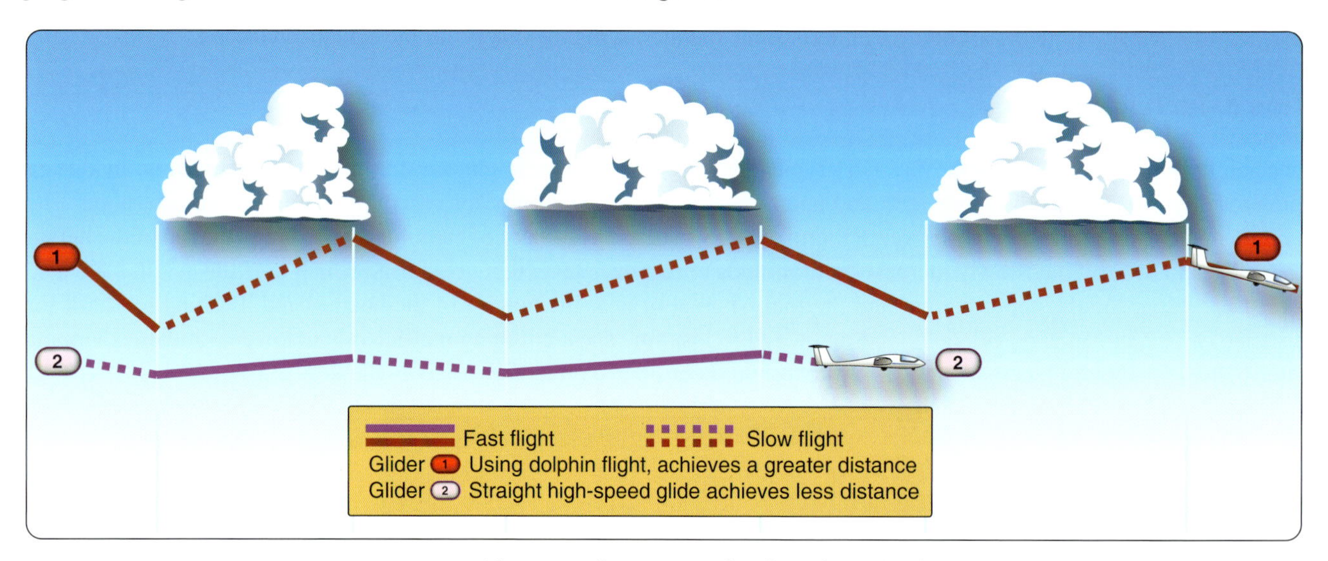

Figure 11-15. *Advantage of proper speed to fly under a cloud street.*

On an actual cross-country flight, the pilot can use a combination of dolphin flight and classic climb and glide. In the example above the pilots decided not to stop and circle in the weak closely spaced thermals, but they could stop and circle if finding areas of strong lift.

Special Situations

Course Deviations

Pilots might consider deviation on a soaring cross-country flight as the norm rather than the exception. Even with fair weather cumulus evenly spaced across all quadrants, the pilot might decide to deviate toward stronger lift. Deviations of 10° or less add little to the total distance and pilots should not hesitate to fly toward such better lift. Even deviations up to 30° may work well if they lead toward better lift or avoid suspected sink ahead. The sooner the deviation starts, the less extra distance the glider will cover during the deviation. [*Figure 11-16*]

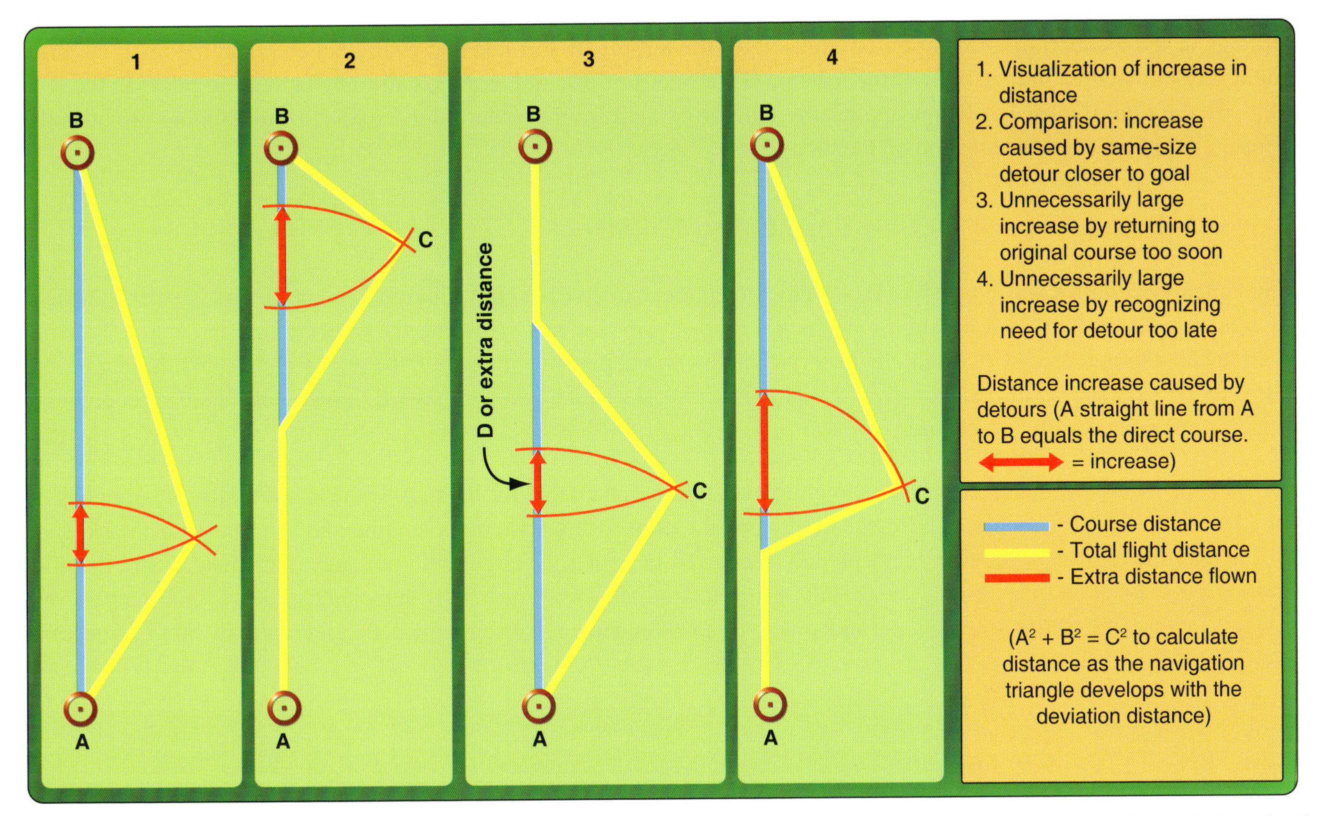

Figure 11-16. *Effects of starting course deviations at different times. Red arrows show extra course distance and indicate the benefit of early course deviations.*

The pilot might use deviations of 45° or even 90° to avoid poor conditions ahead. Areas to avoid could include a large cloudless area or a shaded area where cumulus have spread out into stratus clouds. Cloud development can shade the earth, decrease surface heating, and decrease lift. A pilot could deviate more than 90° to return to active thermals after venturing into stable air.

Generally, glider pilots encountering a large area of stable air or sinking air will end up landing before VFR conditions disappear. Thermalling ceases as the sky becomes cloudy and shades the earth's surface. However, pilots should deviate or have an option for landing rather than fly into lowering ceilings or rain showers ahead. Ridge lift might remain, but cloud bases can obscure that source of lift as well. Thunderstorms along the course can generate outflows and affect surface winds for many miles surrounding the storm. The pilot should avoid landing anywhere near a thunderstorm. Thunderstorms ahead often warrant large course deviations of up to 180° (i.e., retreat to safety).

Lost Procedures

A working GPS system makes it almost impossible to get lost. However, GPS systems can fail, and pilots should have a backup plan if that should occur. The following actions increase the likelihood of reorientation without using GPS.

If lost after some initial searching, the pilot should remain calm and make certain a suitable landing area exists within gliding distance. The pilot should then try to find a source of lift and climb. Even weak lift can provide more time to look for landmarks. Thermalling and climbing while searching has the added advantage of allowing a wider area to scan while circling. The pilot should estimate the last known position, the course flown, and any possible differences in wind at altitude. Maybe the headwind was stronger than anticipated, and the flight did not progress as far along the course as expected. The pilot can try to pinpoint the present position from an estimate of the distance traveled for a given time from a known point and confirm it with visible landmarks on an electronic or paper chart.

Once locating a known landmark on the chart and on the ground, the pilot can confirm the location by finding a few other nearby landmarks. For instance, if seeing a specific town below, an adjacent highway should curve like the one shown on the chart. Airports and airport runways provide valuable clues if sighting an airport, which include runway orientation and markings or the location of a town or a city relative to the airport. If still lost and near a suitable landing area, the pilot should stay in that area until certain of the location. If previous efforts fail, a radio call to other soaring pilots in the area with a description of what lies below and nearby may bring help from another pilot. Finally, a safe landing provides the opportunity to figure out the location on the ground.

Cross-Country Flight in a Self-Launching Glider

Although more complex and expensive, a self-launching glider can give the pilot additional freedom. First, a self- launching glider allows the pilot to fly from airports without a towplane or tow pilot. Second, the pilot can use the engine to avoid off-field landings and extend the flight. In theory, when low in a self-launching glider, the pilot simply starts the engine and climbs to the next source of lift. However, this practice has risks of its own and has led to many accidents due to engine failure or improper starting procedures.

Overreliance on the engine can lead pilots to glide over terrain unsuitable for landing. If the engine fails, the pilot has no safe place to land. Some accidents have occurred when the pilot rushed to avoid a landing and forgot a critical task, such as switching the ignition on. Other accidents have occurred in which the engine did not start immediately, and the pilot flew too far from a suitable landing area while trying to troubleshoot. For a self- launching glider with an engine that stows in the fuselage behind the pilot, the added drag of an extended engine can reduce the glide ratio by 50 to 75 percent. [*Figure 11-17*]

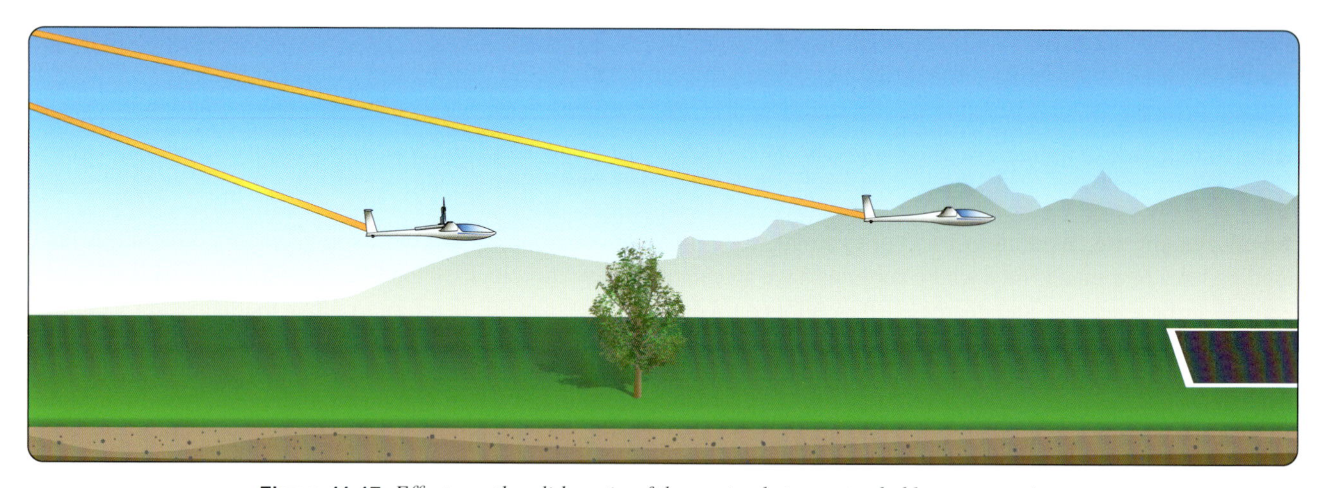

Figure 11-17. *Effects on the glide ratio of the engine being extended but not running.*

The engine starting procedure in a self-launching glider increases the critical decision height to commit to an off- field landing. The time needed to start the engine and extra drag during the starting process can add anywhere from 200 feet to 500 feet of altitude lost to extend and start the engine. Whereas a pure glider may commit to landing at 1,000 feet AGL, the pilot of a self-launching glider probably opts for 1,500 feet AGL, depending on the glider and landing options should the engine fail to start. In this sense, the self-launching glider becomes more restrictive during cross-country flying.

Cross-country flight can also be done under power with a self-launching glider, or a combination of powered and soaring flight. For some self-launching gliders, the most efficient distance per gallon of fuel uses a maximum climb under power followed by a power-off glide. Pilot should check the GFM/POH for recommendations.

Another type of glider features a sustainer engine. Although this type of engine does not have enough power for self-launching, it can keep the glider airborne if lift fails. However, sustainer engines can only produce enough power to overcome the glider's sink rate. Higher sink rates can overwhelm the capability of many sustainer powerplants. Sustainer engines typically eliminate the need for a time-consuming retrieval at the end of a cross country, and they operate with less complexity than their self-launching counterparts. Pilots flying a glider with a sustainer engine have similar concerns as pilots flying self-launching gliders regarding unsuitable terrain and altitude decision heights.

High-Performance Glider Operations & Considerations

Extended cross-country flights have been made in relatively low-performance gliders. However, on any given soaring day, the same pilot can fly a glider with a 40:1 glide ratio farther and faster than one with 20:1 ratio.

Glider Complexity

The experience required to fly a high-performance glider does not necessarily depend on a pilot's total glider hours. Pilots should consider the types of gliders flown (low and high performance). Most high-performance gliders have a single seat. The pilot should obtain some instruction from an authorized flight instructor in a two-seat high performance glider, if available, before attempting to fly a single seat high-performance glider for the first time. Before flying any single-seat glider, pilots should thoroughly familiarize themselves with the GFM/POH, including important speeds, weight and balance issues, and all systems in the glider.

Many high-performance gliders have flaps. Pilots can use a few degrees of positive flap when thermalling, 0° for relatively low-speed glides, and negative flap settings for glides at higher speeds. The GFM/POH and glider polar provide recommended flap settings for different speeds, as well as maximum speeds allowed for different flap settings. A few high-performance gliders have no air brakes and use large positive flap settings for landing.

Water Ballast

Water ballast can maximize average cross-country speed on a day with strong thermals. The gain in speed between thermals outweighs the lost time due to slightly slower climbs with water ballast. Pilots should not plan to use ballast on a day with weak thermals. If strong thermals become weak, the pilot can dump the water ballast.

Water expands when going from the liquid state to the solid state. The force of the water ballast freezing can split composite wing skins. If anticipating flying on a cross-country flight and spending time at levels with temperatures below 0 °C, the pilot should add the GFM/POH recommended antifreeze to the ballast before departure.

Cross-Country Flight Using Other Lift Sources

Ridge or wave lift provides more consistent lift than thermals, allowing long, straight stretches of hundreds of miles at high speed.

Pilots with little experience using ridge lift on a cross-country flight should plan for a day with ideal wind and weather. On relatively low ridges, (e.g., in the eastern United States), ridge lift may not extend very high. The best lift often sits very close to the ridge crest in turbulent air, and pilots might end up flying close to terrain in rough conditions for hours at a time. If the pilot loses lift, maneuvering for an off-field landing might occur with little notice. In milder conditions, the cross country might require thermalling to gain enough height to cross any ridge gap. If cumulus spread out to form a stratus layer shading the ground and eliminate thermals, the pilot might need to use a wind facing slope to maintain soaring flight while waiting for the shade to dissipate. Unless the sun returns and generates the thermal needed to gain sufficient altitude, the pilot would need an alternate plan.

Wave lift can also provide opportunities for long or fast cross-country flights. Most record flights have been along mountain ranges; flights more than 2,000 kilometers having been flown in New Zealand and along the Andes. In the United States, speed records have been set using the wave in the lee of the Sierra Nevada Mountains. In theory, long-distance flights could also occur by climbing high in a wave, then gliding with a strong tailwind to the next range downwind for another climb. The pilot needs to consider physiology (cold, oxygen, etc.) and airspace restrictions for cross-country flights using wave lift.

Convergence zones can also enhance cross-country speed. Even if the convergence does not form a consistent line and acts only as a focus for thermals, dolphin flight could occur. When flying low, awareness of local, small-scale convergence can help the pilot find thermal triggers and enable a climb back to a comfortable cruising height.

A pilot might find a combination of ridge, thermal, wave, and even convergence lift during one cross-country flight. Optimum use of the various lift sources requires mental agility and makes for an exciting and rewarding flight.

Chapter Summary

Pilots should begin planning cross-country flights carefully and within their capability and that of the glider. Planning includes a consideration of weather forecasts, possible routes, what to carry, airspace, and potential landing sites. On the day of a planned flight, the pilot can choose from plans and routes considering the existing environmental conditions. The pilot should brief any ground crew and complete all preflight checks without rushing. Pilots who make cross-country flights know how to maximize the available lift to increase the average speed of the flight. They also understand and follow sensible height restrictions to stay safe. They can divert when necessary and know when to commit to an off-airport landing if lift becomes less than needed to continue. Pilots who fly powered gliders know that they commit to landing procedures at safe altitudes and that they could encounter a situation where their engine does not start.

Chapter 12: Aerotow

Introduction

Glider pilots in the United States commonly launch using an aerotow. [*Figure 12-1*] An aerotow takes time but can offer flexibility regarding energy, location, and altitude. Chapter 7: Launch, Flight Maneuvers, Landings, and Recovery Procedures presents information from the glider pilot's perspective. This chapter provides much of the same information but from the perspective of the tow pilot.

Figure 12-1. *Gliders commonly use tow planes to launch and obtain altitude.*

While all pilots follow Title 14 of the Code of Federal Regulations (14 CFR) part 61, Certification: Pilots, Flight Instructors, and Ground Instructors, and 14 CFR Part 91, General Operating and Flight Rules, the following sections sample some of the requirements of interest to tow pilots:

- 14 CFR part 61, section 61.23—Medical certificates: Requirement and duration.
- 14 CFR part 61, section 61.69—Gliders and Unpowered Ultralight Vehicle Towing: Experience and Training Requirements.
- 14 CFR part 91, section 91.15—Dropping Objects
- 14 CFR part 91, section 91.309—Towing: Gliders and Unpowered Ultralight Vehicles

Equipment Inspections & Operational Checks

Tow Hook

Tow plane equipment in the United States typically uses one of two types of tow hooks: Tost or Schweizer. [*Figure 12-2*] The tow pilot should inspect the tow hook for proper operation daily and prior to any tow activity.

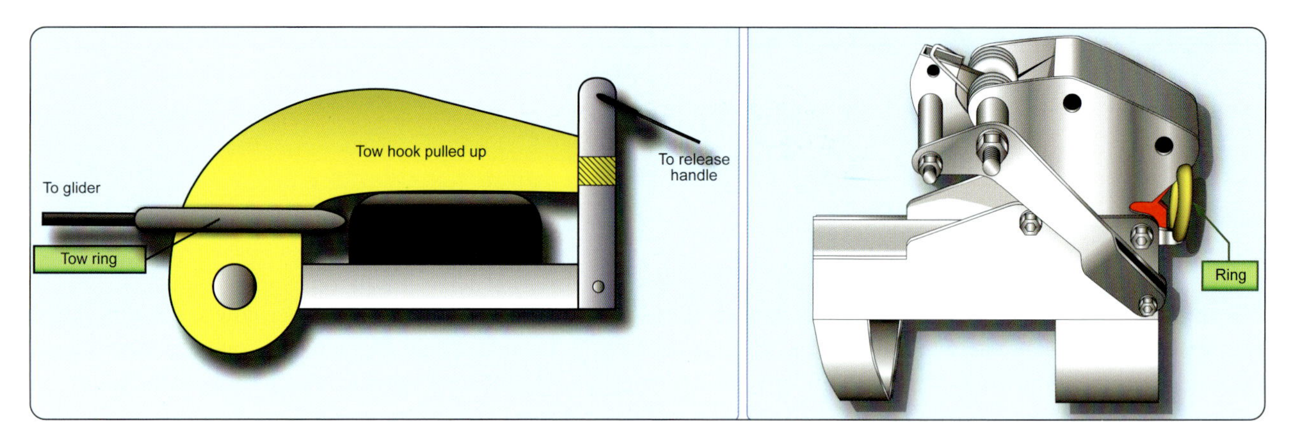

Figure 12-2. *Schweizer tow hook (left) and a Tost tow hook (right).*

Schweizer Tow Hook

Prior to use, the tow pilot should visually inspect the tow hook and release arm for damage, cracks, deformation, and freedom of movement on the pivot bolt. The inspection should ensure that the hook properly engages the release arm. The pilot should also inspect the rubber spacer for general condition and check the condition of the release cable. Inside the airplane, the pilot should verify that the manual release lever does not rub against the aircraft seat or any other obstructions and check the security of the release handle assembly and the cable attachment. The pilot should also perform a functional inspection. If the visual or functional inspection reveals an issue, the pilot should restrict the tow plane from towing duties and repair the tow assembly.

The following functional checks should be performed:

- Attach the tow line to the tow hook and apply tension back on the line.
- With tension on the tow line, have another person pull the release control in the tow plane and check for proper release of the tow line.
- Reattach the tow line and apply a moderate rearward tug.
- Inspect the release assembly to ensure it remains completely closed.

Tost Tow Hook

Before use, the tow pilot should ensure that the release hook opens completely when a helper pulls the release to its fullest extent from inside the airplane. The release hooks should touch the tow hook ring. Inside the airplane, the pilot should check to see that the manual release lever does not rub against the aircraft seat or any other obstructions. The pilot should check the security of the release handle assembly and the cable attachment. When not pulling the manual release lever, the tow hook should return to the fully closed position. The pilot should check the hook for dirt or corrosion and confirm that the airplane end of the tow rope has a Tost ring. If the release mechanism does not work correctly, factory repair should occur before any tow.

The following functional checks should be performed:

- Attach the tow line to the tow hook and apply tension back on the line.
- With tension on the tow line, have another person pull the release control in the tow plane and check for proper release of the tow line.
- Reattach the tow line and apply a moderate rearward tug.
- Inspect the release assembly to ensure it remains completely closed.

Tow Ring Inspection

Tow ring inspection begins with a check for wear and tear. Tow pilots should take rings out of service if they have scratches or dents. The tow ring design corresponds to the type of tow hitch assembly: Schweizer or Tost. [*Figure 12-3*] The Schweizer tow ring uses a single two-inch diameter high-grade, one-quarter inch steel ring magnafluxed with good weld. The Tost tow ring consists of a pair of interconnected rings made from high-grade steel.

Figure 12-3. *Schweizer tow ring and a Tost tow ring.*

The tow pilot should only attach a compatible tow line ring to the tow hitch. While a Schweizer tow ring will not fit into a Tost tow hitch, the tow pilot could mistakenly attach a Tost tow ring to a Schweizer tow hitch. If that happens, the Tost ring might not release when the tow pilot actuates the release mechanism.

Tow Rope Inspection

Although the pilot of the glider has primary responsibility for the selection and inspection of the proper tow line, the tow pilot should confirm the tow line and any required weak links meet the strength requirements of the Federal Aviation Administration (FAA) Regulations and are acceptable for use. See chapter 6, Preflight and Ground Operations, regarding tow line selection and inspection. Pilots should consider replacing the tow rope after a specific period of usage and exposure to the sun or to the elements.

Density altitude affects airplane performance. An increase in air temperature or humidity, or decrease in atmospheric pressure significantly decreases power output and propeller efficiency.

Abort Briefing

Using the computed takeoff data or actual takeoff point for the given conditions, the tow pilot should choose a physical abort point on the runway [*Figure 12-4*] and brief the glider pilot on the abort point and abort procedures. If the tow plane is not off the ground by the chosen abort point, the glider should release, or be released, allowing the tow plane to accomplish a normal takeoff.

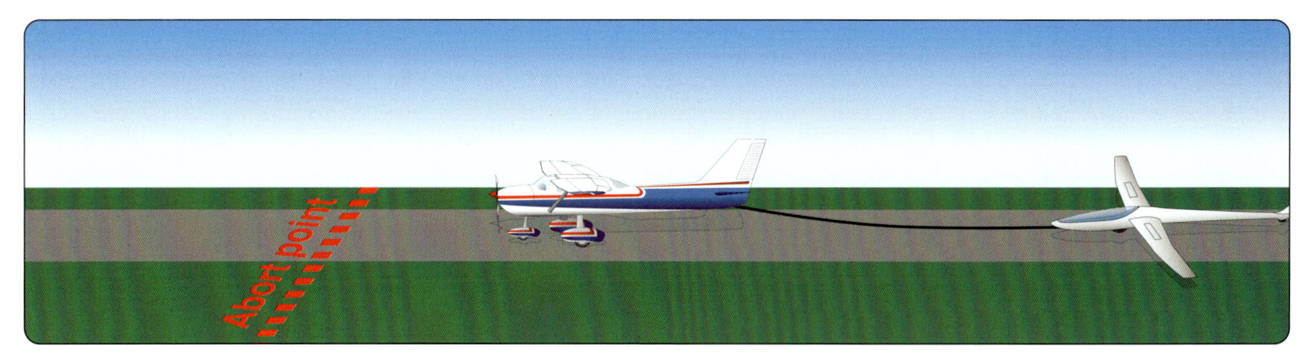

Figure 12-4. *A physical abort point on the runway should be determined and briefed to the glider pilot before starting the towing procedures.*

14 CFR part 91, section 91.309 (a)(5) states, "The pilot of the towing aircraft and the glider have agreed upon a general course of action, including takeoff and release signals, airspeeds, and emergency procedures for each pilot." If any doubt exits about compliance with this rule through briefings or published and agreed upon standard operating procedures, the tow pilot and glider pilot should make certain to clarify all aspects of the upcoming tow.

On the Airport

The tow pilot should maintain an awareness of the direction of the tow plane's prop blast. Blasting launch personnel and glider canopies with wind and debris creates an annoyance and danger. Whenever possible, the tow pilot should angle prop blast away from any ground operations. Prior to taking the active runway for tow line hook-up and takeoff, if applicable, the tow pilot can monitor and announce intentions on the Common Traffic Advisory Frequency (CTAF).

Ground Signals

In most cases, tow pilots use the Standard American Soaring Signals to communicate between the launch crew and tow plane. In some cases, however, specific local procedures take precedence. The tow pilot should receive a briefing on any specific local signals or procedures. The tow pilot might need to observe these signals through the mirror or through an additional signal relay person positioned safely on the side of the runway adjacent to the tow plane. Chapter 7 of this handbook, Launch, Flight Maneuvers, Landing, and Recovery Procedures has figures for the ground signals.

Takeoff & Climb

When ready for takeoff and if applicable, the tow pilot should broadcast the commencement of a glider launch. For example, "Tallahasee traffic, N12345 taking off Runway 33, glider in tow, Tallahasee." 14 CFR part 91, section 91.309(a)(4) states, "Before conducting any towing operation within the lateral boundaries of the surface areas of Class B, Class C, Class D, or Class E airspace designated for an airport, or before making each towing flight within such controlled airspace if required by ATC, the pilot in command notifies the control tower. If a control tower does not exist or is not in operation, the pilot in command must notify the FAA flight service station serving that controlled airspace before conducting any towing operations in that airspace;"

When ready for takeoff and if applicable, the tow pilot should broadcast the commencement of a glider launch. For example, "Tallahasee traffic, N12345 taking off Runway 33, glider in tow, Tallahasee." 14 CFR part 91, section 91.309(a)(4) states, "Before conducting any towing operation within the lateral boundaries of the surface areas of Class B, Class C, Class D, or Class E airspace designated for an airport, or before making each towing flight within such controlled airspace if required by ATC, the pilot in command notifies the control tower. If a control tower does not exist or is not in operation, the pilot in command must notify the FAA flight service station serving that controlled airspace before conducting any towing operations in that airspace;"

On takeoff, the tow pilot advances the throttle smoothly and quickly in one motion. [Figure 12-5] If the tow plane accelerates and then slows down, the glider could overrun the tow line and snag the line in the landing gear. The glider pilot might not have release capability if that occurs.

Figure 12-5. *The tow pilot advances the throttle smoothly and quickly in one motion so that the glider does not overrun the tow line.*

The tow pilot accelerates to liftoff speed keeping in mind that what will occur during the transition out of ground effect. The tow plane leaving ground effect will:

- Require an increase in the angle of attack (AOA) to maintain the same lift coefficient,
- Experience an increase in induced drag and thrust required,
- Experience a decrease in stability and a nose-up change in moment, and
- Experience a reduction in static source pressure and increase in indicated airspeed.

These general effects point out the hazards associated with attempting takeoff prior to achieving the recommended lift-off speed. Due to the reduced drag in ground effect, the tow plane may seem capable of takeoff well below the recommended speed; however, lifting out of ground effect with a lower-than-normal lift off speed may result in poor initial climb performance.

The glider will normally lift off first. The tow pilot should expect the glider pilot to correct for crosswind until the tow plane becomes airborne. At this point, the tow pilot should remain alert to a glider climbing too high and lifting the tow plane tail. Should this happen, the application of full-up elevator on the tow plane may not be sufficient to prevent an accident, and the tow pilot should pull the release handle, release the glider, and regain control. As a rule of thumb, a glider climbing 20 feet or more above a tow plane when using a 200-foot rope presents the danger of an upset.

After liftoff, the tow pilot should pitch up to establish a constant airspeed climb and expect the glider pilot to establish a position directly behind the tow plane. Upon reaching a safe altitude, the tow pilot can initiate a turn to maintain the desired departure path using bank angles limited to 15°–20°.

The tow pilot climbs at full throttle unless otherwise required by the POH. The fuel/air mixture should be leaned in accordance with the POH for maximum power. Each specific model of glider has a published maximum aerotow speed, and the tow pilot should know this speed, which may be very close to the minimum safe speed of the tow plane. [*Figure 12-6*]

Maximum Aero Tow Speed	MPH	Knots
Blanik L-23	93 mph	81 knots
Blanik L-13	87 mph	76 knots
SGS 2-33	98 mph	85 knots
SGS 1-26	95 mph	83 knots
ASK 21	108 mph	94 knots

Figure 12-6. *Sample glider maximum aerotow speeds.*

The towing speed results from consideration of several variables. The tow pilot determines the minimum towing speed after considering minimum speed for proper engine cooling and stall speed of the tow plane, flies at a speed slower than the maximum glider aerotow speed, and conducts the aerotow at the slowest safe speed sufficient for the glider to remain under control. When towing a model of glider for the first time, the tow pilot should obtain a briefing from the glider pilot to ensure compliance with minimum, maximum, and optimal towing speeds. An experienced tow pilot can determine if the speed is sufficient for the glider by observing the glider wing profile in the tow plane mirrors. A tow pilot viewing a larger than normal portion of the underside of the glider wing for a given tow position may indicate a high glider angle of attack and the need for a slight increase in towing speed.

Because of the potential for low altitude emergencies, the initial climb should remain within gliding distance of the airport. The tow should not climb in a direction that would prevent the glider from returning to the traffic pattern with the existing headwind component. If the tow pilot allows the departure path to drift with a crosswind, it reduces the radius of a turn back to the runway in the event of a low altitude emergency.

A thoughtful glider pilot communicates the intention to perform any maneuvers behind the tow plane. However, the tow pilot should remain alert for any unannounced maneuvering. Glider pilots often practice maneuvers such as "Boxing the Wake," explained and illustrated in Chapter 7 of this handbook, Launch, Flight Maneuvers, Landing, and Recovery Procedures. Once detecting this maneuver, which normally begins with the glider descending vertically from high tow position to low tow position, the tow pilot should maintain a constant heading and a wings level attitude. After the "Boxing the Wake" maneuver ends and the glider returns to a stabilized high-tow position, the tow pilot can maneuver as needed.

During tow, a glider instructor may demonstrate and teach slack rope recovery procedures. This maneuver normally involves a climb to one side of the tow plane followed by a small dive to create slack in the tow line. The instructor will then ask the student to take the controls and remove the slack out of the tow line without breaking the rope. The tow pilot should not confuse this maneuver with a release. Chapter 8 of this handbook, Abnormal and Emergency Procedures, contains more information regarding slack line procedures.

Tow Positions, Turns, & Release

Glider Tow Positions

Glider tow operations normally use the high tow position. However, the glider pilot could use a low-tow position in some instances such as a cross-country tow. Either position places the glider outside the wake of the tow plane. [*Figure 12-7*]

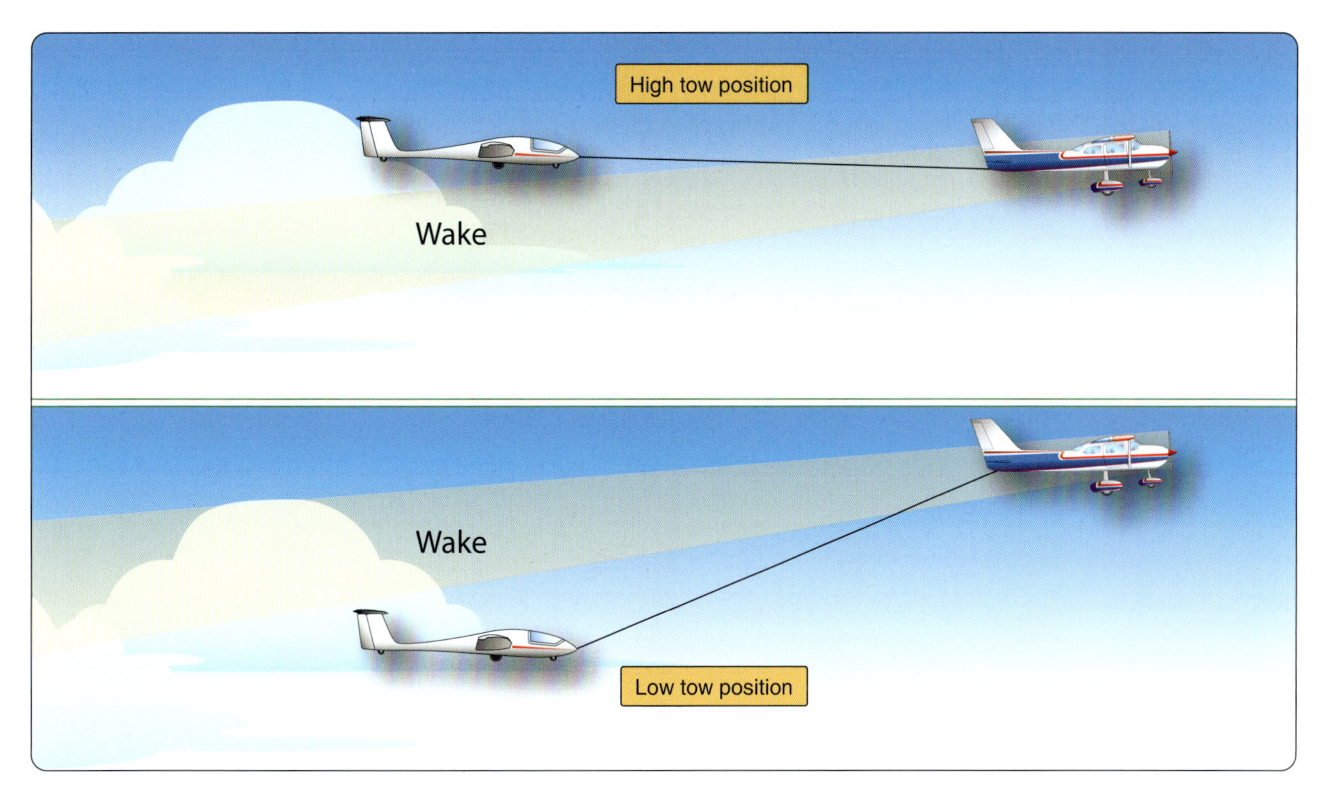

Figure 12-7. *Aerotow climb-out.*

Turns on Tow

The tow pilot can initiate turns upon reaching a safe altitude. The pilot should consider clearance, terrain, and wind gradient. Turbulence and differential wing speed of the glider during turns are potential hazards. Tow pilots should make all turns gently and gradually. The tow pilot expects the glider pilot will attempt to match the flightpath of the tow plane. Due to the length of the wingspan, the glider typically rolls more slowly than the tow plane. Since the bank angles of the tow plane and glider must match to fly the same path, the tow pilot normally uses a maximum of 15–20° of bank.

Approaching a Thermal

When approaching a thermal and planning a release, both pilots should look for other gliders and expect inbound gliders from different directions. Since the first glider in the thermal establishes the direction of turn, any glider joining the thermal should circle in the same direction as the first glider, and the tow pilot should position the flight in a manner that allows the glider proper and safe entry to the thermal. The presence of several gliders in a thermal presents significant risk and the tow pilot should remain clear of any crowded thermal activity.

Release

The winds aloft should be continuously evaluated to help determine the glider release area, and the tow pilot should always attempt to release the glider within gliding distance of the airport considering the winds. The tow pilot should discuss the planned release point with the glider pilot. Use of a consistent tow pattern allows the glider pilot to plan a release point.

Standard glider release procedures include [Figure 12-8]:

1. Some tension on the tow line prior to release allows the tow pilot to feel the release of the glider. A tow pilot looking in the mirror may see a wrinkle in the tow rope after the glider releases.
2. The tow pilot expects the glider to turn right after release but should remain alert for non-standard maneuvering by the glider.

3. Once confirming that the glider has released and cleared, the tow pilot should clear the airspace to the left and start a medium bank, descending left turn.

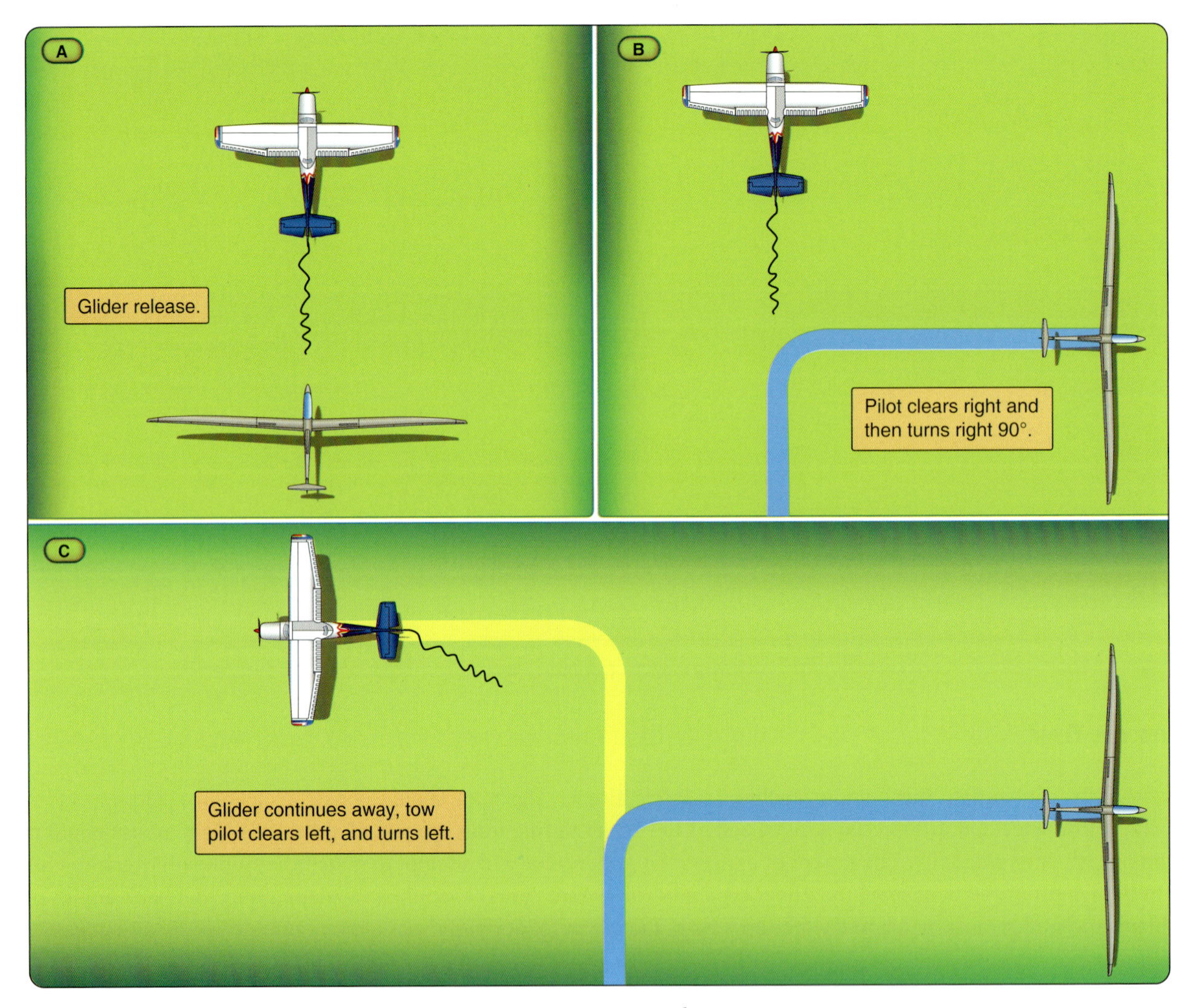

Figure 12-8. *Aerotow release.*

If the glider pilot releases with slack in the tow line, the tow pilot might not detect the release. The tow pilot should only begin a left turn and initiate a descent after observing and confirming the release of the tow line. If there any doubt exists regarding release status, the tow pilot should continue the tow and confirm the release via radio or visually.

Descent, Approach, & Landing

Descent

During the descent, a tow pilot should exercise proper engine management. Good engine conservation practices include both a gradual power reduction and conservative descent airspeeds. Studies indicate that high airspeed may affect the engine more than power reduction. The pilot can use a steep spiral, full flaps, or slipping turns to obtain a suitable rate of descent while reducing power in increments and without a large increase in airspeed. Closing cowl flaps, if equipped, further slows the rate of engine cooling. Each airplane model requires slightly different techniques to keep the engine from cooling too quickly while descending.

Descent and descending flight attitudes increase the potential of a mid-air collision. The tow pilot should consider developing and using specific descent corridors that avoid other glider and powered traffic.

Approach & Landing

A 200-foot tow line hangs down behind the tow plane at a 30- to 40-degree angle. The tow pilot should adjust the altitude of the tow plane to ensure the tow line does not strike obstructions close to the ground.

The tow pilot should know or receive a briefing regarding the location of obstructions around the airport, especially obstructions on the approach end of the runway planned for use. Briefings should include a minimum above ground level (AGL) obstruction crossing height and any factors that may influence altitude judgment, such as visual illusions or other airport distractions.

Regulation does not prohibit landing with the tow line attached; however, the tow pilot should consider the following:

1. Clearing obstructions by more than the tow line length as indicated on the altimeter accounts for instrument lag.
2. Landing with the tow line attached should not occur unless using a field at least 2,500 feet in length with clear approaches.
3. Dragging the tow line on a turf field usually does not cause excessive abrasion. Abrasion from hard ground or paved runways invites early tow line failure.

If releasing the tow line before landing, the pilot normally drops the tow line near the glider launch area during a short approach to the runway. The tow line drop area should have defined dimensions. Ground personnel should receive a drop area briefing before the operation takes place and stay clear of the drop area during towing operations. If seeing an individual in the drop area, the tow pilot should go-around without dropping the tow line.

Cross-Country Aerotow

A safe and successful cross-country tow requires planning. The tow pilot should plan for the maximum fuel consumption for the tow plane used. The pilot should study the route of flight on sectional charts, plan for any potential diversion, and comply with any airspace requirements.

Figure 12-9. *Cross-country tow.*

Since a tow line break can occur without notice, the tow pilot should plan the route over terrain suitable for a glider landing. The tow pilot should consider the physical and mental readiness of both pilots to take the flight. On a particularly long flight, the pilots can plan for rest stops along the way. Both pilots should remain hydrated and have a relief system, if needed. Appropriate use of aircraft trim helps keep the flight within the maximum tow speed of the glider and helps reduce pilot fatigue.

Two-way radio provides for essential communication between the glider and the tow plane during cross-country tows. The pilots should ensure portable radios (if used) or glider batteries have a sufficient charge prior to the flight and conduct a radio check as part of pre-flight activities. [*Figure 12-9*]

Figure 12-9. *On a cross-country tow, the tow pilot and glider pilot should have two-way communication using either panel mounted or portable radios.*

Emergencies

Takeoff Emergencies

Development of an emergency plan ensures successful emergency management. Before the tow and as stated previously, the pilots should select an emergency release point somewhere along the takeoff runway. This release point should leave sufficient room for the glider to land straight ahead, using normal stopping techniques, in the event conditions prevent a safe takeoff with the glider in tow.

During the takeoff and initial climb out, the pilots should remember that position and altitude determine their actions in any low-altitude emergency.

Tow Plane Power Failure on the Runway during Takeoff Roll

The following plan applies in the event the tow plane has a power failure on the runway during the takeoff roll:

- Either pilot should release, and the glider should maneuver to the right side of the runway, if possible.
- The tow plane should maneuver to the left of the runway if space permits. An individual airfield layout and obstacles might dictate an alternate procedure, and the pilots should plan to follow any alternate plan in effect.
- The tow pilot should survey the abort area carefully and know if and where the airplane can roll off the runway (grass or taxiway) without creating a hazard.
- The glider usually lifts off before the tow plane, and the tow pilot should give the glider as much space as possible to land and brake to a stop on the remaining runway.

- The tow pilot should know the stopping characteristics of the glider. Some models have very effective brakes and others do not.

Glider Releases during Takeoff with Tow Plane Operation Normal

The pilot of the tow plane should continue the takeoff to eliminate risk of collision with the glider.

Tow Plane Power Failure while below 200 Feet AGL

Because of airport obstructions near the airport, limited options may exist for a tow plane land out. The pilot of either or both aircraft will normally release to provide landing options for both aircraft. The tow pilot could make slight turns or land straight ahead. Since the tow plane requires considerably more altitude to return to the field in the event of a power failure, the tow pilot should have a specific plan in mind that includes pre-selected landing areas for each runway. Tow pilots should discuss these options during pilot briefings or safety meetings.

Glider Climbs Excessively High during Takeoff

A glider climbing excessively high during takeoff will lift the tail of the tow plane. Should this happen, the application of full-up elevator on the tow plane may not prevent an accident. The tow pilot should pull the release handle immediately to regain control of the tow plane. Any time the glider pulls the nose of the tow plane to a dangerously high or low pitch attitude, the tow pilot should pull the release. An excessively high glider position could jam a Schweizer tow hitch release mechanism.

Airborne Emergencies

Glider Release Failure

If the pilot of the glider cannot release, the glider pilot should inform the tow pilot by means of the aircraft radio or with the following airborne signal. The glider will move out to the left side of the tow plane and rock its wings. [*Figure 12-10*] The tow pilot should wait a few seconds to ensure the wing rocking signals a release failure. Once the tow pilot determines the glider cannot release, the tow plane should return to the airfield and release the glider at a safe altitude over the field.

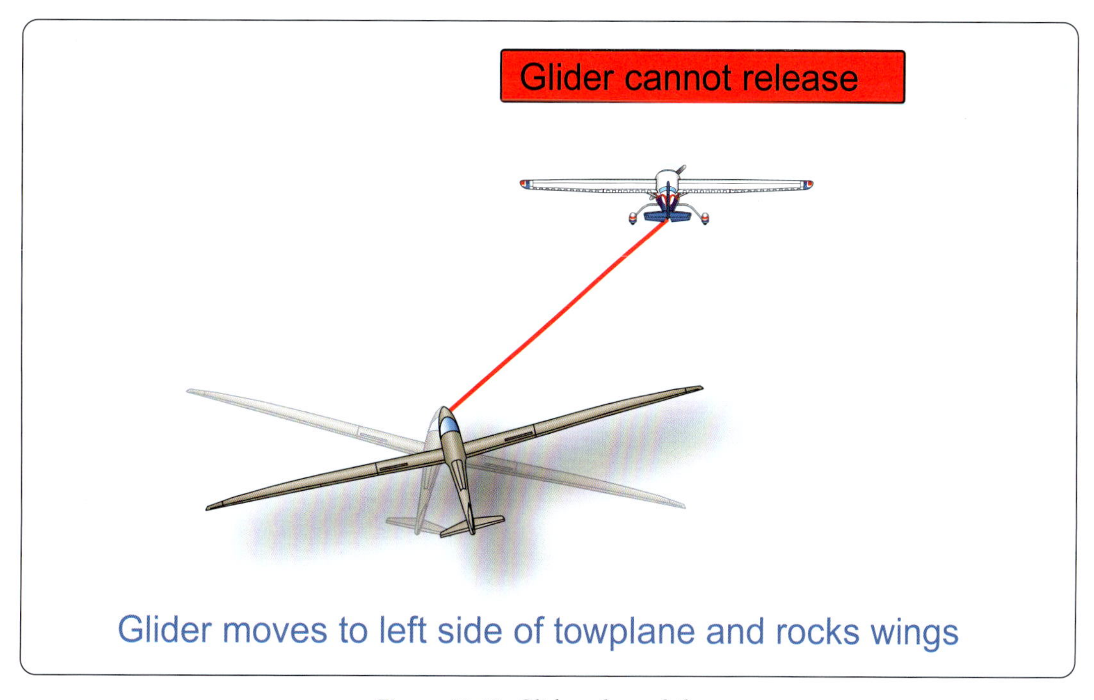

Figure 12-10. *Glider release failure.*

Neither the Tow Plane or Glider Can Release

The pilot of the tow plane informs the pilot of the glider by aircraft radio or airborne signal of this rare occurrence. The tow pilot signals by yawing the tail of the tow plane.

The tow pilot expects the glider to move to the low tow position. Then the tow plane should begin a slow descent toward an airfield of suitable length. The tow pilot flies a wide pattern with a long final approach and sets up a stabilized and gradual 200 to 300 foot per minute descent. The tow pilot should plan on landing long to
allow sufficient altitude for the glider to avoid obstacles on short final.

The glider in the low tow position should land first and the glider pilot should not apply brakes until after the tow plane touches down. The tow pilot should apply brakes gently or not at all to leave room for the glider to stop.

While not well defined in soaring literature as discussed in Chapter 8, Abnormal & Emergency Procedures, some glider pilots may attempt to break the tow rope rather than land behind the tow plane. If the glider does attempt to break the rope, the tow pilot maintains the tow plane in a straight and level attitude to reduce the total forces acting on the tow plane.

Glider Problem

The tow pilot might notice the glider has a problem not yet known to the glider pilot. The most common example involves unintended deployment of the glider spoilers as the glider accelerates on takeoff. The tow pilot can inform the glider pilot via radio or visual signal. The tow pilot waggles or fans the rudder when reaching a safe altitude as the visual signal for "Glider Problem."

Immediate Release

The tow pilot rocks the tow plane wings to indicate the glider pilot should release immediately. This might occur during a critical tow plane emergency such as engine-failure or fire. [*Figure 12-11*]

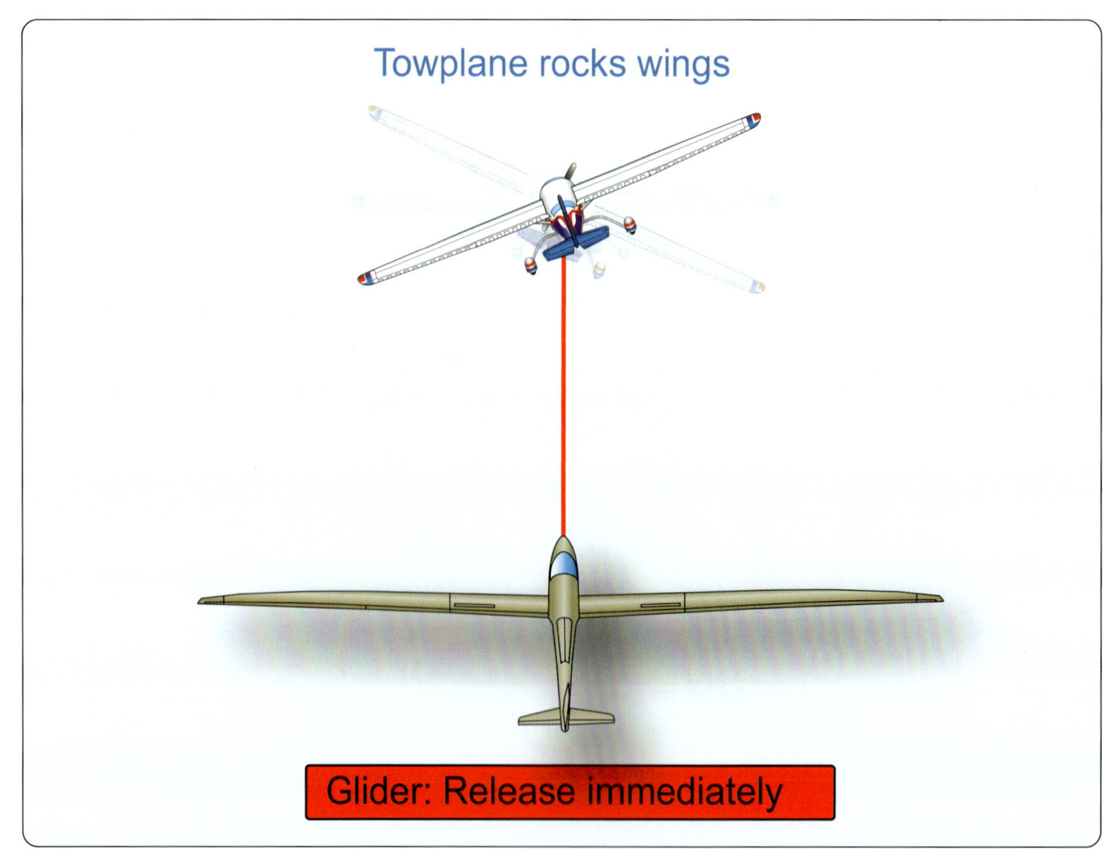

Figure 12-11. *The towplane indicates the glider pilot should release immediately.*

Chapter Summary

The duties and perspectives of a tow pilot differ from those of a glider pilot. However, both pilots have a regulatory responsibility for the safe execution of an agreed to aerotow operation, and both pilots should communicate and cooperate to ensure the safety of both aircraft during an aerotow. Having two-way radio makes communication easier, but no legal requirement exists for it, and pilots may use appropriate visual signals. The tow pilot should maneuver the tow plane accounting for glider characteristics and any glider speed limitations. An aerotow should not preclude the glider from returning to the airport due to wind conditions at release and not put either aircraft in conflict with other aircraft. Each pilot should know the standard procedures for potential emergencies. The tow pilot should manage post-release return to the airport to prevent shock cooling of the tow plane engine. For any long-distance tow such as might occur after a recovery from an airport after a cross-country, both pilots should have the mental and physical readiness to conduct the flight safely.

Chapter 13: Human Factors

Introduction

The study of human factors involves different disciplines. [*Figure 13-1*] When referring to human factors, engineers sometimes refer to the 3Ds: design, development, and deployment of systems that improve the system/human interface. Study of human factors also involves understanding and preventing human errors engineers cannot prevent using the 3Ds. This chapter focuses on the human element—pilot attitudes, pilot error, physiological issues related to soaring safety, pilot management of a glider and its systems, and pilot decision-making as a process that can mitigate glider flight risk and prevent many common types of accidents. For more information on human factors and risk, see the Risk Management Handbook (FAA-H-8083-2) or the Pilot's Handbook of Aeronautical Knowledge (FAA-H-8083-25).

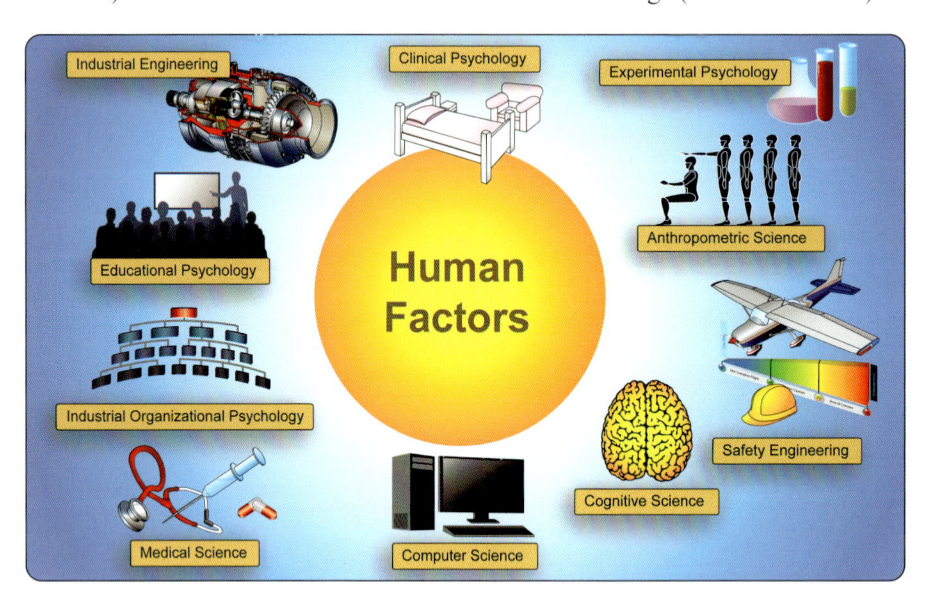

Figure 13-1. *Human factor disciplines.*

Recognizing Hazardous Attitudes

Hazardous attitudes lead to hazardous behaviors that include complacency, indiscipline, and overconfidence, which all increase the risk of a glider accident.

Complacency

Complacency can occur if a pilot feels a false sense of security about the surroundings. This could affect a glider pilot who just wants to fly and does not comprehend the hazards and associated risks. Complacency also affects glider pilots who no longer feel obligated to adhere to standard safety precautions (e.g., "I've done this a million times and don't need to refer to a checklist.").

A few countermeasures include:

- Setting aside sufficient time to prepare for each flight.
- Examining hazards and addressing the associated risk of each flight.
- Challenging oneself to meet a standard of excellence on each flight.

Indiscipline

Aviation accidents sometimes result from pilot failure to comply with regulatory standards. Having the discipline to follow the rules reduces the chance of an accident.

The regulatory requirements set a standard for safety. Pilots should consider developing more stringent personal limitations that may be modified as their experience and proficiency grow. For example, the pilot could establish minimum visibility and wind conditions for flight. In that case, the pilot would not fly if conditions exceeded established personal minimums. A disciplined pilot would only lower personal minimums based on training and rational decision making and not based on a desire to make a particular flight. In addition, a pilot might raise the minimums if under additional stress from personal or work-related issues or if flying infrequently.

Sometimes pilots feel their experience has taught them an easier or faster way to do certain tasks. They should ask themselves if their attitude and procedures align with guidelines set forth by disciplined aviators. If not, these pilots should consider that standardized procedures, rules, and formal risk mitigation strategies offer better protection from accidents.

Pilots interested in more information about a disciplined approach to aviation safety can refer to the FAA Risk Management Handbook (FAA-H-8083-2), which describes structured techniques pilots can use that include how to set or revise personal minimums and how to conduct a disciplined and thorough safety analysis before flight.

Overconfidence

A realistic level of confidence enables a pilot to feel good about a particular flight operation. That confidence comes from experience, training, proficiency, adequate preparation for a flight, and ongoing discipline.

Overconfidence reflects a lack of understanding about the pilot's own limitations, not understanding the aircraft or conditions that could threaten the safety of flight, denial of the reality of pilot shortcomings, or a desire to prove something. Whatever the cause, unjustified confidence can lead to an accident. Pilots should carefully consider their limitations when attempting to fly in an unfamiliar glider or in an unfamiliar environment and resist letting overconfidence or pride interfere with good judgment. Glider pilots who do not fly regularly or who have not flown for several months should recognize the different levels of safety that come from having a current flight review every two years, meeting take-off and landing currency requirements, and having the level of proficiency necessary given the current conditions.

Pilot Error

All pilots make errors, and many pilots may have gaps in knowledge. Training assessments, flight examinations (written or oral), operational checks, critiques, and post-flight pilot self-assessments can highlight what a pilot can do better. An honest and fair assessment can lead to correction of any shortcomings and reduce the potential for accidents.

Types of Errors

One type of unintentional error involves failure to perform an intended action within acceptable tolerances. On the other hand, an incorrect opinion, bad judgment, poor reasoning, careless attitude, or insufficient knowledge can cause a more serious mistake. For example, a pilot with good stick and rudder skills but new to ridge soaring might fly into strong sink on the downwind side of a ridge and narrowly escape hitting terrain. The pilot's lack of knowledge and bad judgment could lead to this kind of mistake.

Intentional

If a pilot knowingly does something wrong, that pilot intentionally deviated from safe practices, procedures, standards, or regulations. In aviation, an intentional error suggests a serious and underlying lack of concern for safety. The pilot should reflect on the reason for the error, seek advice or counseling, and not fly unless able to prevent that behavior.

Fatigue

Fatigue involves a reduction or impairment in any of the following: cognitive ability, decision-making, reaction time, coordination, speed, strength, and balance. Fatigue reduces alertness and often reduces a person's ability to focus and hold attention on a task. [*Figure 13-2*] Emotional fatigue exists and can affect mental and physical performance. Lack of sleep, stress, and overwork can all cause or aggravate fatigue.

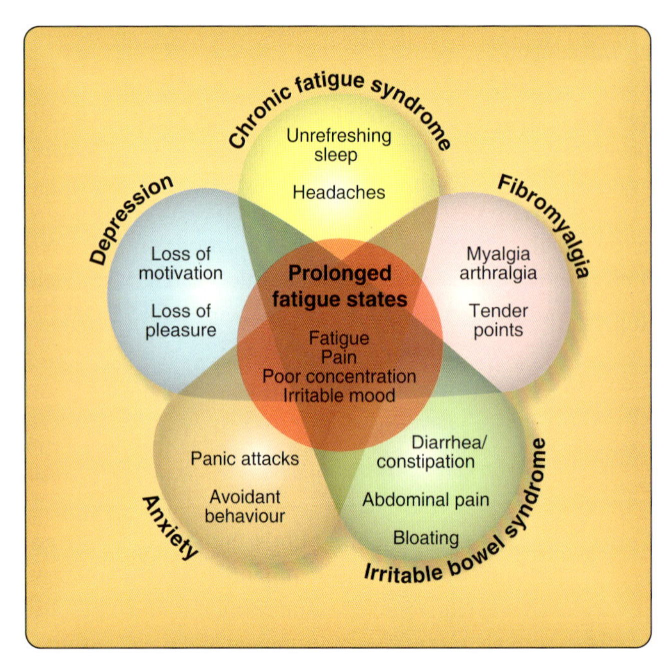

Figure 13-2. *Sampling of factors that interact with fatigue.*

A person's mental and physical state naturally cycles through various levels of performance each day. Variables such as body temperature, blood pressure, heart rate, blood chemistry, alertness, and attention rise and fall in a daily pattern known as a circadian rhythm. [*Figure 13-3*] A person's ability to work and rest rises and falls during this cycle. Activity contrary to a person's circadian rhythm can cause subtle difficulties and fatigue. An affected person might not recognize this situation. Since another person might alert a pilot to the signs of fatigue, flying a glider alone when fatigued creates a particularly dangerous situation. For example, a fatigued glider pilot might not actively see and avoid other traffic. Pilots should avoid flying when not having full night's rest, after working excessive hours, or after an especially exhausting or stressful day.

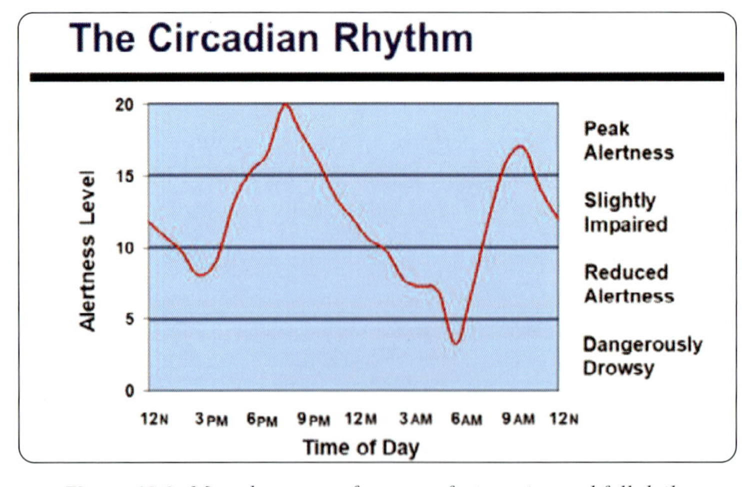

Figure 13-3. *Many human performance factors rise and fall daily.*

The best remedy for fatigue involves getting enough sleep on a regular basis. Pilots should track their hours and quality of sleep for fatigue awareness. Countermeasures to fatigue such as caffeine, may work for a short duration, but many countermeasures may make fatigue worse over the long term. Pilots should exercise caution if using medication to fight fatigue since a fatigued person may have trouble getting needed rest after using medication. A pilot experiencing ongoing fatigue issues or chronic fatigue, should stop flying a glider, consult a physician, and resolve the issue before flying again.

Hyperventilation

Hyperventilation results when a person breathes at an increased rate or breathes more deeply, which reduces the level of carbon dioxide in the blood. Reduced carbon dioxide in the blood can raise blood pH and lead to undesirable health effects.

Hyperventilation might occur as result of emotional stress, fright, or pain. Glider pilots who encounter extreme or unexpected turbulence or strong areas of sink over rough terrain or water may unconsciously increase their breathing rate or breathing volume. When flying at higher altitudes, either with or without oxygen, a tendency to breathe more rapidly than normal may occur.

Figure 13-4 lists the common symptoms of hyperventilation. Treatment for hyperventilation involves restoring the proper carbon dioxide level. Consciously slowing the breathing rate or talking aloud can reverse the effects of hyperventilation. Recovery usually occurs rapidly once the breathing rate returns to normal. In rare cases, hyperventilation can cause unconsciousness.

Common Symptoms of Hyperventilation
Headache
Decreased reaction time
Impaired judgment
Euphoria
Visual impairment
Drowsiness
Lightheaded or dizzy sensation
Tingling in fingers and toes
Numbness
Pale, clammy appearance
Muscle spasms

Figure 13-4. *Common symptoms of hyperventilation.*

Hypoxia

Hypoxia results from reduced oxygen or not enough oxygen. Although human cell tissue will die if deprived of oxygen long enough, the principal concern for pilots is lack of oxygen to the brain, which can reduce cognitive ability and result in life-threatening errors. Hypoxia has several causal factors, including an insufficient supply of oxygen, inadequate transportation of oxygen, or the inability of the body tissues to use oxygen. The forms of hypoxia are based on their causes:

Hypoxic Hypoxia

Hypoxic hypoxia results from insufficient oxygen available to the body as a whole. The reduction in partial pressure of oxygen at high altitude can cause a pilot to experience this type of hypoxia. As an unpressurized aircraft ascends during flight, the percentage of atmospheric oxygen remains constant, but a reduced number of oxygen molecules enter the lungs and pass between the membranes within the respiratory system.

Hypemic Hypoxia

Hypemic hypoxia occurs when the blood cannot take up sufficient oxygen. Causes of this form of hypoxia include reduced blood volume from severe bleeding or certain blood diseases, such as anemia. Carbon monoxide poisoning causes this type of hypoxia. Hemoglobin, the blood molecule that transports oxygen, becomes chemically unable to bind oxygen molecules if exposed to carbon monoxide. Hypemic hypoxia can also occur after a blood donation. While blood volume normalizes quickly following a donation, restoring the lost hemoglobin can take several weeks. Although the effects of the blood loss seem slight at ground level, blood donation can create a flight risk during the recovery period.

Stagnant Hypoxia

Stagnant hypoxia or ischemia results when the oxygen-rich blood in the lungs does not move to the tissues that need it. An arm or leg going to sleep because of restricted blood flow is one form of stagnant hypoxia. This kind of hypoxia can also result from shock, the heart failing to pump blood effectively, or a constricted artery. During flight, excessive G forces can cause stagnant hypoxia. Cold temperatures can also reduce circulation and decrease blood supply to extremities.

Histotoxic Hypoxia

When histotoxic hypoxia occurs, the body transports enough oxygen to the cells, but they cannot make use of it. This impairment of respiration at the cellular level may result from alcohol consumption, exposure to certain drugs or narcotics, or exposure to poisons.

Symptoms of Hypoxia

Oxygen starvation causes impairment of the brain and other vital organs. The first symptoms of hypoxia can include euphoria and a carefree feeling. With increased oxygen deprivation, the extremities become less responsive and flying becomes less coordinated. The symptoms of hypoxia vary with the individual, but common symptoms include:

- Cyanosis (blue fingernails and lips).
- Headache
- Decreased response to stimuli and increased reaction time,
- Impaired judgment
- Euphoria
- Visual impairment
- Drowsiness
- Lightheaded or dizzy sensation
- Tingling in fingers and toes
- Numbness

As hypoxia worsens, the pilot's field of vision begins to narrow, and instrument interpretation can become difficult. Even with all these symptoms, the effects of hypoxia can give a pilot a false sense of security.

Treatment of Hypoxia

Treatment for hypoxia involves increasing the amount of available oxygen. Pilots commonly descend to lower altitudes or use supplemental oxygen to counteract the effects of hypoxia. Time of useful consciousness gives the maximum time available for the pilot to make rational, life-saving decisions and carry them out at a specific altitude without supplemental oxygen. As altitude increases above 10,000 feet, the symptoms of hypoxia increase, and the time of useful consciousness diminishes significantly. [*Figure 13-5*]

Altitude	Time of useful consciousness
45,000 feet MSL	9 to 15 seconds
40,000 feet MSL	15 to 20 seconds
35,000 feet MSL	30 to 60 seconds
30,000 feet MSL	1 to 2 minutes
28,000 feet MSL	2½ to 3 minutes
25,000 feet MSL	3 to 5 minutes
22,000 feet MSL	5 to 10 minutes
20,000 feet MSL	30 minutes or more

Figure 13-5. *Time of useful consciousness.*

Since individuals experience hypoxia differently, experiencing the effects of hypoxia in an altitude chamber [*Figure 13-6*] can improve an individual's recognition of their own symptoms. The Federal Aviation Administration (FAA) provides this opportunity through aviation physiology training, which occurs at the FAA Civil Aerospace Medical Institute (CAMI) in Oklahoma City, Oklahoma, and at many military facilities across the United States. For information about the FAA's one-day physiological training course with altitude chamber and vertigo demonstrations, visit the FAA website.

Figure 13-6. *CAMI altitude chamber.*

Inner Ear Discomfort

The internal pressure of the middle ear cavity changes more slowly than the pressure outside the ear. A pressure difference can develop during an ascent or descent, which may result in temporary hearing loss, discomfort, pain, and distraction from flight tasks.

When a glider ascends, middle ear air pressure may exceed the pressure of the air in the external ear canal, causing the eardrum to bulge outward. This condition usually resolves itself, and the pilot may notice a popping sound as pressure equalizes and hearing sensitivity returns to normal. During a descent, pressure of the air in the external ear canal may increase above that of the middle ear cavity, which causes the eardrum to bulge inward. The condition during a descent seems more painful and unpleasant to most people. Nasal congestion or a cold can make pressure equalization difficult or impossible. Chewing gum, sucking on hard candy, or swallowing might assist pressure equalization during a descent.

Discomfort during a descent ends when enough air flows into the middle ear through the eustachian tube. However, the lower pressure in the middle ear tends to constrict the walls of the eustachian tube and prevent the flow of air to the middle ear. A pilot can try the Valsalva maneuver to correct this situation. The pilot should pinch the nostrils, close the mouth and lips, and blow slowly and gently in the mouth and nose to force air up the eustachian tube into the middle ear.

Scuba Diving

Scuba diving subjects the body to increased pressure, which dissolves more nitrogen in body tissues and fluids. The reduction of atmospheric pressure that accompanies flying can produce physical problems for scuba divers when small bubbles of nitrogen to form inside the body as the gas comes out of solution. These bubbles can cause a painful and potentially incapacitating condition called the bends.

Scuba training emphasizes how to prevent the bends when rising to the surface of a body of water. However, excess nitrogen can remain in tissue fluids for several hours after finishing a dive. A pilot who SCUBA dives can experience the bends from as low as 8,000 feet MSL, with increasing severity as altitude increases. As noted in the Aeronautical Information Manual (AIM), the minimum recommended time between scuba diving on non- decompression stop dives and flying is 12 hours, while the minimum time recommended between decompression stop diving and flying is 24 hours.

Spatial Disorientation

Most gliders used for primary training have basic instrumentation only, and glider pilots do not normally train for flight solely by reference to instruments. Glider pilots should avoid flight in low visibility or in any condition that makes discerning the horizon difficult. These conditions increase the likelihood that the pilot will succumb to an illusion and experience a loss of control. [*Figure 13-7*] Glider pilots who fly in marginal visibility should establish and abide by personal minimums for visibility.

Figure 13-7. *Flying in haze or other restrictions to visibility increases the likelihood of spatial disorientation.*

Flight in a powered glider may occur at night or in instrument conditions provided the glider meets the requirements of 14 CFR part 91, section 91.205, and the pilot meets the applicable requirements of 14 CFR part 61, section 61.57. Pilots have fewer visual cues available to judge flight attitude at night or in instrument conditions. Both regimes present additional potential for illusions, navigation errors, collisions, and loss of control. In addition, any emergency at night becomes much more difficult to handle. Glider pilots who fly in these conditions normally have ratings in other aircraft categories that include training and testing for night or instrument conditions. A glider pilot without additional training, certification, recency, and proficiency should avoid night or instrument operations.

Dehydration

Glider pilots often fly for long periods of time in hot summer temperatures or at high altitudes that can cause dehydration. Although the effects of dehydration may develop slowly, fluid loss from perspiration and breathing can result in fatigue and progress to dizziness, weakness, nausea, tingling of hands and feet, abdominal cramps, and extreme thirst. [*Figure 13-8*]

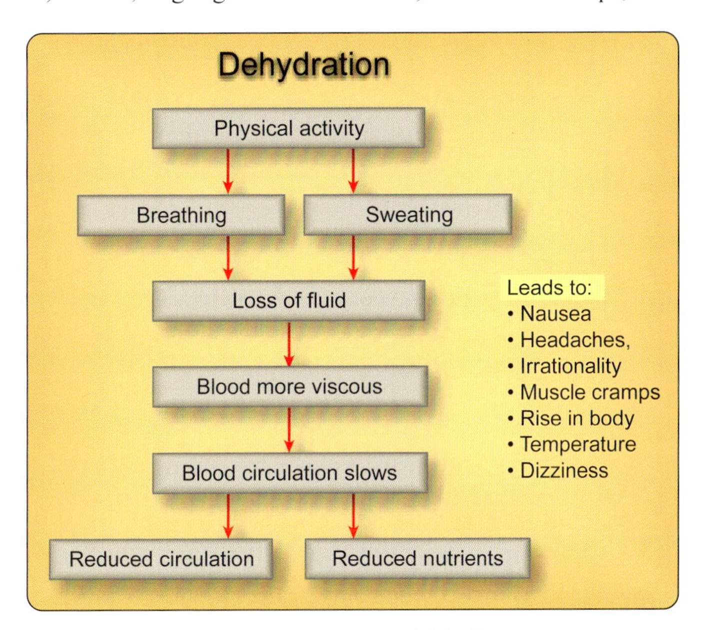

Figure 13-8. *Symptoms of dehydration.*

Pilots should take water on every flight to prevent dehydration. Some glider pilots wear a hat with a rim for shade and to keep a cool head. Pilots should ensure that the brim of the hat does not interfere with the ability to scan for other gliders and air traffic.

Heatstroke

Heatstroke results when the body cannot control excessive high temperature. Onset of this condition may mimic dehydration, but it may also lead to collapse. To prevent these symptoms, the pilot should carry an ample supply of water and use it at frequent intervals on any long flight, even if not thirsty. Wearing light-colored, porous clothing and a hat provides protection from the sun. Ventilating the cabin also helps remove excess heat.

Cold Weather

Preparing for extreme cold may seem odd when comfortable temperatures exist at ground level on wave soaring days. However, when flying at high altitudes, the inside of the glider can get cold. A glider at a high altitude can encounter temperatures of -30° to -60 °C. When soaring, sunshine through the canopy can keep the pilot's upper body warm for a time, but shaded legs and feet can quickly chill or suffer frostbite. After an hour or two at such temperatures, even the upper body can become quite cold. Layered, loose-fitting clothing helps insulate body heat. Either wool gloves or fitted gloves with mittens over them can protect the hands. Two or three pairs of socks in layers with silk on the inside and wool on the outside plus an insulated boot can help keep feet comfortable. Clothing manufacturers produce clothing and socks with internal heating elements and rechargeable lithium batteries that the pilot can turn on and regulate. These make a great addition to a glider pilot's wardrobe.

Low temperatures can cause other unpleasant or hazardous conditions. Pilot or passenger exhaled moisture can condense and then freeze or deposit directly as frost on the inside of the canopy. The pilot can use a clean piece of cloth that does not damage the canopy to wipe off condensation or light frost. Allowing fresh air through a vent can clear condensation and stop frost formation. Unfortunately, this also quickly lowers the inside temperature, and may require adding a layer of clothing.

The body dehydrates more rapidly in extreme cold and refraining from drinking water can cause dehydration even in cold conditions. Because cold weather causes the kidneys to excrete liquid at a faster rate, the pilot should make a bathroom stop before takeoff and consider a relief system for the flight.

Cabin Management & Equipment

Prior to launch, the pilot should brief any passengers on use of safety belts, shoulder harnesses, and emergency procedures. The pilot should check the security of any trim ballast, organize the items carried onboard, and properly stow and secure all other items. Placement of charts, tablets, and cross-country aids should allow the pilot to reach them easily.

Parachute

Pilots may use a parachute for emergencies. A certificated and appropriately rated parachute rigger must repack a nylon parachute within the preceding 180 days. The packing date information is usually found on a card contained in a small pocket on the body of the parachute. Refer to 14 CFR, part 91, section 91.307 for more information.

Supplemental Oxygen

High-altitude soaring flights require the use of supplemental oxygen. In some parts of the country, soaring routinely occurs to a 16,000- to 18,000-foot cloud base in thermals. Flight using mountain waves may lead to flight at altitudes more than 30,000 feet in the United States.

Breathing supplies oxygen to the blood and removes carbon dioxide. In each breath at 18,000 feet, the pilot breathes in only half as much oxygen as at sea level. The pilot's automatic reaction without an adequate supply of oxygen would involve breathing twice as fast. This hyperventilation, or over-breathing, would result in eliminating too much carbon dioxide from the blood.

14 CFR, part 91, section 91.211, dictates time and altitude requirements for use of supplemental oxygen. Prior to use of supplemental oxygen, the system should be checked for oxygen availability and flow. The pilot can use the PRICE checklist:

- P = Pressure
- R = Regulator
- I = Indicator
- C = Connections
- E = Emergency bail-out bottle

Aviation Oxygen Systems

A portable aviation oxygen system delivers oxygen using lightweight and compact components. It delivers a calibrated amount of oxygen based on extensive research in human flight physiology. Prior to purchasing any type of oxygen system, pilots should consider the type of flying they do. Two different common types of systems used today include the Continuous-Flow System and the Electronic Pulse Demand Oxygen System (EDS). Pilots should only use aviator's oxygen supplied in a green bottle with these systems and never use medical oxygen. Medical oxygen may contain water and could render a pilot's oxygen system unworkable.

Continuous-Flow System

The continuous-flow system uses a high-pressure storage tank and a pressure-reducing regulating valve that reduces the pressure in the cylinder to approximately atmospheric pressure at the mask. [*Figure 13-9*] The oxygen flows continuously when the system is turned on provided the bottle contains sufficient oxygen. In some installations, the pilot can adjust the amount of oxygen flow manually for low, intermediate, and high altitudes; automatic regulators adjust the oxygen flow by means of a bellows, which varies the flow according to altitude. When using the continuous-flow oxygen system, the pilot can use either an oxygen mask or a nasal cannula. [*Figures 13-10* and *13-11*]

Figure 13-9. *Continuous-flow oxygen system.*

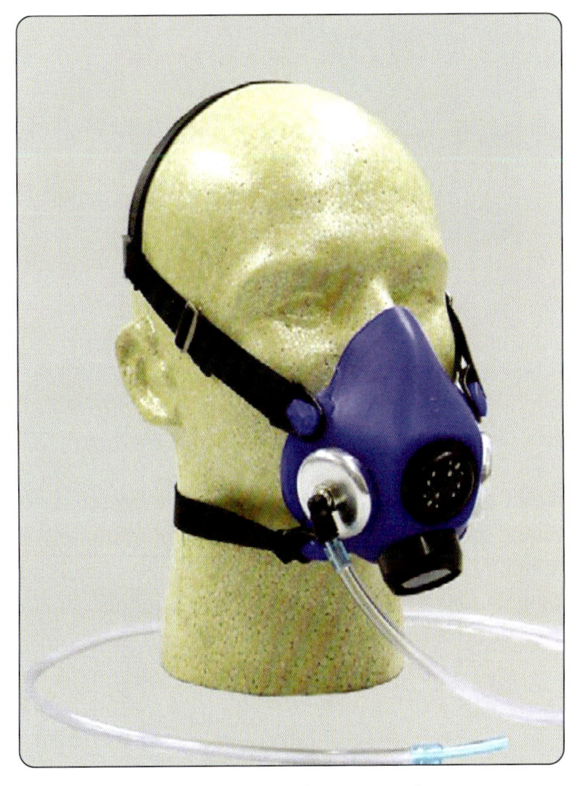

Figure 13-10. *Oxygen mask.*

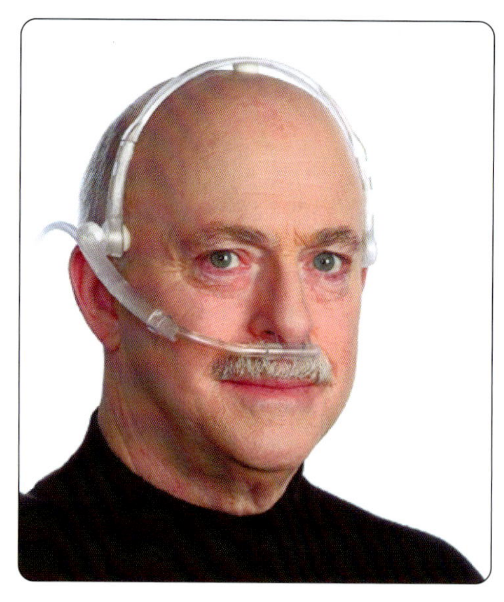

Figure 13-11. *Nasal cannula.*

Electronic Pulse Demand Oxygen System (EDS)

The EDS delivers altitude-compensated pulses of oxygen only when the pilot inhales. It typically uses 1⁄6 the amount of oxygen at 1⁄4 the weight and volume of conventional constant-flow systems. [*Figure 13-12*] The EDS has a micro-electronic pressure altitude barometer that automatically determines the volume for each oxygen pulse up to pressure altitudes of 32,000 feet. The EDS automatically goes to a 100 percent pulse-demand mode at pressure altitudes above 32,000 feet.

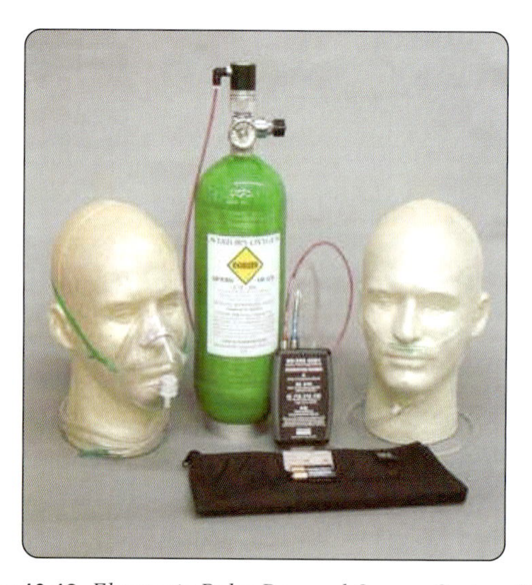

Figure 13-12. *Electronic Pulse Demand Oxygen System (EDS).*

The pilot can set an EDS to different modes and delays. For example, it can respond with oxygen at altitudes where needed and conserve oxygen at lower altitudes. It can also be set to night or now mode where it responds from sea-level and up. The EDS limits its response to a maximum respiration rate of about 20 breaths per minute, virtually eliminating any hyperventilation. The pilot turns it on and does not need to read scales or adjust any knobs when climbing or descending. These devices need no local altimeter setting since they respond directly to pressure altitude, just as the body does.

Risk Management

Risk management for pilots includes identification of hazards that pose a flight risk, assessment of the level of risk associated with each hazard, and decision making to manage and mitigate any unacceptable risks.

Hazards include any condition that can foreseeably cause or contribute to an aircraft accident. Typical hazards include pilot condition or lack of proficiency, aircraft or equipment malfunctions or shortcomings, and environmental conditions that include weather, mountains, obstacles, other aircraft, wires, and high-altitude flight. Any external pressure to take a particular flight can create an additional hazard. These items form part of the PAVE checklist (Pilot, Aircraft, enVironment, External pressure) that pilots can use to consider and manage common flight risks.

The composite of predicted severity and likelihood of the potential effect of a hazard constitutes the risk. The level of risk associated with a hazard depends on the likelihood of an accident and the severity of damage or injury that could occur. Risk mitigation involves an analysis and implementation of changes the pilot can make to lower the level of risk. For example, a glider pilot recognizes the risk associated with a tow line break. The pilot lowers the level of this risk by training for this possibility. Before each tow, the pilot should also consider a plan of action in the event of a rope break. If that consideration does not adequately mitigate that risk for the given set of conditions on a particular day, the pilot might decide to wait for the winds to change, ask for a different direction of tow, or postpone or cancel the operation.

Safety Management System (SMS)

This Handbook focuses on the individual and not on safety management systems (SMS). SMS addresses risk management from an organizational perspective as an ongoing activity. Interested persons may obtain information about SMS from the Risk Management Handbook (FAA-H-8083-2).

Aeronautical Decision-Making (ADM)

A pilot makes numerous aeronautical decisions involving risk management before a flight begins. One structured means involves using a Flight Risk Assessment Tool or FRAT. Pilots can check the FAA Risk Management Handbook (FAA-H-8083-2) for a sample FRAT and can use one to enhance the safety of flight.

FAA regulations set up specific minimum safety requirements for some conditions, but that does not mean every pilot has the capability to fly in those conditions. As previously mentioned, pilots can establish their own more stringent rules for the personal, equipment, and environmental conditions that might lead them to decide not to fly on a given day. The FAA Risk Management Handbook (FAA-H-8083-2) has a chapter explaining the rationale for personal minimums in more detail and how to establish and maintain these minimums.

Aeronautical decision-making (ADM) in flight usually involves a systematic mental process pilots use to determine a course of action in response to a given set of circumstances. Pilots should perceive any hazards that threaten the safety of flight, determine a course of action that will lead to a successful outcome, and then perform the steps expected to lead to that successful outcome. The process of hazard perception, situation processing, and performance should repeat as the flight progresses. At the conclusion of a flight the pilot can self-assess on the quality of decision making.

Despite advancement in training methods, airplane equipment and systems, and services for pilots, incidents and accidents still occur. Despite all the changes in technology to improve flight safety, the human factor plays a role in a high percentage of all aviation accidents.

Human factor-related accidents usually do not involve a single decision but result from a chain of decisions and factors that might lead to an accident. An error chain describes the sequence of several events in a human factors-related accident. Breaking one link in the chain would normally change the outcome of the sequence of events.

This list presents different hazards and pilot responses. Did these pilots carefully consider the consequences of their decisions?

- Circumstance: My oxygen system has a slow leak. Decision: Soaring conditions are perfect, and I will not need oxygen for today.

- Circumstance: High winds are forecast later today. Decision: I can fly and make it back before the wind changes.

- Circumstance: My aircraft or radio batteries are low. Decision: I am only planning a short flight, so I'll go.

Circumstances as mundane as a slow oxygen leak, a high wind forecast, or low batteries become part of a decision chain that can lead to an incident or accident. In the previous circumstances, the pilot might interrupt an accident chain by having the slow oxygen leak repaired, respecting the high wind forecast and postponing the flight, or recharging the low batteries before the flight.

Advisory Circular (AC) 60-22, Aeronautical Decision Making, provides introductory material, background information, and reference material on ADM. The material in this AC provides a systematic approach to risk assessment and stress management in aviation, illustrates how personal attitudes can influence decision-making, and how those attitudes can be modified to enhance safety. This AC also provides instructors with methods for teaching ADM techniques and skills in conjunction with conventional flight instruction. Individuals learning to fly gliders should seek out instructors who integrate ADM training. The FAA Risk Management Handbook (FAA-H-8083-2) provides an overview and examples of what pilots do to make their flights safe and enjoyable.

Analysis of Previous Accidents

The National Transportation Safety Board (NTSB) compiles an accident report any time a reportable glider accident occurs. Interested persons can find this public information at www.ntsb.gov.

An individual using the NTSB's accident database, can use their supplied query tool to perform a simple search using the term "glider" (and other optional elements) to retrieve accident reports involving gliders. The detail in these reports provides a narrative of circumstances and events that led to each accident. Since pilot decisions before and during flight may have prevented these accidents, a review can illustrate the concept of an accident chain. That understanding should prompt pilots to make safety and risk mitigation a high priority. *Figure 13-13* contains a summary of several glider accident final reports from the past several years.

Event Date	Glider Type	Injury	Probable Cause
2/9/2020	Aviastroitel AC 4C	Fatal	The pilot's exceedance of the glider's critical angle of attack while maneuvering for landing, which resulted in an aerodynamic stall and subsequent loss of control.
3/1/2020	Let L 23 SUPER BLANIK	Serious	The pilot's failure to maintain glider control and his exceedance of the glider's critical angle of attack while maneuvering in gusting wind conditions, which resulted in an aerodynamic stall.
4/7/2020	Schempp Hirth Standard Cirrus	Fatal	The pilot's exceedance of the glider's critical angle of attack following his premature termination of the tow for reasons that could not be determined, which resulted in an aerodynamic stall/spin during a turn back to the departure airport.
5/11/2020	Schempp Hirth VENTUS 2CT	Minor	A loss of thermal lift during a motor glider flight, which resulted in an off-airport landing. Contributing to the accident was the pilot's delayed attempted engine start.
6/4/2020	Gilasflugel Mosquito	None	The pilot's decision to divert to the private airport and his subsequent failure to maintain directional control while landing on a turf runway that contained tall grass.
6/13/2020	Pipistrel PIPISTREL SINUS 912	None	The pilot's failure to maintain airspeed while landing with a quartering tailwind, which resulted in a loss of control and a hard landing.
7/11/2020	Schleicher ASW27	Fatal	The pilot's loss of glider control while maneuvering near a mountain ridge in downdrafts and dry microbursts at an altitude that precluded recovery.
8/16/2020	Evektor Aerotechnik L13	Serious	The pilot's failure to stow the speed brake prior to attempting takeoff.
9/19/2020	PIK PIK-200	Minor	The pilot's failure to maintain directional control during takeoff that resulted in a collision with another glider that was parked close to the departure runway, and the pilot's improper decision to attempt a takeoff without ensuring he had safe clearance from the parked glider.
9/29/2020	Schleicher ASK21	None	The pilot's improper control input during a bounced landing that resulted in the glider impacting terrain.
11/7/2020	Schleicher ASW20C (A1); Schleicher ASW27 (A2)	Serious	The failure of both pilots of each glider to see an avoid one another while maneuvering, which resulted in a mid-air collision.
12/27/2020	Glasflugel CLUB LIBELLE 205	Serious	The pilot's failure to maintain adequate clearance from trees during an off-airport landing.
4/22/2021	Schempp Hirth Ventus C	Serious	Impact with trees during a forced landing in atmospheric lift conditions that were insufficient to maintain flight. Contributing was the pilot's delayed decision to return to the airport.
5/16/2021	I.C.A. BRASOV (ROMANIA) IS- 2882	Serious	The flight instructor's failure to maintain aircraft control resulting in the exceedance of the glider's critical angle of attack following the breakage of the weak/safety link during a winch launch, which resulted in an aerodynamic stall and spin, and subsequent impact with trees and terrain.
5/20/2021	Pilatus 84-PC11	Minor	The pilot's misjudged approach angle which resulted in impact with trees and terrain.
6/5/2021	Schweizer SGS 2-33A	None	The pilot's failure to maintain directional control while landing in gusting wind conditions.
6/6/2021	Schweizer SGS 1-35	Fatal	The pilot's low-altitude release from tow for reasons that could not be determined, and his subsequent exceedance of the glider's critical angle of attack while returning to the runway, which resulted in an aerodynamic stall and impact with terrain.
6/13/2021	Schweizer SGS 2-33A	Serious	The student pilot's failure to maintain an appropriate glide path to the runway.
6/20/2021	Schweizer SGS 1-35	Minor	The glider pilot's loss of visual references during the landing approach resulting in an offairport landing on rough sloping terrain.
7/15/2021	ALEXANDER SCHLEICHER GMBH & CO ASW 27-18	None	The pilot's failure to maintain the glider's stability in the roll axis during the takeoff roll, which resulted in a dragged wingtip and ground loop.
7/25/2021	BURKHART GROB G 103 TWIN	Minor	The glider's encounter with atmospheric conditions where the lift was not sufficient to maintain flight which resulted in an off-airport landing and a collision with a fence.
8/15/2021	Schweizer SGS 2-33A	None	The check pilot's failure to account for the extended departure distance from the airport during a simulated tow rope break and recovery.
9/7/2021	Aeromot AMT-100	None	The pilot's failure to extend the landing gear.
10/21/2021	Pipistrel Apis-Bee	Serious	The pilot's failure to maintain control of the glider during the landing approach, which resulted in an aerodynamic stall and subsequent impact with the runway.
11/17/2021	PHOENIX AIR U-15 PHOENIX	None	The pilot's failure to maintain distance with an airport sign while taxiing.
4/8/2022	ALEXANDER SCHLEICHER GMBH & CO ASK 21	None	The gliders encounter with atmospheric conditions where the lift was not sufficient to maintain flight and subsequent impact with mountainous terrain.
6/19/2022	ROLLADEN-SCHNEIDER 15-6	None	The glider's encounter with atmospheric conditions where the lift was not sufficient to maintain flight and subsequent water ditching.
6/22/2022	LET L-23 SUPER BLANIK	Serious	The pilot's encounter with sinking air conditions that resulted in a loss of lift and a subsequent loss of control.
6/22/2022	DG FLUGZEUGBAU GMBH DG 10005	None	The glider's encounter with atmospheric conditions where the lift was not sufficient to maintain flight. Contributing to the accident was the pilot's decision to overfly a suitable landing site which resulted in an off-field landing in a lake.
7/10/2022	Schleicher ASW-198	None	The pilot's misidentification of the runway during the visual approach which resulted in an off runway landing and impact with a trailer.

Figure 13-13. *Glider accident data from the NTSB database.*

Chapter Summary

This chapter focuses on the subset of human factors that pilots can control to prevent accidents. Hazardous attitudes play a role in accidents, and pilot should recognize and avoid them and the types of errors they lead to. The chapter discusses human physiology related to glider flight. Pilots should know how to take care of their physical needs before and during flight. The chapter also discusses systems pilots might use during glider flight including oxygen systems. The chapter discusses risk management and reducing the level of risk to avoid accidents. The NTSB database contains accident reports that illustrate the concept of an accident chain. While glider flight has inherent associated risk, the principles discussed in this chapter and throughout this handbook can reduce a glider pilot's level of exposure to that risk.

Glossary

A

Advection. The transport of an atmospheric variable due to mass motion by the wind. Usually the term as used in meteorology refers only to horizontal transport.

Ailerons. The hinged portion of the trailing edge of the outer wing used to bank or roll around the longitudinal axis.

Air density. The mass of air per unit volume.

Airfoil. The surfaces on a glider that produce lift.

Air mass. A widespread mass of air having similar characteristics (e.g., temperature), that usually helps to identify the source region of the air. Fronts are distinct boundaries between air masses.

Amplitude. In wave motion, one half the distance between the wave crest and the wave trough.

Angle of attack. The angle formed between the relative wind and the chord line of the wing.

Angle of incidence. The angle between the chord line of the wing and the longitudinal axis of the glider. The angle of incidence is built into the glider by the manufacturer and cannot be adjusted by the pilot's movements of the controls.

Aspect ration. The ratio between the wing span and the mean chord of the wing.

Asymmetrical airfoil. One in which the upper camber differs from the lower camber.

Atmospheric sounding. A measure of atmospheric variables aloft, usually pressure, temperature, humidity, and wind.

Atmospheric stability. Describes a state in which an air parcel resists vertical displacement or, once displaced (for instance by flow over a hill), tends to return to its original level.

B

Bailout bottle. Small oxygen cylinder connected to the oxygen mask supplying several minutes of oxygen. It can be used in case of primary oxygen system failure or if an emergency bailout at high altitude became necessary.

Ballast. Term used to describe any system that adds weight to the glider. Performance ballast employed in some gliders increases wing loading using releasable water in the wings (via integral tanks or water bags). This allows faster average cross-country speeds. Trim ballast is used to adjust the flying CG, often necessary for light-weight pilots. Some gliders also have a small water ballast tank in the tail for optimizing flying CG.

Barograph. Instrument for recording pressure as a function of time. Used by glider pilots to verify flight performance for badge or record flights.

Best glide speed (best L/D speed). The airspeed that results in the least amount of altitude loss over a given distance. This speed is determined from the performance polar. The manufacturer publishes the best glide (L/D) airspeed for specified weights and the resulting glide ratio. For example, a glide ratio of 36:1 means that the glider will lose 1 foot of altitude for every 36 feet of forward movement in still air at this airspeed.

C

Camber. The curvature of a wing when looking at a cross section. A wing has upper camber on its top surface and lower camber on its bottom surface.

Cap cloud. Also called a foehn cloud. These are clouds forming on mountain or ridge tops by cooling of moist air rising on the upwind side followed by warming and drying by downdrafts on the lee side.

Centering. Adjusting circles while thermalling to provide the greatest average climb.

Center of pressure. The point along the wing chord line where lift is considered to be concentrated.

Centrifugal force. The apparent force occurring in curvilinear motion acting to deflect objects outward from the axis of rotation. For instance, when pulling out of a dive, it is the force pushing you down in your seat.

Centripetal force. The force in curvilinear motion acting toward the axis of rotation. For instance, when pulling out of a dive, it is the force that the seat exerts on the pilot to offset the centrifugal force.

Chord line. An imaginary straight line drawn from the leading edge of an airfoil to the trailing edge.

Cloud streets. Parallel rows of cumulus clouds. Each row can be as short as 10 miles or as long as a 100 miles or more.

Cold soaked. Condition of a self-launch or sustainer engine making it difficult or impossible to start in flight due to long-time exposure to cold temperatures. Usually occurs after a long soaring flight at altitudes with cold temperatures (e.g., a wave flight).

Convection. Transport and mixing of an atmospheric variable due to vertical mass motions (e.g., updrafts).

Convective condensation level (CCL). The level at which cumulus forms from surface-based convection. Under this level, the air is dry adiabatic and the mixing ratio is constant.

Conventional tail. A glider design with the horizontal stabilizer mounted at the bottom of the vertical stabilizer.

Convergence. A net increase in the mass of air over a specified area due to horizontal wind speed and/or direction changes. When convergence occurs in lower levels, it is usually associated with upward air motions.

Convergence zone. An area of convergence, sometimes several miles wide, at other times very narrow. These zones often provide organized lift for many miles (e.g., a sea-breeze front).

Critical angle of attack. Angle of attack, typically around 18°, beyond which a stall occurs. The critical angle of attack can be exceeded at any airspeed and at any nose attitude.

Cross-country. In soaring, any flight out of gliding range of the takeoff airfield. Note that this is different from the definitions in the 14 CFR for meeting the experience requirements for various pilot certificates and/or ratings.

Cumulus congestus. A cumulus cloud of significant vertical extent and usually displaying sharp edges. In warm climates, these sometimes produce precipitation. Also called towering cumulus, these clouds indicate that thunderstorm activity may soon occur.

Cumulonimbus (CB). Also called thunderclouds, these are deep convective clouds with a cirrus anvil and may contain any of the characteristics of a thunderstorm: thunder, lightning, heavy rain, hail, strong winds, turbulence, and even tornadoes.

D

Dead reckoning. Navigation by computing a heading from true airspeed and wind, then estimating time needed to fly to a destination.

Density altitude. Pressure altitude corrected for nonstandard temperature variations. Performance charts for many older gliders are based on this value.

Dewpoint (or dewpoint temperature). The temperature to which a sample of air must be cooled, while the amount of water vapor and barometric pressure remain constant, in order to attain saturation with respect to water.

Dihedral. The angle at which the wings are slanted upward from the root to the tip.

Diurnal effects. A variation (may be in temperature, moisture, wind, cloud cover, etc.) that recurs every 24 hours.

Downburst. A strong, concentrated downdraft, often associated with a thunderstorm. When these reach the ground, they spread out, leading to strong and even damaging surface winds.

Drag. The force that resists the movement of the glider through the air.

Dry adiabat. A line on a thermodynamic chart representing a rate of temperature change at the dry adiabatic lapse rate.

Dry adiabatic lapse rate (DALR). The rate of decrease of temperature with height of unsaturated air lifted adiabatically (not heat exchange). Numerically the value is 3 °C or 5.4 °F per 1,000 feet.

Dust devil. A small vigorous circulation that can pick up dust or other debris near the surface to form a column hundreds or even thousands of feet deep. At the ground, winds can be strong enough to flip an unattended glider over on its back. Dust devils mark the location where a thermal is leaving the ground.

Dynamic stability. A glider's motion and time required for a response to static stability.

E

Elevator. Attached to the back of the horizontal stabilizer, if controls movement around the lateral axis.

Empennage. The tail group of the aircraft usually supporting the vertical stabilizer and rudder, as well as the horizontal stabilizer and elevator, or on some aircraft, the V-Tail.

F

Flaps. Hinged portion of the trailing edge between the ailerons and fuselage. In some gliders, ailerons and flaps are interconnected to produce full-span flaperons. In either case, flap change the lift and drag on the wing.

Flutter. Resonant condition leading to rapid, unstable oscillations of part of the glider structure (e.g., the wing) or a control surface (e.g., elevator or aileron). Flutter usually occurs at high speeds and can quickly lead to structural failure.

Forward slip. A slide used to dissipate altitude without increasing the glider's speed, particularly in gliders without flaps or with inoperative spoilers.

G

Glider. A heavier-than-air aircraft that is supported in flight by the dynamic reaction of the air against its lifting surfaces, and whose free flight does not depend on an engine.

Graupel. Also called soft hail or snow pellets, these are white, round or conical ice particles ⅛ to ¼ inch in diameter. They often form as a thunderstorm matures and indicate the likelihood of lightning.

Ground effect. A reduction in induced drag for the same amount of lift produced. Within one wingspan above the ground, the decrease in induced drag enables the glider to fly at a lower airspeed. In ground effect, a lower angle of attack is required to produce the same amount of lift.

H

Height band. The altitude range in which the thermals are strongest on any given day. Remaining with the height band on a cross-country flight should allow the fastest average speed.

House thermal. A thermal that forms frequently in the same or similar location.

Human factors. The study of how people interact with their environments. In the case of general aviation, it is the study of how pilot performance is influenced by such issues as the design of cockpits, the function of the organs of the body, the effects of emotions, and the interaction and communication with the other participants of the aviation community, such as other crewmembers and air traffic control personnel.

I

Induced drag. Drag that is the consequence of developing lift with a finite-span wing. It can be represented by a vector that results from the difference between total and vertical lift.

Inertia. The tendency of a mass at rest to remain at rest or, if in motion, to remain in motion unless acted upon by some external force.

Instrument meteorological conditions (IMC). Meteorological conditions expressed in terms of visibility, distance from cloud, and ceiling less than the minimum specified for visual meteorological conditions (VMC). Gliders rarely fly in IMC due to instrumentation and air traffic control requirements.

Inversion. Usually refers to an increase in temperature with height, but may also be used for other atmospheric variables.

Isohumes. Lines of equal relative humidity.

Isopleth. A line connecting points of constant or equal value.

Isotherm. A contour line of equal temperature.

K

Katabatic. Used to describe any wind blowing down slope.

Kinetic energy. Energy due to motion, defined as one half mass times velocity squared.

L

Lapse rate. The decrease with height of an atmospheric variable, usually referring to temperature, but can also apply to pressure or density.

Lateral axis. An imaginary straight line drawn perpendicularly (laterally) across the fuselage and through the center of gravity. Pitch movement occurs around the lateral axis, and is controlled by the elevator.

Lenticular cloud. Smooth, lens-shaped clouds marking mountain-wave crests. They may extend the entire length of the mountain range producing the wave and are also called wave clouds or lennies by glider pilots.

Lift. Produced by the dynamic effects of the airstream acting on the wing, lift opposes the downward force of weight.

Limit load. The maximum load, expressed as multiples of positive and negative G (force of gravity), that an aircraft can sustain before structural damage becomes possible. The load limit varies from aircraft to aircraft.

Load factor. The ratio of the load supported by the glider's wings to the actual weight of the aircraft and its contents.

Longitudinal axis. An imaginary straight line running through the fuselage from nose to tail. Roll movement occurs around the longitudinal axis, and is controlled by the ailerons.

M

Mesoscale convection system (MCS). A large cluster of thunderstorms with horizontal dimensions on the order of 100 miles. MCSs are sometimes organized in a long line of thunderstorms (e.g., a squall line) or as a random grouping of thunderstorms. Individual thunderstorms within the MCS may be severe.

Microburst. A small-sized downburst of 2.2 nautical mile or less horizontal dimension.

Minimum sink airspeed. Airspeed, as determined by the performance polar, at which the glider achieves the lowest sink rate. That is, the glider loses the least amount of altitude per unit of time at minimum sink airspeed.

Mixing ration. The ratio of the mass of water vapor to the mass of dry air.

Multicell thunderstorm. A group or cluster of individual thunderstorm cells with varying stages of development. These storms are often self propagating and may last for several hours.

O

Olphin flight. Straight flight following speed-to-fly theory. Glides can often be extended and average cross-country speeds increased by flying faster in sink and slower in lift without stopping to circle.

P

Parasite drag. Drag caused by any aircraft surface that deflects or interferes with the smooth airflow around the airplane.

Pilotage. Navigational technique based on flight by reference to ground landmarks.

Pilot-induced oscillation (PIO). Rapid oscillations caused by the pilot's overcontrolled motions. PIOs usually occur on takeoff or landings with pitch-sensitive gliders and in severe cases can lead to loss of control or damage.

Pitch attitude. The angle of the longitudinal axis relative to the horizon. Pitch attitude serves as a visual reference for the pilot to maintain or change airspeed.

Pitot-static system. System that powers the airspeed altimeter and variometer by relying on air pressure differences to measure glider speed, altitude, and climb or sink rate.

Placards. Small statements or pictorial signs permanently fixed in the cockpit and visible to the pilot. Placards are used for operating limitations (e.g., weight or speeds) or to indicate the position of an operating lever (e.g., landing gear retracted or down and locked).

Precipitable water. The amount of liquid precipitation that would result if all water vapor were condensed.

Pressure altitude. The height above the standard pressure level of 29.92 "Hg. It is obtained by setting 29.92 in the barometric pressure window and reading the altimeter.

R

Radiant energy. Energy due to any form of electromagnetic radiation (e.g., from the sun).

Radius of turn. The horizontal distance an aircraft uses to complete a turn.

Rate of turn. The amount of time it takes for a glider to turn a specified number of degrees.

Relative wind. The airflow caused by the motion of the aircraft through the air. Relative wind, also called relative airflow, is opposite and parallel to the direction of flight.

Rotor. A turbulent circulation under mountain-wave crests, to the lee side and parallel to the mountains creating the wave. Glider pilots use the term rotor to describe any low-level turbulent flow associated with mountain waves.

Rotor streaming. A phenomenon that occurs when the air flow at mountain levels may be sufficient for wave formation, but begins to decrease with altitude above the mountain. In this case, the air downstream of the mountain breaks up and becomes turbulent, similar to rotor, with no lee waves above.

Rudder. Attached to the back of the vertical stabilizer, the rudder controls movement about the vertical axis.

S

Sailplane. A glider used for traveling long distances and remaining aloft for extended periods of time.

Saturated Adiabatic Lapse Rate (SALR). The rate of temperature decrease with height of saturated air. Unlike the dry adiabatic lapse rate (DALR), the SALR is not a constant numerical value but varies with temperature.

Self-launching glider. A glider equipped with an engine, allowing it to be launched under its own power. When the engine is shut down, a self-launching glider displays the same characteristics as a non-powered glider.

Side slip. A slip in which the glider's longitudinal axis remains parallel to the original flightpath but in which the flightpath changes direction according to the steepness of the bank.

Slip. A descent with one wing lowered and the glider's longitudinal axis at an angle to the flightpath. A slip is used to steepen the approach path without increasing the airspeed, or to make the glider move sideways through the air, counteracting the drift resulting from a crosswind.

Speed to fly. Optimum speed through the (sinking or rising) air mass to achieve either the furthest glide or fastest average cross-country speed depending on the objectives during a flight.

Spin. An aggravated stall that results in the glider descending in a helical, or corkscrew, path.

Spoilers. Devices on the tops of wings to disturb (spoil) part of the airflow over the wing. The resulting decrease in lift creates a higher sink rate and allows for a steeper approach.

Squall line. A line of thunderstorms often located along or ahead of a vigorous cold front. Squall lines may contain severe thunderstorms. The term is also used to describe a line of heavy precipitation with an abrupt wind shift but no thunderstorms, as sometimes occurs in association with fronts.

Stabilator. A one-piece horizontal stabilizer used in lieu of an elevator.

Stability. The glider's ability to maintain a uniform flight condition and return to that condition after being disturbed.

Stall. Condition that occurs when the critical angle of attack is reached and exceeded. Airflow begins to separate from the top of the wing, leading to a loss of lift. A stall can occur at any pitch attitude or airspeed.

Standard atmosphere. A theoretical vertical distribution of pressure, temperature and density agreed upon by international convention. It is the standard used, for instance, for aircraft performance calculations. At sea level, the standard atmosphere consists of a barometric pressure of 29.92 inches of mercury ("Hg) or 1013.2 millibars, and a temperature of 15 °C (59 °F). Pressure and temperature normally decrease as altitude increases. The standard lapse rate in the lower atmosphere for each 1,000 feet of altitude is approximately 1 "Hg. and 2 °C (3.5 °F). For example, the standard pressure and temperature at 3,000 feet mean sea level (MSL) is 26.92 "Hg. (29.92 – 3) and 9 °C (15 °C – 6 °C).

Static stability. The initial tendency to return to a state of equilibrium when disturbed from that state.

Supercell thunderstorm. A large, powerful type of thunderstorm that forms in very unstable environments with vertical and horizontal wind shear. These are almost always associated with severe weather, strong surface winds, large hail, and/or tornadoes.

T

T-tail. A type of glider with the horizontal stabilizer mounted on the top of the vertical stabilizer forming a T.

Thermal. A buoyant plume or bubble of rising air.

Thermal index (TI). For any given level is the temperature of the air parcel having risen at the dry adiabatic lapse rate (DALR) subtracted from the ambient temperature. Experience has shown that a TI should be –2 for thermals to form and be sufficiently strong for soaring flight.

Thermal wave. Waves, often but not always marked by cloud streets, that are excited by convection disturbing an overlying stable layer. Also called convection waves.

Thermodynamic diagram. A chart presenting isopleths of pressure, temperature, water vapor content, as well as dry and saturated adiabats. Various forms exist, the most commonly used in the United States being the Skew-T/Log-P.

Thrust. The forward force that propels a powered glider through the air.

Total drag. The sum of parasite and induced drag.

Towhook. A mechanism allowing the attachment and release of a towrope on the glider or towplane. On gliders, it is located near the nose or directly ahead of the main wheel. Two types oftowhooks commonly used in gliders are manufactured by Tost and Schweizer.

Trim devices. Any device designed to reduce or eliminate pressure on the control stick. When properly trimmed, the glider should fly at the desired airspeed with no control pressure from the pilot (i.e., hands off). Trim mechanisms are either external tabs on the elevator (or stabilator) or a simple spring-tension system connected to the control stick.

True altitude. The actual height of an object above mean sea level.

V

V-tail. A type of glider with two tail surfaces mounted to form a V. V-tails combine elevator and rudder movements.

Variometer. Sensitive rate of climb or descent indicator that measures static pressure between the static ports and an external capacity. Variometers can be mechanical or electrical and can be compensated to eliminate unrealistic indications of lift and sink due to rapid speed changes.

Vertical axis. An imaginary straight line drawn through the center of gravity and perpendicular to the lateral and longitudinal axes. Yaw movement occurs around the vertical axis and is controlled by the rudder.

Visual meteorological conditions (VMC). Meteorological conditions expressed in terms of visibility, distance from cloud, and ceiling equal to or better than a specified minimum. VMC represents minimum conditions for safe flight using visual reference for navigation and traffic separation. Ceilings and visibility below VMC constitutes instrument meteorological conditions (IMC).

W

Washout. Slight twist built in towards the wingtips, designed to improve the stall characteristics of the wing.

Water vapor. Water present in the air while in its vapor form. It is one of the most important of atmospheric constituents.

Wave length. The distance between two wave crests or wave troughs.

Wave window. Special areas arranged by Letter of Agreement (LOA) with the controlling ATC wherein gliders may be allowed to fly under VFR in Class A Airspace at certain times and to certain specified altitudes.

Weight. Acting vertically through the glider's center of gravity, weight opposes lift.

Wind triangle. Navigational calculation allowing determination of true heading with a correction for crosswinds on course.

NOTES

NOTES

NOTES